Quantum Nonlocality

Quantum Nonlocality

Special Issue Editor

Lev Vaidman

MDPI • Basel • Beijing • Wuhan • Barcelona • Belgrade

MDPI

Special Issue Editor
Lev Vaidman
Tel Aviv University
Israel

Editorial Office
MDPI
St. Alban-Anlage 66
4052 Basel, Switzerland

This is a reprint of articles from the Special Issue published online in the open access journal *Entropy* (ISSN 1099-4300) from 2018 to 2019 (available at: https://www.mdpi.com/journal/entropy/special_issues/Quantum_Nonlocality)

For citation purposes, cite each article independently as indicated on the article page online and as indicated below:

LastName, A.A.; LastName, B.B.; LastName, C.C. Article Title. *Journal Name* **Year**, *Article Number*, Page Range.

ISBN 978-3-03897-948-7 (Pbk)
ISBN 978-3-03897-949-4 (PDF)

Contents

About the Special Issue Editor

Lev Vaidman was born in Leningrad and studied physics in Israel. He received his Ph.D. from Tel Aviv University under the guidance of Yakir Aharonov, with whom he continues to collaborate until today. After three years at University of South Carolina, he returned to Tel Aviv, where he became head of a quantum research group. His research is centered upon the foundations of quantum mechanics and quantum information. Vaidman is a theoretical physicist, and many of his proposals have been implemented in laboratories around the world, though he himself only became recently involved in the experimental realizations of his ideas. Vaidman is mainly known for introducing teleportation of continuous variables, cryptography with orthogonal states, novel types of quantum measurement (nonlocal, weak, protective, interaction-free), and introducing numerous quantum paradoxes. His analyses of quantum mechanical interpretations are centered in the development of the many-worlds interpretation, for which he is apparently the strongest proponent.

MDPI

Editorial
Quantum Nonlocality

Lev Vaidman

Raymond and Beverly Sackler School of Physics and Astronomy, Tel-Aviv University, Tel-Aviv 69978, Israel; vaidman@post.tau.ac.il; Tel.: +972-545908806

Received: 23 April 2019; Accepted: 24 April 2019; Published: 29 April 2019

Keywords: nonlocality; entanglement; quantum

The role of physics is to explain observed phenomena. Explanation in physics began as a causal chain of local actions. The first nonlocal action was Newton's law of gravity, but Newton himself considered the nonlocal action to be something completely absurd which could not be true—and indeed, gravity today is explained through local action of the gravitational field. It is the quantum theory which made physicists believe that there was nonlocality in Nature. It also led to the acceptance of randomness in Nature, the existence of which is considered as another weakness of science today. In fact, I hope that it is possible to remove randomness and nonlocality from our description of Nature [1]. Accepting the existence of parallel worlds [2] eliminates randomness and avoids action at a distance, but it still does not remove nonlocality. This special issue of Entropy is an attempt to more deeply understand the nonlocality of the quantum theory. I am interested to explore the chances of removing nonlocality from the quantum theory, and such an attempt is the most desirable contribution to this special issue; however, other works presented here which characterize the quantum nonlocality and investigate the role of nonlocality as an explanation of observed phenomena also shed light on this question.

It is important to understand what the meaning of nonlocality is in quantum theory. Quantum theory does not have the strongest and simplest concept of nonlocality, which is the possibility of making an instantaneous observable local change at a distance. However, all single-world interpretations do have actions at a distance. The quantum nonlocality also has an operational meaning for us, local observers, who can live only in a single world. Given entangled particles placed at a distance, a measurement on one of the particles instantaneously changes the quantum state of the other, from a density matrix to a pure state. It is only in the framework of the many-worlds interpretation, considering all worlds together, where the measurement causes no change in the remote particle, and it remains to be described by a density matrix. Another apparent nonlocality aspect is the existence of global topological features, such as the Aharonov-Bohm effect [3]. I believe I succeeded in removing this kind of nonlocality from quantum mechanics [4], but the issue is still controversial [5–8]. Unfortunately, no contributions clarifying this problem appear in this issue.

It is of interest to analyze nonlocal properties of composite quantum systems, the properties of systems in separate locations [9]. These properties are nonlocal by definition, and the nonlocality of their description does not necessarily tell us that the Nature is nonlocal. It is not surprising that nonlocal properties obey nonlocal dynamical equations. Although unrelated to the question of nonlocality in Nature, it is a useful tool for quantum information which, due to quantum technology revolution, becomes not just the future, but the present of practical applications. See the discussion of this aspect of quantum nonlolcality in this issue and note the recent first experimental realization of measurements of nonlocal variables [10].

For the problem of nonlocality of Nature, the important question is: which of the nonlocal features of composite systems cannot be specified by local measurements of its parts? More precisely, this is the question of nonlocality of a single world, would it be one of the worlds of the many-worlds theory or the only world of one of the single-world interpretations. Even if it does not answer the question of nonlocality of the physical universe incorporating all the worlds, this is the question relevant for harnessing the quantum advantage for tasks which cannot be accomplished classically.

What seems to be an unavoidable aspect of nonlocality of the quantum theory—which is present even in the framework of all worlds together—is entanglement. Measurement on one system does not change the state of the other system in the physical universe, but in each world created by the measurement, the state of the remote system is different. The entanglement, that is, the nonlocal connection between the outcomes of measurements shown to be unremovable using local hidden variables, is the ultimate nonlocality of quantum systems.

Very subjectively—I find the most interesting contribution to be the work by Brassard and Raymond-Robichaud [11], "Parallel Lives: A Local-Realistic Interpretation of 'Nonlocal' Boxes". The work challenges the ultimate question of nonlocality of entanglement. It is part of the ongoing program which was introduced by Deutsch and Hayden [12] to completely eliminate nonlocality from quantum mechanics. The present authors promise to complete it in a future publication. The current paper, instead, provides a wider picture, considering, in a local way, different theories that are currently viewed as nonlocal. The analysis of Popescu Rohrlich (PR) boxes [13], the Einstein–Podolsky–Rosen argument, and Bell's theorem puts the picture in proper and clear perspective. I am optimistic that Brassard and Raymond-Robichaud will succeed in building their fully local picture as they promise. However, I am also pretty sure that they will have to pay a very high price for removing all aspects of nonlocality by carrying a huge amount of local information in order to reconstruct the consequences of entanglement. Currently, I feel that I will not adopt the "parallel lives" picture, and will stay with the many-worlds interpretation [2], an elegant economical interpretation that has no randomness and action at a distance, but still has nonlocality in the concept of a world. However, I am very curious to see the quantum theory of the parallel lives. The possibility of the construction of a fully local theory, even if it is not economical, is of great importance.

The main test bed for considering nonlocal theories has been the example of PR boxes. It is the topic of the contribution by Rohrlich and Hetzroni [14], "GHZ States as Tripartite PR Boxes: Classical Limit and Retrocausality". The starting point of this work is Rohrlich's questioning of his own discovery: can we obtain a classical limit for PR boxes [15]? I am not sure that we have to worry about a classical limit for PR boxes; there is no compelling reason to assume the existence of such a hypothetical construction, as well as the existence of its classical limit. The message of Rohrlich and Hetzroni is that even if the lack of a classical limit for PR boxes represents a conceptual difficulty, there is no difficulty in the case of a quantum-mechanical setup—namely the Greenberger–Horne–Zeilinger setup—which is structurally similar to PR boxes but sufficiently different to have a classical limit. Their paper has also a nice analysis of how retrodiction might solve nonlocality paradoxes.

Retrodiction is also discussed in the contribution by Parks and Spence [16], "Capacity and Entropy of a Retro-Causal Channel Observed in a Twin Mach–Zehnder Interferometer During Measurements of Pre- and Post-Selected Quantum Systems". The test bed is now a peculiar interferometer considered as a retro-causal channel, analyzed in terms of weak and strong measurements performed on a pre- and post-selected particle. Experimental data collected from an optical experiment performed in 2010 was analyzed. The entropy of this retro-causal structure was considered, making it very relevant for the journal hosting the special issue. The developed formalism is capable of quantitative analysis of other interference experiments.

The level of complexity goes up in the contribution by Bharti, Ray, and Kwek [17], "Non-Classical Correlations in *n*-Cycle Setting". The compatibility relation among the observables is represented by graphs, where edges indicate compatibility. PR boxes and other nonlocal boxes such as Kochen–Specker–Klyachko boxes are considered for the *n*-cycle case. Non-contextuality is brought up, and extensive analysis of various inequalities characterizing the nonlocality is performed. The work holds the potential to be valuable for the future of quantum computation, as it provides a tight quantitative comparison of efficiency for several tasks of classical methods, quantum methods, and those built on PR boxes.

Another approach for characterizing the nonlocality of quantum theory and some general classes of nonlocal theories (e.g., PR boxes) can be found in the contribution by Carmi and Cohen [18], "On the Significance of the Quantum Mechanical Covariance Matrix". It also has a direct connection to the journal through the suggestion that the Tsallis entropy quantifies the extent of nonlocality. The key element in this new approach is the connection between nonlocality and a subtle form of uncertainty applicable to general covariance matrices. The most interesting result is that the nonlocality originating from these new characteristics can be measured using feasible weak and strong measurements.

A new approach to harnessing entropic uncertainty relations for investigating quantum nonlocality was presented in the contribution by Costa, Uola, and Gühne [19], "Entropic Steering Criteria: Applications to Bipartite and Tripartite Systems". Steering may be seen as an action at a distance in one-world interpretations, and thus a robust manifestation of quantum nonlocality. The authors introduce entropic steering criteria, and derive several strong bounds using modest numerical calculations.

A general review of basic techniques for certification of EPR steering was presented by Zhen, Xu, Liu, and Chen [20], "The Einstein–Podolsky–Rosen Steering and Its Certification". It specified the remaining open problem of how much entanglement is sufficient for EPR steering, and how much EPR steering is sufficient for nonlocality. Solving this problem will advance the realization of nonlocality-based quantum protocols.

Montina and Wolf, in their paper [21] "Discrimination of Non-Local Correlations", presented a surprisingly efficient algorithm which allowed to answer a very complex problem of characterization of nonlocality using numerical tractable computation. The method shows its validity by successfully reproducing known results, and provides a direction for dealing with difficult, unsolved problems.

Several "loophole-free" Bell-type experiments performed in recent years led to a strong consensus that Nature, or at least the world we live in, has Bell-type nonlocality, but does not have the strong nonlocality of superluminal signalling. Nevertheless, some statistical results of locality testing experiments showed apparently incompatible results. Liang and Zhang, in their paper [22] "Bounding the Plausibility of Physical Theories in a Device-Independent Setting via Hypothesis Testing", adapted the prediction-based-ratio method (which was originally designed for testing Bell-locality) for testing non-superluminal signaling, the quantum hypothesis, as well as some other natural hypotheses. Their method has provided a unified platform for testing all these different hypotheses at the same time, and is thus a means to evaluate the strength and correctness of various Bell-type experiments.

A paper by Podoshvedov [23], "Efficient Quantum Teleportation of Unknown Qubit Based on DV-CV Interaction Mechanism", analyzes a novel scheme of qubit teleportation based on continuous variables, arguing that the method is optimal under some realistic constraints. Quantum teleportation is arguably the most spectacular application of quantum nonlocality, as it cannot be explained in the framework of the hidden variables theory.

The question of information transfer in teleportation is, in my view, the key issue in understanding quantum nonlocality [12]. Some light on this question was shed by Cruzeiro and Gisin in their paper [24], "Bell Inequalities with One Bit of Communication". Their results are based on the development of recent years which showed that the Bell-Type correlations can be simulated by classical means with the help of transmitting a surprisingly small number of bits. They derived a large class of new Bell-type inequalities, and presented a way in which to generate many others.

The formalism of quantum theory allows for the analysis of nonlocal properties which cannot be considered in the classical domain. Classically, a property is either true or false, while in quantum theory, we have the new concept of superposition which has no classical analogue. In the paper [25], "Non-Local Parity Measurements and the Quantum Pigeonhole Effect", Paraoanu extended the gedanken experiment proposed by Aharonov et al. [26], proposing two constructions of measurement of parity, a manifestly nonlocal variable. This adds a new conceptual twist in the paradox by exposing, in an unexpected way, the tension between quantum physics and local realism.

Quantum nonlocality is not just a peculiar feature which can be harnessed in quantum information tasks—it is also present in many situations. Martínez, Rodríguez, Fierro, Otero, and Aguilar, in their paper [27] "Quantum Nonlocality and Quantum Correlations in the Stern–Gerlach Experiment", showed the presence of quantum nonlocality in the iconic quantum measurement performed on a single atom.

Quantum nonlocality is an important element for explaining observed quantum effects of organic molecules. Summhammer, Sulyok, and Bernroider analyzed such a situation in their paper [28], "Quantum Dynamics and Non-Local Effects Behind Ion Transition States during Permeation in Membrane Channel Proteins". The analyzed system is very complex, and some approximations are required, but the observed behaviour was satisfactorily explained only after taking into account quantum nonlocality.

Another work showing the need for quantum nonlocality to explain observed behavior was presented by Iotti and Rossi in [29] "Microscopic Theory of Energy Dissipation and Decoherence in Solid-State Quantum Devices: Need for Nonlocal Scattering Models". Here, nonlocal generalization of semiclassical (local) scattering models [30] was successful, whereas numeral calculations based on local models failed.

Even if the current special issue does not provide complete answers to all questions about quantum nonlocality, I do see significant progress and am confident that the questions posed here bring us closer to understanding this bizarre feature of quantum mechanics.

Acknowledgments: I thank all authors of submitted contributions and I sincerely believe that our efforts deepen our understanding of Nature. The special issue could not get it current form without support of very professional staff of the journal Entropy and MDPI. This work has been supported in part by the Israel Science Foundation Grant No. 1311/14.

Conflicts of Interest: The author declares no conflict of interest.

References

1. Vaidman, L. Quantum theory and determinism. *Quantum Stud. Math. Found.* **2014**, *1*, 5–38. [CrossRef]
2. Vaidman, L. Many-Worlds Interpretation of Quantum Mechanics. In *The Stanford Encyclopedia of Philosophy*, Fall 2018 ed.; Zalta, E.N., Ed.; Metaphysics Research Lab: Stanford, CA, USA, 2018.
3. Aharonov, Y.; Bohm, D. Significance of Electromagnetic Potentials in the Quantum Theory. *Phys. Rev.* **1959**, *115*, 485–491. [CrossRef]
4. Vaidman, L. Role of potentials in the Aharonov-Bohm effect. *Phys. Rev. A* **2012**, *86*, 040101. [CrossRef]
5. Aharonov, Y.; Cohen, E.; Rohrlich, D. Comment on "Role of potentials in the Aharonov-Bohm effect". *Phys. Rev. A* **2015**, *92*, 026101. [CrossRef]
6. Vaidman, L. Reply to "Comment on 'Role of potentials in the Aharonov-Bohm effect'". *Phys. Rev. A* **2015**, *92*, 026102. [CrossRef]
7. Aharonov, Y.; Cohen, E.; Rohrlich, D. Nonlocality of the Aharonov-Bohm effect. *Phys. Rev. A* **2016**, *93*, 042110. [CrossRef]
8. Pearle, P.; Rizzi, A. Quantum-mechanical inclusion of the source in the Aharonov-Bohm effects. *Phys. Rev. A* **2017**, *95*, 052123. [CrossRef]
9. Aharonov, Y.; Albert, D.Z.; Vaidman, L. Measurement process in relativistic quantum theory. *Phys. Rev. D* **1986**, *34*, 1805. [CrossRef]
10. Xu, X.-Y.; Pan, W.-W.; Wang, Q.-Q.; Dziewior, J.; Knips, L.; Kedem, Y.; Sun, K.; Xu, J.-S.; Han, Y.-J.; Li, C.-F.; et al. Measurements of nonlocal variables and demonstration of the failure of the product rule for a pre- and postselected pair of photons. *Phys. Rev. Lett.* **2019**, *122*, 100405. [CrossRef]
11. Brassard, G.; Raymond-Robichaud, P. Parallel lives: A local-realistic interpretation of "nonlocal" boxes. *Entropy* **2019**, *21*, 87. [CrossRef]
12. Deutsch, D.; Hayden, P. Information flow in entangled quantum systems. *Proc. R. Soc. A Math. Phys. Eng. Sci.* **2000**, *456*, 1759–1774. [CrossRef]
13. Popescu, S.; Rohrlich, D. Quantum nonlocality as an axiom. *Found. Phys.* **1994**, *24*, 379–385. [CrossRef]

14. Rohrlich, D.; Hetzroni, G. GHZ States as Tripartite PR Boxes: Classical Limit and Retrocausality. *Entropy* **2018**, *20*, 478. [CrossRef]
15. Rohrlich, D. PR-box correlations have no classical limit. In *Quantum Theory: A Two-Time Success Story*; Springer: Basel, Switzerland, 2014; pp. 205–211.
16. Parks, A.; Spence, S. Capacity and Entropy of a Retro-Causal Channel Observed in a Twin Mach-Zehnder Interferometer During Measurements of Pre-and Post-Selected Quantum Systems. *Entropy* **2018**, *20*, 411. [CrossRef]
17. Bharti, K.; Ray, M.; Kwek, L.C. Non-Classical Correlations in n-Cycle Setting. *Entropy* **2019**, *21*, 134. [CrossRef]
18. Carmi, A.; Cohen, E. On the significance of the quantum mechanical covariance matrix. *Entropy* **2018**, *20*, 500. [CrossRef]
19. Costa, A.; Uola, R.; Gühne, O. Entropic Steering Criteria: Applications to Bipartite and Tripartite Systems. *Entropy* **2018**, *20*, 763. [CrossRef]
20. Zhen, Y.Z.; Xu, X.Y.; Liu, N.L.; Chen, K. The Einstein–Podolsky–Rosen Steering and Its Certification. *Entropy* **2019**, *21*, 422. [CrossRef]
21. Montina, A.; Wolf, S. Discrimination of Non-Local Correlations. *Entropy* **2019**, *21*, 104. [CrossRef]
22. Liang, Y.C.; Zhang, Y. Bounding the Plausibility of Physical Theories in a Device-Independent Setting via Hypothesis Testing. *Entropy* **2019**, *21*, 185. [CrossRef]
23. Podoshvedov, S.A. Efficient quantum teleportation of unknown qubit based on DV-CV interaction mechanism. *Entropy* **2019**, *21*, 150. [CrossRef]
24. Zambrini Cruzeiro, E.; Gisin, N. Bell inequalities with one bit of communication. *Entropy* **2019**, *21*, 171. [CrossRef]
25. Paraoanu, G. Non-Local Parity Measurements and the Quantum Pigeonhole Effect. *Entropy* **2018**, *20*, 606. [CrossRef]
26. Aharonov, Y.; Colombo, F.; Popescu, S.; Sabadini, I.; Struppa, D.C.; Tollaksen, J. Quantum violation of the pigeonhole principle and the nature of quantum correlations. *Proc. Natl. Acad. Sci. USA* **2016**, *113*, 532–535. [CrossRef]
27. Piceno Martínez, A.; Benítez Rodríguez, E.; Mendoza Fierro, J.; Méndez Otero, M.; Arévalo Aguilar, L. Quantum Nonlocality and Quantum Correlations in the Stern–Gerlach Experiment. *Entropy* **2018**, *20*, 299. [CrossRef]
28. Summhammer, J.; Sulyok, G.; Bernroider, G. Quantum dynamics and non-local effects behind ion transition states during permeation in membrane channel proteins. *Entropy* **2018**, *20*, 558. [CrossRef]
29. Iotti, R.; Rossi, F. Microscopic Theory of Energy Dissipation and Decoherence in Solid-State Quantum Devices: Need for Nonlocal Scattering Models. *Entropy* **2018**, *20*, 726. [CrossRef]
30. Iotti, R.C.; Dolcini, F.; Rossi, F. Wigner-function formalism applied to semiconductor quantum devices: Need for nonlocal scattering models. *Phys. Rev. B* **2017**, *96*, 115420. [CrossRef]

entropy

MDPI

Article

Parallel Lives: A Local-Realistic Interpretation of "Nonlocal" Boxes

Gilles Brassard [1,2,*] and Paul Raymond-Robichaud [1,*]

1   Département d'informatique et de recherche opérationnelle, Université de Montréal, Montréal, QC H3C 3J7, Canada
2   Canadian Institute for Advanced Research, Toronto, ON M5G 1M1, Canada
*   Correspondence: brassard@iro.umontreal.ca (G.B.); paul.r.robichaud@gmail.com (P.R.-R.)

Received: 1 July 2018; Accepted: 11 January 2019; Published: 18 January 2019

Abstract: We carry out a thought experiment in an imaginary world. Our world is both local and realistic, yet it violates a Bell inequality more than does quantum theory. This serves to debunk the myth that equates local realism with local hidden variables in the simplest possible manner. Along the way, we reinterpret the celebrated 1935 argument of Einstein, Podolsky and Rosen, and come to the conclusion that they were right in their questioning the completeness of the Copenhagen version of quantum theory, provided one believes in a local-realistic universe. Throughout our journey, we strive to explain our views from first principles, without expecting mathematical sophistication nor specialized prior knowledge from the reader.

Keywords: Bell's theorem; Einstein–Podolsky–Rosen argument; local hidden variables; local realism; no-signalling; parallel lives

---

1. Introduction

Quantum theory is often claimed to be nonlocal, or more precisely that it cannot satisfy simultaneously the principles of locality and realism. These principles can be informally stated as follows:

- Principle of realism: There is a real world whose state determines the outcome of all observations.
- Principle of locality: No action taken at some point can have any effect at some remote point at a speed faster than light.

We give a formal definition of local realism in a companion paper [1]; here, we strive to remain at the intuitive level and explain all our concepts, results and reasonings without expecting mathematical sophistication nor specialized prior knowledge from the reader.

The belief that quantum theory is nonlocal stems from the correct fact proved by John Bell [2] that it cannot be described by a *local hidden variable theory*, as we shall explain later. However, the claim of nonlocality for quantum theory is also based on the incorrect equivocation of local hidden variable theories with local realism, leading to the following fallacious argument:

1. Any local-realistic world must be described by local hidden variables.
2. Quantum theory cannot be described by local hidden variables.
3. *Ergo*, quantum theory cannot be both local and realistic.

The first statement is false, as we explain at length in this paper; the second is true; the third is a legitimate application of *modus tollens* (if *p* implies *q* but *q* is false, then *p* must be false as well), but the argument is unsound since it is based on a false premise. As such, our reasoning does not imply that quantum theory can be both local and realistic, but it establishes decisively that the usual reasoning against the local realism of quantum theory is fundamentally flawed.

In a companion paper, we go further and explicitly derive a full and complete local-realistic interpretation for finite-dimensional unitary quantum theory [3], which had already been discovered by David Deutsch and Patrick Hayden [4]. See also Refs. [5,6]. Going further, we show in another companion paper [1] that the local realism of quantum theory is but a particular case of the following more general statement: Any reversible-dynamics theory that does not allow instantaneous signalling admits a local-realistic interpretation.

In order to invalidate statement (1) above, we exhibit an imaginary world that is both local and realistic, yet that cannot be described by local hidden variables. Our world is based on the so-called *nonlocal box*, also known as the *PR box*, introduced by Sandu Popescu and Daniel Rohrlich [7], which is already known to violate a Bell inequality even more than quantum theory (more on this later), which indeed implies that it cannot be explained by local hidden variables (more on this later also). Nevertheless, we provide a full local-realistic explanation for our imaginary world. Even though this world is not the one in which we live, its mathematical consistency suffices to debunk the myth that equates local realism with local hidden variables. In conclusion, the correct implication of Bell's theorem is that quantum theory cannot be described by local hidden variables, *not* that it is not local-realistic. *That's different!*

Given that quantum theory has a local-realistic interpretation, why bother with nonlocal boxes, which only exist in a fantasy world? The main virtue of the current paper, compared to Refs. [1,3–6], is to invalidate the fallacious, yet ubiquitous, argument sketched above in the simplest and easiest possible way, without needing to resort to sophisticated mathematics. The benefit of working with nonlocal boxes, rather than dealing with all the intricacies of quantum theory, was best said by Jeffrey Bub in his book on *Quantum Mechanics for Primates*: "The conceptual puzzles of quantum correlations arise without the distractions of the mathematical formalism of quantum mechanics, and you can see what is at stake—where the clash lies with the usual presuppositions about the physical world" [8].

The current paper is an expansion of an informal self-contained 2012 poster [9] reproduced in Appendix A with small corrections, which explains our key ideas in the style of a graphic novel, as well as of a brief account in a subsequent paper [10]. A similar concept had already been formulated by Mark A. Rubin ([5], p. 318) in the context of two distant observers measuring their shares of a Bell state in the same basis, as well as Colin Bruce in his popular-science book on *Schrödinger's Rabbits* ([11], pp. 130–132). To the best of our knowledge, the latter was the first local-realistic description of an imaginary world that cannot be described by local hidden variables.

After this introduction, we describe the Popescu–Rohrlich nonlocal boxes, perfect as well as imperfect, in Section 2. We elaborate on no-signalling, local-realistic and local hidden variable theories in Section 3, which we illustrate with the Einstein–Podolsky–Rosen argument [12] and the nonlocal boxes. Bell's Theorem is reviewed in Section 4 in the context of nonlocal boxes, and we explain why they cannot be described by local hidden variables. The paper culminates with Section 5, in which we expound our theory of *parallel lives* and how it allows us to show that "nonlocal" boxes are perfectly compatible with both locality and realism. Having provided a solution to our conundrum, we revisit Bell's Theorem and the Einstein–Podolsky–Rosen argument in Section 6 in order to understand how they relate to our imaginary world. There, we argue that our theory of parallel lives is an unavoidable consequence of postulating that the so-called nonlocal boxes are in fact local and realistic. We conclude with a discussion of our results in Section 7. Finally, we reproduce in Appendix A an updated version of our 2012 poster [9], which illustrates the main concepts. Throughout our journey, we strive to illustrate how the arguments formulated in terms of nonlocal boxes and the more complex quantum theory are interlinked.

2. The Imaginary World

We now proceed to describe how our imaginary world is perceived by its two inhabitants, Alice and Bob. We postpone to Section 5 a description of what is really going on in that world. The main ingredient that makes our world interesting is the presence of perfect nonlocal boxes, a theoretical idea invented by Popescu and Rohrlich [7].

2.1. The Nonlocal Box

Nonlocal boxes always come in pairs: one box is given to Alice and the other to Bob. Some people prefer to define the nonlocal box as consisting of both boxes, so that the pair of boxes that we describe here would constitute a single nonlocal box; it's a matter of taste. One can think of a nonlocal box as an ordinary-looking box with two buttons labelled 0 and 1. Whenever a button is pushed, the box instantaneously flashes either a red or green light, with each outcome being equally likely. This concept is illustrated in Figure 1 and in Appendix A.

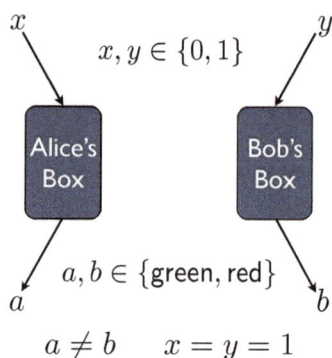

Figure 1. Nonlocal boxes.

If Alice and Bob meet to compare their results after they have pushed buttons, they will find that each pair of boxes produced outputs that are correlated in the following way: Whenever they had both pushed input button 1, their boxes flashed different colours, but if at least one of them had pushed input button 0, their boxes flashed the same colour. See Table 1.

Table 1. Behaviour of nonlocal boxes.

Alice's Input	Bob's Input	Output Colours
0	0	Identical
0	1	Identical
1	0	Identical
1	1	Different

For example, if Alice pushes 1 and sees green, whereas Bob pushes 0, she will discover when she meets Bob that he has also seen green. However, if Alice pushes 1 and sees green (as before), whereas Bob pushes 1 instead, she will discover when they meet that he has seen red.

A nonlocal box is designed for one-time use: once a button has been pushed and a colour flashed, the box will forever flash that colour and is no longer responsive to new inputs. However, Alice and Bob have an unlimited supply of such pairs of disposable nonlocal boxes.

2.2. Testing the Boxes

Our two inhabitants, Alice and Bob, would like to verify that their nonlocal boxes behave indeed according to Table 1. Here is how they proceed:

1. Alice and Bob travel far apart from each other with a large supply of numbered unused boxes, so that Alice's box number i is the one that is paired with Bob's box bearing the same number.
2. They flip independent unbiased coins labelled 0 and 1 and push the corresponding input buttons on their nonlocal boxes. For each box number, they record the randomly-chosen input and the observed resulting colour. Because they are sufficiently far apart, the experiment can be performed with sufficient simultaneity that Alice's box cannot know the result of Bob's coin flip (hence the input to Bob's box) before it has to flash its own light, and vice versa.
3. After many trials, Alice and Bob come back together and verify that the boxes work perfectly: no matter how far they were from each other and how simultaneously the experiment is conducted, the correlations promised in Table 1 are realized for each and every pair of boxes.

Note that neither Alice nor Bob can confirm that the promised correlations are established until they meet, or at least send a signal to each other. In other words, data collected locally at Alice's and at Bob's need to be brought together before any conclusion can be drawn. This detail may seem insignificant at first, but it will turn out to be crucial in order to give a local-realistic explanation for "nonlocal" boxes.

2.3. Imperfect Nonlocal Boxes

So far, we have talked about perfect nonlocal boxes, but we could consider nonlocal boxes that are sometimes allowed to give incorrectly correlated outputs. We say that a pair of nonlocal boxes *works with probability p* if it behaves according to Table 1 with probability p. With complementary probability $1 - p$, the opposite correlation is obtained.

2.3.1. Quantum Theory and Nonlocal Boxes

Although we shall concentrate on perfect nonlocal boxes in this paper, quantum theory makes it possible to implement nonlocal boxes that work with probability

$$p_{\text{quant}} = \cos^2\left(\tfrac{\pi}{8}\right) = \tfrac{2+\sqrt{2}}{4} \approx 85\%$$

but no better according to Cirel'son's theorem [13]. It follows that our imaginary world is distinct from the world in which we live since perfect nonlocal boxes cannot exist according to quantum theory.

For our purposes, the precise mathematics and physics that are needed to understand how it is possible for quantum theory to implement nonlocal boxes that work with probability p_{quant} do not matter. Let us simply say that it is made possible by harnessing entanglement in a clever way. Entanglement, which is the most nonclassical of all quantum resources, is at the heart of quantum information science.It was discovered by Einstein, Podolsky and Rosen in 1935 in Einstein's most cited paper [12], although there is some evidence that Erwin Schrödinger had discovered it earlier. It is also because of entanglement that the quantum world in which we live is often thought to be nonlocal.

3. The Many Faces of Locality

Recall that the principle of locality claims that no action taken at some point can have any effect at some remote point at a speed faster than light. An apparently weaker principle would allow such effects under condition that they cannot be observed at the remote point. This is the Principle of no-signalling, which we now explain.

3.1. No-Signalling

It is important to realize that nonlocal boxes do not enable instantaneous communication between Alice and Bob. Indeed, no matter which button Alice pushes (or if she does not push any button at all), Bob has an equal chance of seeing red or green flashing from his box whenever he pushes either of his buttons. Said otherwise, any action taken by Alice has no effect whatsoever on the probabilities of events that Bob can observe.

It follows that our imaginary world shares an important property with the quantum world: it obeys the principle of no-signalling.

- Principle of no-signalling: No action taken at some point can have any observable effect at some remote point at a speed faster than light.

Among observable effects, we include anything that would affect the probability distribution of outputs from any device. The principle of no-signalling implies in general that, for any pair of devices shared by Alice and Bob (not only PR boxes), Bob's output distribution depends only on Bob's input, and not on Alice's input, provided they are sufficiently far from each other and Alice does not provide her input to her device too long before Bob's device must produce its output.

3.2. Local Realism Implies No-Signalling

The principle of no-signalling follows from the principles of locality and realism: any local-realistic world is automatically no-signalling, as shown by the following informal argument:

1. By the principle of locality, no action taken at point A can have any effect on the state of the world at point B faster than at the speed of light.
2. By the principle of realism, anything observable at point B is a function of the state of the world at that point.
3. It follows that no action at point A can have an observable effect at point B faster than at the speed of light.

Here, we have relied on the tacit assumption that, in a local-realistic world, what is observable at some point is a function of the state of the world at that same point. The above argument is fully formalized, with all hypotheses made explicit, in Ref. [1].

3.3. Local Hidden Variable Theories

The most common type of local-realistic theories, which were studied in particular by John Bell [2], is based on local hidden variables (explained below). The misconception according to which all local-realistic theories have to be of that type has led to the widespread misguided belief that quantum theory cannot be local-realistic because it cannot be based on local hidden variables according to Bell's theorem.

In the idealized context of nonlocal boxes, a local hidden variable theory would consider arbitrarily sophisticated pairs of devices that are allowed to share randomness for the purpose of explaining the observed behaviour. The individual boxes would also be allowed independent sources of internal randomness. The initial shared randomness, along with the internal randomness and the inputs provided by Alice and Bob would be used to determine which colours to flash. However, what is not allowed is for the output of one of Alice's boxes to depend on the input of Bob's corresponding box, or vice versa. This can be enforced by the principle of locality, provided both input buttons are pushed with sufficient simultaneity to prevent a signal from one box to reach the other in time, even at the speed of light, to influence its outcome.

In any such theory (not just those pertaining to PR boxes), it is always possible to remove internal sources of randomness and replace them by parts of the initial source of shared randomness that would be used by one side only (provided we allow an infinite amount of shared randomness). However, the following section shows that, in the case of *perfect* nonlocal boxes, internal randomness should *never* be used to influence the behaviour of PR boxes.

3.4. The Einstein–Podolsky–Rosen Argument

Even though they were obviously not talking about Popescu–Rohrlich nonlocal boxes, the original 1935 argument of Einstein, Podolsky and Rosen (EPR) applies *mutatis mutandis* to prove that, in the context of local hidden variable theories, the output of Bob's nonlocal box should be completely determined by the initial randomness shared between Alice's and Bob's boxes and by Bob's input (and vice versa, with Alice and Bob interchanged).

1. Suppose Alice pushes her input button first. Note that for simplicity, we ignore the fact that there would be no such thing as absolute time if we took account of relativity, so that the notion of who pushes the button first may be ill-defined. This has no impact on the current reasoning because it is well-defined whether the effect of a button push can reach the other side before the other button is pushed.

2. When she pushes her button, this cannot have any instantaneous effect on Bob's box, by the principle of locality.

3. After seeing her output, Alice can know with certainty what colour Bob will see as a function of his input (even though she does not know which input he will choose). For example, if Alice had pushed 1 and seen green, she knows that if Bob chooses to push 0 he will also see green, whereas, if he chooses to push 1, he will see red.

4. Since it is possible for Alice to know with certainty what colour Bob will see when he pushes either button, and she can obtain this knowledge without influencing his system, it must be that his colour was predetermined as a function of which button he would push. This predetermination can only come from the initial source of shared randomness, and errors could occur if it were influenced by local randomness at Bob's.

This argument was used in the original Einstein–Podolsky–Rosen paper [12] to prove, under the implicit assumption of local hidden variables, that there are instances in quantum theory in which both the position and the momentum of a particle must be simultaneously defined. This clashed with the Copenhagen vision of quantum theory, according to which Heisenberg's uncertainty principle arises not merely from the fact that measuring one of those properties necessarily disturbs the other, but because they can never be fully defined simultaneously. The conclusion of Einstein, Podolsky and Rosen was that (the Copenhagen) quantum-mechanical description of physical reality can*not* be considered complete. After Niels Bohr's response [14], the physics community consensus was largely in his favour, asserting that the EPR argument was unsound and that the Copenhagen interpretation is indeed complete. In a companion paper [3], we prove that, under the metaphysical principle of local realism, it is Einstein, Podolsky and Rosen who were correct after all in arguing that the usual formulation of quantum theory cannot be a complete description of physical reality, and furthermore we provide a solution to make it complete along lines similar to those already discovered by Deutsch and Hayden [4].

However, let us come back to the imaginary world of nonlocal boxes. . . .

3.5. Local Hidden Variable Theory for Nonlocal Boxes

In a local hidden variable theory for nonlocal boxes, we have seen that all correlations should be explained by the initial shared randomness. Since each box implements a simple one-bit to one-colour-out-of-two function, it suffices to use only two bits of the randomness shared with its twin box to do so. It is natural to call those bits A_0 and A_1 for Alice, and B_0 and B_1 for Bob. If we define function

$$c : \{0,1\} \rightarrow \{\text{green}, \text{red}\}$$

by $c(0) =$ green and $c(1) =$ red, then Alice's box would flash colour $a = c(A_x)$ when input button x is pushed by Alice, whereas Bob's box would flash colour $b = c(B_y)$ when input button y is pushed by Bob. See Figure 1 again.

In order to fulfil the requirements of nonlocal boxes given in Table 1, it is easy to verify that the four local hidden variables must satisfy the condition

$$A_x \oplus B_y = x \cdot y \qquad (1)$$

for all $x, y \in \{0, 1\}$ simultaneously, where "$\oplus$" and "$\cdot$" denote the sum and the product modulo 2. For example, if Alice and Bob select $x = 0$ and $y = 1$, respectively, their boxes must flash the same colour $a = c(A_0) = c(B_1) = b$, according to Table 1, and therefore the hidden variables A_0 and B_1 must be equal since function c is one-to-one. In symbols, $A_0 = B_1$, which is equivalent to $A_0 \oplus B_1 = 0$, which indeed is equal to $x \cdot y$ in this case.

Is this possible?

4. Bell's Theorem

Theorem 1 (Bell's Theorem). *No local hidden variable theory can explain a nonlocal box that would work with a probability better than 75%. In particular, no local hidden variable theory can explain perfect nonlocal boxes.*

Proof. We have just seen that any local hidden variable theory that enables perfect nonlocal boxes would have to satisfy Equation (1) for all $x, y \in \{0, 1\}$. This gives rise to the following four explicit equations:

$$A_0 \oplus B_0 = 0,$$
$$A_0 \oplus B_1 = 0,$$
$$A_1 \oplus B_0 = 0,$$
$$A_1 \oplus B_1 = 1.$$

If we sum modulo 2 the equations on both sides and rearrange the terms using the associativity and commutativity of addition modulo 2, as well as the fact that any bit added modulo 2 to itself gives 0, we obtain:

$$(A_0 \oplus B_0) \oplus (A_0 \oplus B_1) \oplus (A_1 \oplus B_0) \oplus (A_1 \oplus B_1) = 0 \oplus 0 \oplus 0 \oplus 1,$$
$$(A_0 \oplus A_0) \oplus (A_1 \oplus A_1) \oplus (B_0 \oplus B_0) \oplus (B_1 \oplus B_1) = 1,$$
$$0 \oplus 0 \oplus 0 \oplus 0 = 1,$$
$$0 = 1,$$

which is a contradiction. Therefore, it is not possible for all four equations to hold simultaneously. At least one of the four possible choices of buttons pushed by Alice and Bob is bound to give incorrect results. It follows that any attempt at creating a nonlocal box that works with probability better than $3/4 = 75\%$ is doomed to fail in any theory based on local hidden variables. □

The reader can easily verify from the proof of Theorem 1 that any three of the four equations can be satisfied by a proper choice of local hidden variables. For example, setting $A_0 = B_0 = A_1 = B_1 = 0$ results in the first three equations being satisfied, but not the fourth. A more interesting strategy would be for Alice's box to produce $A_x = x$ and for Bob's box to produce $B_y = 1 - y$. In this case, the last three equations are satisfied but not the first. For each equation, there is a simple strategy that satisfies the other three but not that one (more than one such strategy in fact). More interestingly, it is possible to create a pair of nonlocal boxes that works with probability 75% *regardless of the chosen input* if the boxes share three bits of randomness. The first two bits determine which one of the four equations is jettisoned, thus defining an arbitrary pre-agreed strategy that fulfils the other three. If the third random bit is 1, both boxes will in fact produce the complement of the output specified in their strategy (which has no effect on which equations are satisfied). The purpose of this third shared random bit is that a properly functioning pair of PR boxes should produce an unbiased random output on each side if we only consider marginal probabilities.

We say of any world in which nonlocal boxes exist that work with a probability better than 75% that it *violates a Bell inequality* in honour of John Bell, who established the first result along the lines of Theorem 1, albeit not explicitly the one described here [2].

Quantum Theory and Bell's Theorem

The usual conclusion from Theorem 1 is that any world containing nonlocal boxes that work with a probability better than 75% cannot be both local and realistic. Since quantum theory enables boxes that work $\approx$ 85% of the time, as we have seen in Section 2.3.1, it seems inescapable that the quantum world cannot be local-realistic.

Similarly, it is tempting to assert that the more a Bell inequality is violated by a theory, the more nonlocal it is. In particular, our imaginary world would be more nonlocal than the quantum world itself. As we shall now see—and this is the main point of this paper—all these conclusions are unsound because local realism and local hidden variables should not be equated.

5. A Local Realistic Solution—Parallel Lives

Here is how the seemingly impossible can be accomplished. Let us assume for simplicity that Alice and Bob have a single pair of "nonlocal" boxes at their disposal, which is sufficient to rule out local hidden variable explanations. When Alice pushes a button on her box, she splits in two, together with her box. One Alice sees the red light flash on her box, whereas the other sees the green light flash. Both Alices are equally real. However, they are now living *parallel lives*: they will never be able to see each other or interact with each other. In fact, neither Alice is aware of the existence of the other, unless they infer it by pure thought as the only local-realistic explanation for what they will experience when they test their boxes according to Section 2.2. From now on, any unsplit object (or person) touched by either Alice or her box splits and inherits this splitting power. This does not have to be direct physical touching: a message sent by Alice has the same splitting effect on anything it reaches. Hence, Alice's splitting ripples through space, but no faster than at the speed of light. It is crucial to understand that it is not the entire universe that splits instantaneously when Alice pushes her button, as this would be a highly nonlocal effect.

The same thing happens to Bob when he pushes a button on his box: he splits and neither copy is aware of the other Bob. One copy sees a red light flash and the other sees a green light flash. If both Alice and Bob push a button at about the same time, we have two independent Alices and two independent Bobs, and for now the Alices and the Bobs are also independent of one another.

It is only when Alice and Bob interact that correlations are established. Let us assume for the moment that both Alice and Bob always push their buttons before interacting. The magical rule is that an Alice is allowed to interact with a Bob if and only if they jointly satisfy the conditions of the nonlocal box set out in Table 1.

For example, if Alice pushes button 1, she splits. Consider the Alice who sees green. Her system can be imagined to carry the following rule: You are allowed to interact with Bob if either he had pushed button 0 on his box and seen green, or pushed button 1 and seen red. Should this Alice ever come in presence of a Bob who had pushed button 1 and seen green, she would simply not become aware of his presence and could walk right through him without either one of them noticing anything. Of course, the other Alice, the one who had seen red after pushing button 1, would be free to shake hands with that Bob.

If Bob had pushed button 0 and seen green, his system can likewise be imagined to carry the following rule: You are allowed to interact with Alice if and only if she sees green, regardless of which button she had pushed. It is easy to generalize this idea to all cases covered by Table 1 because there will always be one green Alice and one red Alice, one green Bob and one red Bob, and whenever green Alice is allowed to interact with one Bob, red Alice is allowed to interact with the other Bob. From their perspective, each Alice and each Bob will observe correlations that seem to "emerge from outside space-time" [15]. However, this interpretation is but an illusion due to their intrinsic inability to perceive some of the actors in the world in which they live.

Our imaginary world is fully local because Alice's state is allowed to depend only on her own input and output at the moment she pushes a button. It is true that the mysterious correlations given in Table 1 would be impossible for any local hidden variable theory. However, Alice and Bob cannot experience those correlations before they actually meet (or at least before they share their data), *and these encounters cannot take place faster than at the speed of light*. When they meet, the correlations they experience are simply due to the matching rule that determines which Alices are allowed to interact with which Bobs, and *not* to a magical (because nonlocal) spukhafte Fernwirkung ("spooky action at a distance"), which was so abhorrent to Einstein, and rightly so.

What if Alice pushes her button, but Bob does not? In the discussion above, we assumed for simplicity that both Alice and Bob had pushed buttons on their boxes before interacting. A full story should include various other scenarios. It could be that Alice pushes a button on her box and travels to interact with a Bob who had not yet touched his box. Or it could be that after pushing a button on her box, only the Alice who had seen green travels to interact with Bob, whereas the Alice who had seen red stays where she is.

For instance, consider the case in which Alice had pushed button 1 on her box, split, and only the Alice who had seen green travels to meet unsplit Bob. At the moment they meet, Bob and his box automatically split. One Bob now owns a box programmed as follows: "if button 0 is pushed, flash green, but if button 1 is pushed, flash red"; the other Bob owns a box containing the complementary program, with "green" and "red" interchanged. As for our travelling Alice, she will see the first one of those Bobs and be completely oblivious of the other, who will not even be aware that an Alice had just made the trip to meet him.

It would be tedious, albeit elementary, to go through an exhaustive list of all possible scenarios. We challenge the interested reader to figure out how to make our imaginary world behave according to Table 1 in all cases. However, rather than get bored at this exercise, why not enjoy Appendix A, which illustrates the concept of parallel lives in the form of a graphic novel [9]?

Quantum Theory, Parallel Lives and Many Worlds

We coined the term "parallel lives" for the idea that a system is allowed to be in a superposition of several states, but that all splittings occur locally. This was directly inspired by the many-worlds interpretation of quantum theory, whose pioneer was Hugh Everett [16] more than six decades ago. However, our parallel lives theory is fundamentally distinct from the highly nonlocal—and very popular ([17], pp. 119–121)—naïve version of its many-worlds counterpart according to which the entire universe would split whenever Alice pushes a button on her box (or makes a measurement that has more than one possible outcome according to standard quantum theory). Later, Deutsch and Hayden provided the first explicitly local formulation of quantum theory, including a very lucid explanation of why Bell's theorem is irrelevant [4]. Even though they did not use the term "parallel lives", their approach was akin to ours. In their solution, the evolution of the quantum world is fully local, and individual systems, including observers, are implicitly allowed to be in superposition. In a companion paper [3], we offer our own local formalism for quantum theory along the lines of this paper, complete with full proofs of our assertions.

6. Revisiting Bell's Theorem and the Einstein–Podolsky–Rosen Argument

Having provided a solution to our conundrum with the explicit construction of a local-realistic imaginary world in which perfect Popescu–Rohrlich "nonlocal" boxes are possible, we revisit the Einstein–Podolsky–Rosen argument in order to understand how it relates to our imaginary world. This leads us to conclude that our theory of parallel lives is an unavoidable consequence of postulating that those boxes are compatible with local realism.

6.1. Parallel Lives versus Hidden Variable Theories

To understand the main difference between parallel lives and local hidden variable theories, consider again the scenario according to which Alice had pushed button 1 and her box flashed a green colour. According to local hidden variable theories, she would know with certainty what colour Bob will see as a function of his choice of input: he will also see green if he pushes button 0, but he will see red if he pushes button 1. This was at the heart of the Einstein–Podolsky–Rosen argument of Section 3.4 to the effect that the colours flashed by Bob's box had to be predetermined as a function of which button he would push since Alice could know this information without interacting with Bob's box. To quote the original argument, "If, without in any way disturbing a system, we can predict with certainty the value of a physical quantity, then there exists an element of physical reality corresponding to this physical quantity" [12]. The "element of physical reality" in question is what we now call local hidden variables and the "physical quantity" is the mapping between input buttons and output colours.

The parallel-lives interpretation is fundamentally different. Whenever Alice pushes a button on her box, she cannot infer anything about Bob's box. Instead, she can predict how her various lives will interact with Bob's in the future, in case they meet. Consider for example a situation in which both Alice and Bob push their input buttons, whose immediate effect is the creation of two Alices and two Bobs. Let us call them Green-Alice, Red-Alice, Green-Bob and Red-Bob, depending on which colour they have seen. If the original Alice had pushed her button 1, Green-Alice may now infer that she will interact with Red-Bob if he had also pushed his button 1, whereas she will interact with Green-Bob if he had pushed his button 0. The opposite statement is true of Red-Alice. As we can see, this is a purely local process since this instantaneous knowledge of both Alices has no influence on whatever the faraway Bobs may observe, which is actually both colours!

6.2. How an Apparent Contradiction Leads to Parallel Lives

Consider the following argument concerning nonlocal boxes, and pretend that you have never heard of parallel lives (nor of many worlds), yet you believe in locality:

1. Let us say that Alice pushes button 1 on her box. Without loss of generality, say that her box flashes the green colour.
2. Now, we know that Bob will see green if he pushes his button 0, whereas he will see red if he pushes his button 1, according to Table 1. By the principle of locality, this conclusion holds regardless of Alice's previous action since she was too far for her choice of button to influence Bob's box.
3. What would have happened had Alice pushed her button 0 instead at step 1? She must see the same colour as Bob, regardless of Bob's choice of button, since her pushing button 0 precludes the possibility that both Alice and Bob will press their button 1, which is the only case yielding different colours, again according to Table 1.
4. Statements 2 and 3 imply together that, when Alice pushes her button 0, she must see both red and green!

Despite appearances, statement 4 is not a contradiction, and indeed it can be resolved. Both results seen by Alice must be equally real by logical necessity. The only way for her to see both colours, *yet be convinced she saw only one*, is that there are in fact two Alices unaware of each other. In other words, the postulated locality of Popescu–Rohrlich "nonlocal" boxes *forces us* into a parallel-lives theory, which, far from being a postulate, is in fact an ineluctability.

If both Alices are indeed mathematically necessary to describe a local-realistic world, then both Alices are real in that world, inasmuch as we accept as a philosophical axiom the claim that, whenever a mathematical quantity is necessary to describe reality, that quantity corresponds to something that is real, and is not a mere artifact of the theory.

The same conclusion applies whenever any theory is shown to be inconsistent with all possible local hidden variable theories. Indeed, such theories carry the rarely-mentioned assumption that, once concluded, any experiment has a single outcome. Other outcomes that could have been possible simply did not occur. The inescapable resolution of any such inconsistency is to accept the conclusion that all possible outcomes occur within parallel lives of the experimenter.

7. Conclusions

We have exhibited a local-realistic imaginary world that violates a Bell inequality. For this purpose, we introduced the concept of *parallel lives*, but argued subsequently that this was an unavoidable consequence of postulating that the so-called nonlocal boxes are in fact local and realistic. The main virtue of our work is to demonstrate in an exceedingly simple way that local reality can produce correlations that are impossible in any theory based on local hidden variables. In particular, it is fallacious to conclude that quantum theory is nonlocal simply because it violates Bell's inequality.

In quantum theory, ideas analogous to ours can be traced back at least to Everett [16]. They were developed further by Deutsch and Hayden [4], and subsequently by Rubin [5] and Tipler [6]. Furthermore, Bruce ([11], pp. 130–132) gave the first local-realistic explanation for a theory that is neither quantum nor classical. In companion papers, we have proven that unitary quantum mechanics is local-realistic [3] (which had already been shown in Ref. [4]) and, more generally, that this is true for *any* reversible-dynamics no-signalling operational theory [1]. The latter paper provides a host of suggestions in its final section for a reader eager to pursue this line of work in yet unexplored directions.

Throughout our journey, we have revisited several times the Einstein–Podolsky–Rosen argument and have come to the conclusion that they were right in questioning the completeness of Bohr's Copenhagen quantum theory. Perhaps Einstein was correct in his belief of a local-realistic universe after all and in wishing for quantum theory to be completed? Perhaps we live parallel lives. . . .

Author Contributions: Both authors contributed significantly to this work but P.R.-R. spearheaded the effort. According to the tradition in our field, the authors are listed in alphabetical order. Conceptualization, P.R.-R.; Investigation, G.B. and P.R.-R.; Methodology, G.B. and P.R.-R.; Supervision, G.B.; Writing—Original Draft, P.R.-R.; Writing—Review and Editing, G.B. and P.R.-R.

Funding: The work of Gilles Brassard is supported in part by the Canadian Institute for Advanced Research, the Canada Research Chair program, Canada's Natural Sciences and Engineering Research Council (NSERC) and Québec's Institut transdisciplinaire d'information quantique. The work of Paul Raymond-Robichaud was supported in part by NSERC and the Fonds de recherche du Québec—Nature et technologies.

Acknowledgments: We acknowledge stimulating discussions with Charles Alexandre Bédard, Charles Bennett, Jeff Bub, Giulio Chiribella, David Deutsch, Stéphane Durand, Marcin Pawłowski, Sandu Popescu, Renato Renner, Alain Tapp and Stefan Wolf, as well as careful proofreading by Yuval Elias. We also acknowledge useful suggestions provided by the anonymous referees. Furthermore, we acknowledge the artwork of Louis Fernet-Leclair, who drew the poster reproduced in Appendix A according to our specifications. Finally, G.B. is grateful to Christopher Fuchs for having dragged him *up* the slope of quantum foundations years ago, despite their subsequent divergent paths.

Conflicts of Interest: The authors declare no conflict of interest. The funding sponsors had no role in the design of the study, in the writing of the manuscript, and in the decision to publish the results.

Abbreviations

The following abbreviations are used in this manuscript:

EPR   Einstein–Podolsky–Rosen
PR   Popescu–Rohrlich

Appendix A. Poster on Parallel Lives

We reproduce below the poster realized by Louis Fernet-Leclair in 2012 (improved in 2018) according to our specifications [9].

References

1. Brassard, G.; Raymond-Robichaud, P. The equivalence of local-realistic and no-signalling theories. *arXiv* **2017**, arXiv:1710.01380.
2. Bell, J.S. On the Einstein–Podolsky–Rosen paradox. *Physics* **1964**, *1*, 195–200. [CrossRef]
3. Brassard, G.; Raymond-Robichaud, P. A local realistic formalism for quantum theory. In preparation.
4. Deutsch, D.; Hayden, P. Information flow in entangled quantum systems. *Proc. R. Soc. Lond.* **2000**, *A456*, 1759–1774. [CrossRef]
5. Rubin, M.A. Locality in the Everett interpretation of Heisenberg-picture quantum mechanics. *Found. Phys. Lett.* **2001**, *14*, 301–322. [CrossRef]
6. Tipler, F.J. Quantum nonlocality does not exist. *Proc. Natl. Acad. Sci. USA* **2014**, *111*, 11281–11286. [CrossRef] [PubMed]
7. Popescu, S.; Rohrlich, D. Quantum nonlocality as an axiom. *Found. Phys.* **1994**, *24*, 379–385. [CrossRef]
8. Bub, J. *Bananaworld: Quantum Mechanics for Primates*; Oxford University Press: Oxford, UK, 2016.

9.   Parallel Lives: A Local-Realistic Interpretation of 'Nonlocal' Boxes. Poster Realized by Louis Fernet-Leclair. Available online: http://www.iro.umontreal.ca/~brassard/ParallelLives (accessed on 22 January 2019).

10.   Brassard, G.; Raymond-Robichaud, P. Can free will emerge from determinism in quantum theory? In *Is Science Compatible with Free Will? Exploring Free Will and Consciousness in the Light of Quantum Physics and Neuroscience*; Suarez, A., Adams, P., Eds.; Springer: Berlin, Germany, 2013; Chapter 4, pp. 41–60.

11.   Bruce, C. *Schrödinger's Rabbits: The Many Worlds of Quantum*; The National Academies Press: Washington, DC, USA, 2004.

12.   Einstein, A.; Podolsky, B.; Rosen, N. Can quantum-mechanical description of physical reality be considered complete? *Phys. Rev.* **1935**, *47*, 777–780. [CrossRef]

13.   Cirel'son, B.S. Quantum generalizations of Bell's inequality. *Lett. Math. Phys.* **1980**, *4*, 93–100. [CrossRef]

14.   Bohr, N. Can quantum-mechanical description of physical reality be considered complete? *Phys. Rev.* **1935**, *48*, 696–702. [CrossRef]

15.   Gisin, N. Are there quantum effects coming from outside space-time? Nonlocality, free will and 'no many-worlds'. In *Is Science Compatible with Free Will? Exploring Free Will and Consciousness in the Light of Quantum Physics and Neuroscience*; Suarez, A., Adams, P., Eds.; Springer: Berlin, Germany, 2013; Chapter 3, pp. 23–40.

16.   Everett, H., III. 'Relative State' formulation of quantum mechanics. *Rev. Mod. Phys.* **1957**, *29*, 454–462. [CrossRef]

17.   Bub, T.; Bub, J. *Totally R∀ndom: Why Nobody Understands Quantum Mechanics—A Serious Comic on Entanglement*; Princeton University Press: Princeton, NJ, USA, 2018.

entropy

MDPI

Article

GHZ States as Tripartite PR Boxes: Classical Limit and Retrocausality

Daniel Rohrlich [1],* and Guy Hetzroni [2]

[1] Department of Physics, Ben-Gurion University of the Negev, Beersheba 84105, Israel
[2] Program in the History and Philosophy of Science, The Hebrew University of Jerusalem, Jerusalem 91905, Israel; guy.hetzroni@mail.huji.ac.il
* Correspondence: rohrlich@bgu.ac.il

Received: 21 March 2018; Accepted: 6 June 2018; Published: 20 June 2018

Abstract: We review an argument that bipartite "PR-box" correlations, though designed to respect relativistic causality, in fact *violate* relativistic causality in the classical limit. As a test of this argument, we consider Greenberger–Horne–Zeilinger (GHZ) correlations as a tripartite version of PR-box correlations, and ask whether the argument extends to GHZ correlations. If it does—i.e., if it shows that GHZ correlations violate relativistic causality in the classical limit—then the argument must be incorrect (since GHZ correlations do respect relativistic causality in the classical limit.) However, we find that the argument does not extend to GHZ correlations. We also show that both PR-box correlations and GHZ correlations can be retrocausal, but the retrocausality of PR-box correlations leads to self-contradictory causal loops, while the retrocausality of GHZ correlations does not.

Keywords: axioms for quantum theory; PR box; nonlocal correlations; classical limit; retrocausality

PACS: 03.65.Ta; 03.65.Ca; 03.30.+p; 03.65.Ud

Quantum mechanics might make more sense to us if we could derive it from simple axioms with clear physical content, instead of opaque axioms about Hilbert space. Aharonov [1,2] and, independently, Shimony [3,4] conjectured that quantum mechanics might follow from the two axioms of nonlocality and relativistic causality (no superluminal signalling). For example, quantum correlations respect relativistic causality, but they are nonlocal: they violate the Bell-CHSH [5–7] inequality. Could quantum mechanics be *unique* in reconciling these axioms, just as the special theory of relativity is unique in reconciling the axioms of relativistic causality and the equivalence of inertial frames? So-called "PR-box" [8] correlations disprove this conjecture. Like quantum correlations, they respect relativistic causality; but unlike quantum correlations, they violate the Bell-CHSH inequality *maximally*. Nevertheless, Ref. [9] argues that the addition of one minimal axiom of clear physical content—namely, the existence of a classical limit—suffices for ruling out PR-box correlations.

The additional axiom is minimal in the following sense: Quantum mechanics has a classical limit in which there are no uncertainty relations; there are only jointly measurable macroscopic observables. This classical limit—our direct experience—is an inherent constraint, a boundary condition, on quantum mechanics and on any generalization of quantum mechanics. Thus PR-box correlations, too, must have a classical limit. Reference [9] argues that in this classical limit, PR-box correlations (and, by extension [10,11], all stronger-than-quantum bipartite correlations) allow observers "Alice" and "Bob" to exchange superluminal signals. (A similar statement appears in Ref. [12] with "macroscopic locality" taking the place of "classical limit". Yet Ref. [12] assumes that Alice and Bob can detect fluctuations of order $\sqrt{N}$ in their measurements, an assumption we do not make.) The argument [9,10] relies on measurement sequences that are observable but exponentially improbable. It is therefore of interest to test the argument by applying it to a different problem. In particular, GHZ correlations [13] are a tripartite version of PR-box correlations in the sense of being all-or-nothing correlations (perfect correlations and anticorrelations). Could Alice, Bob and an additional observer, "Jim", use GHZ correlations, in the classical limit, to exchange superluminal signals? Does the argument of Ref. [9] lead to this conclusion? If so, it is clearly an incorrect argument: quantum mechanics and its classical limit do *not* violate relativistic causality! The first section of this paper reviews the arguments of Ref. [9] and attempts to extend them to show how Alice, Bob and Jim could exchange superluminal signals in the classical limit; but this attempt fails. The second section compares PR-box and GHZ correlations to show how retrocausality is self-contradictory in the first case but not in the second.

1. GHZ and PR-Box Correlations in the Classical Limit

Let Alice and Bob make spacelike separated measurements on pairs of particles. For each pair (indexed by i), one member of the pair is in Alice's laboratory, and she can choose to measure observables a_i or a_i' (but not both) on it; the other member is in Bob's laboratory, and Bob can choose to measure observables b_i or b_i' (but not both) on it. All four observables a_i, a_i', b_i and b_i' take values ± 1 with equal probability. The definition of PR-box correlations,

$$C(a_i, b_i) = C(a_i, b_i') = C(a_i', b_i) = 1 = -C(a_i', b_i'),\tag{1}$$

implies that Alice can manipulate the correlations between the observables b_i, b_i' of Bob's particle by choosing whether to measure a_i or a_i': indeed, b_i and b_i' are perfectly correlated if she measures a_i (as both of them are perfectly correlated with her outcome), and perfectly anticorrelated if she measured a_i' (as b_i is correlated with her outcome and b_i' is anticorrelated with it). Thus, even though Alice's choice of measurement does not affect Bob's distribution of either b_i or b_i', it does affect correlations between these two observables. So can Alice exploit these correlations to signal to Bob? No, she cannot, since, by assumption, b_i and b_i' are incompatible and Bob cannot measure both. But, notably, this assumption cannot apply in the classical limit.

Following Ref. [9], we define the classical limit of PR-box correlations as follows: Macroscopic (classical) quantities are averages over arbitrarily large ensembles of microscopic observables. To see how this definition applies, let us consider an ensemble of N pairs shared by Alice and Bob and obeying Equation (1). Apparently, the N pairs are just as useless for signalling as one pair, since, for each pair, Bob is allowed to measure only b_i or b_i'. But the classical limit as defined means that given a large enough ensemble, Bob can measure quantities which depend upon macroscopic averages such as $B = \sum_{i=1}^N b_i / N$ and $B' = \sum_{i=1}^N b_i' / N$, obtaining *some* information about both of them. There is no fundamental limit on how many times Alice and Bob can repeat their measurements, hence no matter how large they choose N (so as to minimize the variances in B and B'), there is no limit to the strength of the (anti-)correlations that they may observe.

Now let us imagine two possible scenarios. In one scenario, Alice measures a_i consistently on all her N particles. In the other scenario, she measures a_i' consistently on all her N particles. What does Bob obtain from his measurements? The average value of B is $\langle B \rangle = 0$. Even typical deviations of B are small, i.e., of order $1/\sqrt{N}$, so they disappear in the classical limit. Apparently the scenarios lead to the exact same conclusion: Bob cannot read Alice's 1-bit message, encoded in her choice of what to measure.

Yet it will sometimes happen (with probability 2^{-N}) that B will take the value 1. If Alice and Bob repeat either scenario exponentially many times, they can produce arbitrarily many cases of $B = 1$. True, there will be measurement errors in Bob's results, but in the classical limit Bob must obtain at least *some* information about *both B and B'*. Now if Alice consistently measures a_i, Bob can expect to obtain $B = 1$ with probability close to 2^{-N}. And he can also expect to obtain $B = 1 = B'$ with the *same* probability, and not with probability 2^{-2N}, because Alice's choice has correlated $\langle B \rangle$ with $\langle B' \rangle$. Conversely, if Alice consistently measures a_i', then Bob can expect to obtain $B = 1$ with probability close to 2^{-N}, and he can also expect to obtain $B = 1 = -B'$ with the *same* probability, and not with probability 2^{-2N}, because Alice's choice has *anti*correlated $\langle B \rangle$ with $\langle B' \rangle$. Another way for Bob to get Alice's message is to observe the variance in his measurements of $B \pm B'$: if Alice measures a_i consistently, the distribution of $B + B'$ (over repeated trials with N pairs at a time) is binomial, while the distribution of $B - B'$ has zero variance, and vice versa in the other scenario. Thus Alice can send Bob a (superluminal) message in the classical limit.

It does not matter that the price of a one-bit message from Alice to Bob may be astronomical. As long as it is possible, at any price, it constitutes a violation of relativistic causality, which we cannot allow. Hence PR-box correlations violate relativistic causality in the classical limit, as claimed. (Note that we cannot obtain the classical limit $N \rightarrow \infty$ by setting $N = \infty$. Rather, we take N finite but arbitrarily large, and for any N, there is no fundamental bound on the number of times Alice and Bob can repeat their measurements in order to obtain the accuracy they need for B and B', etc.)

Before proceeding to tripartite (GHZ) correlations, let us stop to consider bipartite *quantum* correlations. Does the above argument imply that they, too, allow signalling in the classical limit? If so, it cannot be correct. Most similar to PR-box correlations are quantum correlations that saturate Tsirelson's bound [14] for the Bell-CHSH inequalities. Without loss of generality, we can consider entangled pairs of spin-1/2 particles in the state $[|\uparrow\rangle_A|\uparrow\rangle_B + |\downarrow\rangle_A|\downarrow\rangle_B]/\sqrt{2}$. In this state, Alice and Bob always obtain perfect correlations if they measure spin along the same axes in the xz plane.

Quantum correlations saturate Tsirelson's bound when $a = \sigma_z^A$, $a' = \sigma_x^A$, $b = (\sigma_z^B + \sigma_x^B)/\sqrt{2}$ and $b' = (\sigma_z^B - \sigma_x^B)/\sqrt{2}$, where each of the four observables takes the values ± 1. (We suppress the index i.) Their correlations are

$$C(a,b) = C(a,b') = C(a',b) = \frac{\sqrt{2}}{2} = -C(a',b') . \qquad (2)$$

If Alice measures a, then b and b' are correlated with her results. If she measures a', then b is correlated with her results and b' is anticorrelated. Can Bob thus detect what Alice measures? As in the discussion of PR-box correlations, we can compute and compare the variances of $(b + b')/\sqrt{2}$ vs. $(b - b')/\sqrt{2}$. But, by definition, these observables correspond to σ_z^B and σ_x^B, respectively, i.e., to a and a' on Bob's particle in the pair, which is left in the same state as Alice's. Now if Alice measures a consistently on her particles and Bob measures $(b + b')/\sqrt{2}$, the variance in Bob's results is maximal just because the variance in Alice's results is maximal. (That is, she has equal probability to obtain ± 1). Conversely, if Alice measures a consistently on her particles and Bob measures $(b - b')/\sqrt{2}$, the variance in Bob's results is maximal simply because a measurement of σ_x^B after Alice measures a is equally likely to be ± 1, whatever Alice obtains. We thus find that the correlations in Equation (2) are not strong enough to induce any difference between the variances of the observables $B + B'$ and $B - B'$. Indeed, they are the strongest correlations that do not induce such a difference and therefore do not permit signalling in the classical limit [10,11].

Reference [9] claims that correlations that are too strong violate relativistic causality in the classical limit, and that PR-box correlations are too strong because they provide absolute "all or nothing" correlations. But quantum mechanics, as well, provides "all or nothing" correlations. Consider a triplet of spin-half particles in a GHZ state $|\Psi_{GHZ}\rangle = [|\uparrow\rangle_A|\uparrow\rangle_B|\uparrow\rangle_J - |\downarrow\rangle_A|\downarrow\rangle_B|\downarrow\rangle_J]/\sqrt{2}$ shared by Alice, Bob and Jim in their respective laboratories. Suppose that these observers measure either σ_x or σ_y on their respective particles. Let a_x denote Alice's outcome from a measurement of σ_x^A (the x component of the spin of her particle) and let a_y denote Alice's outcome from a measurement of σ_y^A (the y component of the spin), with analogous notations for Bob and Jim. The state $|\Psi_{GHZ}\rangle$ is an eigenstate of the following four operators, satisfying

$$
\begin{aligned}
|\Psi_{GHZ}\rangle &= \sigma_y^A \sigma_x^B \sigma_y^J |\Psi_{GHZ}\rangle \\
&= \sigma_y^A \sigma_y^B \sigma_x^J |\Psi_{GHZ}\rangle \\
&= \sigma_x^A \sigma_y^B \sigma_y^J |\Psi_{GHZ}\rangle \\
&= -\sigma_x^A \sigma_x^B \sigma_x^J |\Psi_{GHZ}\rangle .
\end{aligned}
\qquad (3)
$$

The implication is that if all three observers measure σ_x on their particles, they will discover that $a_x b_x j_x = -1$. Similarly, if the appropriate measurements are carried out, they will discover that

$a_x b_y j_y = 1 = a_y b_x j_y = a_y b_y j_x$ as in Equation (3). In their famous paper [13], Greenberger, Horne and Zeilinger (GHZ) used these facts to show that there is no way to assign simultaneous values consistently to all six variables a_x, a_y, b_x, b_y, j_x and j_y. This fact rules out any local hidden variable model for the GHZ state.

Can Alice, Bob and Jim use GHZ states to signal? For definiteness, let us assume that Jim tries to send a signal to Alice and Bob via his choice of what to measure, σ_x^J or σ_y^J. Before going to the classical limit, let's ask whether Jim can send Alice and Bob a signal using just a few triplets. Note that if Jim measures σ_x^J and gets $j_x = -1$, then a_x and b_x must be correlated; we write $a_x b_x = 1$. In the same notation, $a_y b_y = -1$. In fact, if Jim measures σ_x^J, we find $a_x b_x = -a_y b_y$ whatever he gets. On the other hand, if Jim measures σ_y^J, we obtain the analogous equation $a_x b_y = a_y b_x$, whatever he gets, and no correlation between a_x and b_x or a_y and b_y. Are these correlations of any use? Alice and Bob *cannot* measure all their observables a_x, a_y, b_x, b_y to infer Jim's choice.

But the commutation relations

$$[\, \sigma_x^A \sigma_x^B, \, \sigma_y^A \sigma_y^B \,] = 0 = [\, \sigma_x^A \sigma_y^B, \, \sigma_y^A \sigma_x^B \,] \ , \tag{4}$$

imply that Alice and Bob *can* obtain $a_x b_x$ and $a_y b_y$ to see if they are anticorrelated or, alternatively, can obtain $a_x b_y$ and $a_y b_x$ to see if they are correlated! In the first case, Jim must have measured σ_x^J and in the second case, he must have measured σ_y^J. Right?

Wrong. This scheme fails. To see why, we first note that if Alice and Bob measure both $\sigma_x^A \sigma_x^B$ and $\sigma_y^A \sigma_y^B$, they will certainly find that $a_x b_x = -a_y b_y$ simply because the product of operators $\sigma_x^A \sigma_x^B \sigma_y^A \sigma_y^B$ equals $-\sigma_z^A \sigma_z^B$, which yields -1 when applied to $|\Psi_{GHZ}\rangle$. Likewise, if Alice and Bob measure both $\sigma_x^A \sigma_y^B$ and $\sigma_y^A \sigma_x^B$, they will verify that $a_x b_y = a_y b_x$, simply because the product of operators $\sigma_x^A \sigma_y^B \sigma_y^A \sigma_x^B$ equals $\sigma_z^A \sigma_z^B$, which yields 1 when applied to $|\Psi_{GHZ}\rangle$. In fact, Alice and Bob can learn nothing about Jim's choice from their measurements.

We are back to square one. So let us try to apply the classical-limit argument of Ref. [9]. By analogy with Ref. [9], let Alice, Bob and Jim make collective measurements on ensembles of N triplets at a time, with Jim measuring either σ_x^J or σ_y^J consistently on his particles. For large enough N, we can define a collective variable $J_x = \sum j_x / N$, if Jim chooses to measure σ_x^J, or alternatively $J_y = \sum j_y / N$, if he chooses to measure σ_y^J, where the j_x and j_y represent Jim's particles in any given ensemble. (As before, we suppress the index i.) We can then define also the collective variables $A_x = \sum a_x / N$, $A_y = \sum a_y / N$, $B_x = \sum b_x / N$ and $B_y = \sum b_y / N$. In some (rare) cases, one or more of these collective variables may even reach ± 1. Above we noted that, for a given triplet of particles, Alice and Bob cannot measure all their observables a_x, a_y, b_x and b_y to infer Jim's choice. But, according to the classical-limit argument, there cannot be such complementary between A_x and A_y, or between B_x and B_y. Alice and Bob must have access to at least *some* information about all these variables. True, their expectation values all vanish, but if Alice, Bob and Jim repeat their measurements exponentially many times, they will find fluctuations as large as ± 1. Since Equation (3) involves products, we cannot directly sum over it to get a relation between A_x or A_y and B_x, B_y, J_x and J_y. Even so, suppose Jim measures σ_x^J and obtains $j_x = -1$ for every particle in his ensemble. Then for each of the other two particles in the triplet, a_x and b_x are correlated and a_y and b_y are anticorrelated. But Alice and Bob will not be able to detect this correlation unless another "miracle" occurs, in addition to the "miracle" that happened in Jim's laboratory. For example, suppose that $A_x = 1$. It follows from Equation (3) that $B_x = 1$ (up to fluctuations due to measurement errors). Then Alice and Bob could compare their results for A_x and B_x to uncover a striking correlation between them and conclude that Jim had measured J_x and not J_y.

But this conclusion can be valid only if the statistics support it. In this scenario, we have assumed rare fluctuations: $J_x = -1$ and $A_x = 1$. Since the two fluctuations are independent, their combined probability is the product of their individual probabilities, namely $2^{-N} \times 2^{-N} = 2^{-2N}$. For this rare scenario, we don't need to assume also that $B_x = 1$; Equation (3) requires it. Thus, with probability 2^{-2N}, Alice and Bob will obtain $A_x = 1 = B_x$. Does this result imply that Jim consistently measured σ_x^J

on his particles? How likely is it that Alice and Bob would have obtained $A_x = 1$ and $B_x = 1$ if Jim had chosen to measure σ_y^J on all his particles, making a_x and b_x uncorrelated? The probability would have been 2^{-2N}, exactly the same. So, once again, Alice and Bob have no way of reading Jim's one-bit message (his choice of what to measure). Likewise, Alice and Bob can try to signal to Jim by, say, measuring $\sigma_x^A = \sigma_x^B$. If they get $A_x = 1 = B_x$, Jim will certainly obtain $J_x = -1$. But the probability that Jim will obtain $J_x = -1$ by chance is 2^{-N}, at least as large as the probability 2^{-2N} that Alice and Bob will obtain $A_x = 1 = B_x$ or even the probability 2^{-N} that Alice and Bob will obtain $\sigma_x^A \sigma_x^B = 1$ for all the N pairs in their ensemble.

The statistics don't work out in the case of GHZ triplets as they do in the case of PR-box pairs. We therefore conclude that despite the similarity between Equations (1) and (3), GHZ correlations do not allow Jim to signal to Alice and Bob by choosing which observable to measure (at least via the above attempts), even if we assume a classical limit in which they can measure the ensemble averages of incompatible observables. The argument of Ref. [9] passes the test we prepared for it.

2. Retrocausality in PR-Box and GHZ Correlations

Instantaneous signalling directly violates relativity theory, opening the door to causal loops and contradictions. In particular, consider the classical limit of a PR-box ensemble, with Alice sending one bit of information $i_A \in \{0, 1\}$ to distant Bob. In an "unprimed" reference frame, Bob receives Alice's message instantaneously (at time $t_B = t_A$); but in an appropriate "primed" reference frame, Alice's bit could be a message into the past, e.g., Bob receives her bit (at time t'_B) before she sends it (at time $t'_A > t'_B$). Applying the principle of relativity, we infer that in the primed reference frame, Bob could send a bit $i_B \in \{0, 1\}$ at time t'_B that Alice would receive instantaneously (at time t'_B) before sending i_A. Then if Alice's device is set to echo whatever message she receives from Bob (so that $i_A = i_B$), and Bob's device is set to yield the inverse of the message he receives from Alice (so that $i_B = 1 - i_A$), together they create a self-contradictory causal loop, as in Figure 1.

From this example it may seem obvious that PR-box correlations and GHZ correlations are distinguished, in that PR-box correlations in the classical limit can be retrocausal, and create self-contradictory causal loops, whereas GHZ correlations cannot be retrocausal. It is therefore of interest to note that this distinction is *not* valid. GHZ correlations can be understood as retrocausal, as well! Yet the predictions implied by Equation (3) do not create causal loops. How can quantum correlations affect distant or past events without creating causal loops?

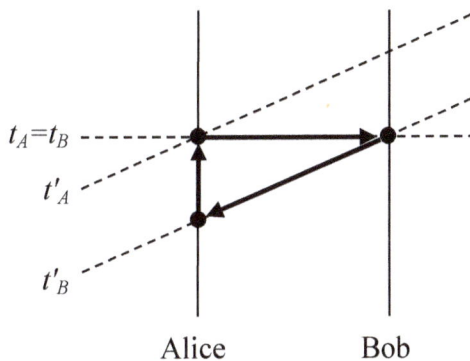

Figure 1. The horizontal dotted line represents an equal-time surface in the unprimed frame, while the tilted dotted lines represent two equal-time surfaces in the primed frame. The arrows, each representing a cause and an effect, form a closed causal loop.

Reference [15] imagines an action called "jamming" in which Jim "the Jammer" can, by pushing a button on a device he holds, decide at any moment whether to turn an ensemble of entangled pairs of particles shared by Alice and Bob into a product state. Although jamming is action at a distance, it is consistent with relativistic causality if two conditions are met. The first condition, the *unary* condition, states that Alice and Bob cannot infer Jim's decision from the results of their *separate* measurements. For example, if—regardless of Jim's decision—Alice measures either a or a', and obtains results ± 1 with equal probability, and likewise Bob measures either b or b', and obtains results ± 1 with equal probability, then the unary condition is fulfilled. The *binary* condition states that if $\hat{a}$ is the spacetime event of Alice's measurements on her ensemble, $\hat{b}$ is the spacetime event of Bob's measurements on *his* ensemble, and $\hat{j}$ is the spacetime of event of Jim pushing the button on his device, then the overlap of the forward light cones of $\hat{a}$ and $\hat{b}$ lies entirely within the forward light cone of $\hat{j}$. (See Figure 2). As shown in Ref. [15], if jamming obeys the unary and binary conditions, then it is consistent with relativistic causality even though $\hat{a}$ and $\hat{b}$ may be *earlier* in time than $\hat{j}$. While jamming is natural in the context of quantum information theory, in Ref. [15] it provides an example of how a nonlocal equation of motion can be consistent with the no-signalling constraint.

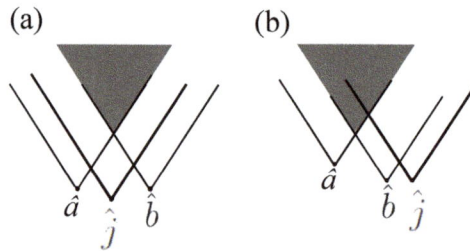

Figure 2. The overlap of the future light cones of $\hat{a}$ and $\hat{b}$ either (**a**) lies or (**b**) does not lie entirely within the future light cone of $\hat{j}$.

We return now to the GHZ correlations of Equation (3) and show that they permit jamming [16]. Suppose Alice, Bob and Jim share an ensemble of particle triplets in the GHZ state. If Jim consistently measures σ_z^J, he disentangles Alice's particles from Bob's, regardless of the outcomes he gets. If he measures σ_x^J, Alice's particles remain entangled with Bob's particles, and their spins are correlated. For example, σ_x^A and σ_x^B are perfectly correlated or perfectly anticorrelated, depending on Jim's outcome. If the information regarding Jim's outcomes is delivered to Alice and Bob, they can bin their σ_x measurements in two ensembles corresponding to Jim's outcomes ± 1. They will find that their results, within each ensemble, are perfectly (anti-)correlated in the case that Jim had chosen to measure σ_x^J, or uncorrelated in case he had measured σ_z^J.

This realization of jamming satisfies the unary condition because, regardless of Jim's decision, Alice's measurements of σ_x^A average to zero, and likewise for Bob's measurements of σ_x^B. It fulfills the binary condition because Jim must report to Alice and Bob the results of his measurements of σ_z^J or σ_x^J for them to determine, from the results of *their* measurements, whether their pairs were entangled or not. Now, Alice and Bob can make their determination *only* in the overlap of the future light cones of $\hat{a}$ and $\hat{b}$, which must lie in the future light cone of $\hat{j}$ for them to receive Jim's input. Thus jamming via GHZ triplets is consistent with relativistic causality. Nevertheless, Jim's decision, whether to leave the pairs shared by Alice and Bob in entangled or product states, can take place even *later* than $\hat{a}$ and $\hat{b}$, and even at a timelike separation from both measurements $\hat{a}$ and $\hat{b}$. (See Figure 3). Even then, it is only in the forward light cone of $\hat{j}$ that Alice and Bob can combine their data and determine whether Jim jammed their measurements.

(a)   (b)

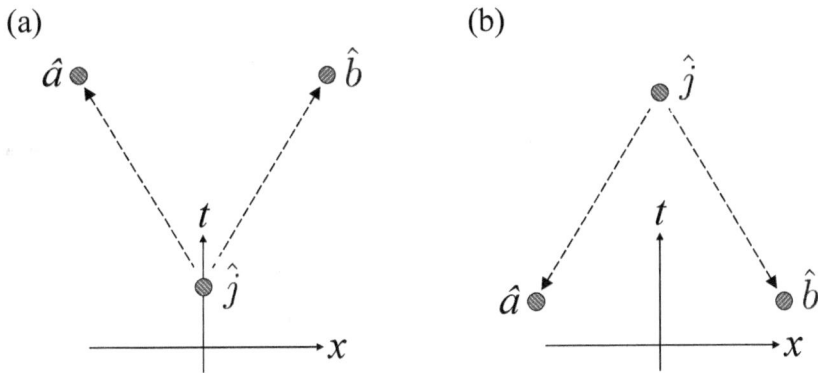

Figure 3. Configurations in which Jim can (**a**) causally and (**b**) retrocausally put pairs of particles shared by Alice and Bob in product or entangled states, as he chooses. The dashed arrows connect cause with effect.

So what makes PR-box correlations different from GHZ correlations, such that the former violate relativistic causality (in the classical limit) while the latter do not? We might have replied, "PR-box correlations are retrocausal whereas GHZ correlations are not". But we have just seen that this distinction fails. So let us return to our comparison, in the first section, of PR-box correlations and bipartite quantum correlations. We noted that even quantum correlations that violate the Bell-CHSH inequality maximally are not strong enough to permit signalling. Are GHZ correlations, which like PR-box correlations can be 0 or 1, strong enough? No! They are indeed stronger, but their strength dissipates over the *two* stages Alice and Bob require in attempting to receive Jim's signal. Relativistic causality in the classical limit is a subtle, but effective, constraint on quantum mechanics.

We introduced this work by stating that three axioms with clear physical meaning, namely nonlocality, relativistic causality, and the existence of a classical limit, might be sufficient for deriving quantum mechanics, or at least an important part of the theory. We can consider reducing these three axioms to two simply by eliminating nonlocality as an axiom. Indeed, axioms in physical theories are, in general, constraints. The constraint of locality could be an axiom, but absence of this constraint need not be an axiom. And it seems from our work that quantum mechanics is just as nonlocal as it can be without violating relativistic causality. The retrocausality we have seen in jamming via GHZ correlations suggests that also retrocausality, like nonlocality, can appear wherever it is not forbidden by relativistic causality.

Author Contributions: Daniel Rohrlich proposed the calculations and the authors contributed equally to the writing and revisions of the paper.

Acknowledgments: We thank all the Referees for helpful comments. This publication was made possible through the support of grants from the John Templeton Foundation (Project ID 43297), from the Israel Science Foundation (grant no. 1190/13). The opinions expressed in this publication are those of the authors and do not necessarily reflect the views of either of these supporting foundations.

Conflicts of Interest: The authors declare no conflict of interest.

References

1. Aharonov, Y. (University of South Carolina, Columbia, SC, USA). Unpublished Lecture Notes.
2. Aharonov, Y.; Rohrlich, D. *Quantum Paradoxes: Quantum Theory for the Perplexed*; Wiley-VCH: Weinheim, Germany, 2005; Chapters 6, 18.
3. Shimony, A. Controllable and uncontrollable nonlocality. In *Foundations of Quantum Mechanics in Light of the New Technology*; Kamefuchi, S., Fujikawa, K., Eds.; Japan Physical Society: Tokyo, Japan, 1983; p. 225.

4. Shimony, A. Events and processes in the quantum world. In *Quantum Concepts of Space and Time*; Penrose, R., Isham, C., Eds.; Clarendon Press: Oxford, UK, 1986; p. 182.
5. Bell, J.S. On the Einstein-Podolsky-Rosen paradox. *Physics* **1964**, *1*, 195–200. [CrossRef]
6. Clauser, J.F.; Horne, M.A.; Shimony, A.; Holt, R.A. Proposed experiment to test local hidden-variable theories. *Phys. Rev. Lett.* **1969**, *23*, 880–884. [CrossRef]
7. Bell, J.S. The theory of local beables. *Epist. Lett.* **1976**, *9*, 11.
8. Popescu, S.; Rohrlich, D. Quantum nonlocality as an axiom. *Found. Phys.* **1994**, *24*, 379–385. [CrossRef]
9. Rohrlich, D. PR-box correlations have no classical limit. In *Quantum Theory: A Two-Time Success Story [Yakir Aharonov Festschrift]*; Struppa, D.C., Tollaksen, J.M., Eds.; Springer: Milan, Italy, 2013; pp. 205–211.
10. Rohrlich, D. Stronger-than-quantum bipartite correlations violate relativistic causality in the classical limit. *arXiv* **2014**, arXiv:1408.3125.
11. Gisin, N. Quantum correlations in Newtonian space and time: Faster than light communication or nonlocality. In *(The Frontiers Collection) Quantum [Un]Speakables II: Half a Century of Bell's Theorem*; Bertlmann, R., Zeilinger, A., Eds.; Springer: Berlin, Germany, 2017; pp. 321–330.
12. Navascués, M.; Wunderlich, H. A glance beyond the quantum model. *Proc. R. Soc. A* **2010**, *466*, 881–890. [CrossRef]
13. Greenberger, D.M.; Horne, M.; Zeilinger, A. Going beyond Bell's theorem. In *Bell's Theorem, Quantum Theory, and Conceptions of the Universe*; Kafatos, M., Ed.; Kluwer Academic Pub.: Dordrecht, The Netherlands, 1989; pp. 69–72.
14. Tsirelson, B.S. Quantum generalizations of Bell's inequality. *Lett. Math. Phys.* **1980**, *4*, 93–100. [CrossRef]
15. Grunhaus, J.; Popescu, S.; Rohrlich, D. Jamming nonlocal quantum correlations. *Phys. Rev. A* **1996**, *53*, 3781–3874. [CrossRef] [PubMed]
16. Rohrlich, D. Three attempts at two axioms for quantum mechanics. In *Probability in Physics*; Ben-Menahem, Y., Hemmo, M., Eds.; Springer: Berlin, Germany, 2012; pp. 187–200.

entropy

MDPI

Article

Capacity and Entropy of a Retro-Causal Channel Observed in a Twin Mach-Zehnder Interferometer During Measurements of Pre- and Post-Selected Quantum Systems

Allen Parks * and Scott Spence

Electromagnetic and Sensor Systems Department, Naval Surface Warfare Center Dahlgren Division, Dahlgren, VA 22448, USA; scott.e.spence@navy.mil
* Correspondence: allen.parks@navy.mil; Tel.: +1-540-653-0582

Received: 20 April 2018; Accepted: 23 May 2018; Published: 27 May 2018

Abstract: Simple intuitive models are presented for the capacity and entropy of retro-causal channels in measured ensembles of quantum systems which can be represented as statistical mixtures of pre-selected only and pre- and post-selected systems. Measurement data from a twin Mach-Zehnder interferometer experiment are used in these models to discuss the capacity and entropy of an apparent retro-causal channel observed in the experimental data. It is noted that low capacity/low entropy retro-causal channels can exist in strong measurement systems.

Keywords: quantum measurement; pre- and post-selected systems; retro-causal channel; channel capacity; channel entropy

---

1. Introduction

The possibility that future events can influence the present has long been argued by physicists and philosophers. In 1964 Aharonov, Bergmann, and Lebowitz proposed a time symmetric theory for non-relativistic quantum mechanics [1] and was further developed by Aharonov et al. in terms of weak measurements and weak values [2,3]. This theory not only employs standard forward in time evolving quantum states, but also the retro-causal property of quantum states evolving backward in time. Although the retro-causal interpretation of weak value theory is controversial, e.g., [4–7], experiments performed in recent years have verified many of the theory's counterintuitive predictions, e.g., [8–12].

Inspired by a weak value gedanken experiment discussed by Tollaksen et al. [13], an optical twin Mach-Zehnder interferometer (MZI) experiment was performed in 2010 which confirmed the predictions made by the gedanken experiment and provided indirect experimental evidence that single particle quantum interference phenomena can be explained in terms of a non-local exchange of modular momentum [14,15]. A recent re-analysis of the reduced 2010 experimental data also suggests that a possible retro-causal channel was observed in the twin MZI apparatus during the associated measurements of ensembles of pre-selected and post-selected (PPS) quantum systems [16,17].

Using the data from the 2010 experiment as a guide, simple models are presented here for the capacity and entropy of this retro-causal channel. The capacity model assumes that the probability distribution for the measurement pointer can be represented as a statistical mixture of the probability distributions for the pointers associated with the PPS systems and the pre-selected only (PSO) systems produced by the measurement. The entropy of the channel is modelled as a classical binary entropy function for a Bernoulli process. Application of the capacity model to the 2010 experimental data shows that even though the capacity of the channel is greatest when the measurement is weak, the channel persists—albeit with increasingly small capacity—as the measurement becomes stronger. It is also shown that—as expected for a binary entropy function—the entropy is smallest when the measurement is extremely weak or extremely strong and reaches its maximum value when the pointer distribution of the mixture is the mean of the PPS and PSO pointer distributions.

An overview of the theory of projector measurement for PPS and PSO systems is given in Section 2 (measurements of a projector were made in the 2010 experiment). The capacity and entropy models for retro-causal channels present in PPS and PSO mixtures (like that assumed for the 2010 experiment) are presented in Section 3. Section 4 reviews the relevant features of the 2010 experiment and discusses—from the perspective of these models—the properties of the retro-causal channel associated with the twin MZI used in the experiment. Concluding remarks comprise the final section of this paper.

2. Theory of Projector Measurement

The measured value of a quantum mechanical observable for a PSO (PPS) system is the statistical result of a standard measurement procedure performed upon an ensemble of identical PSO (PPS) quantum systems. Such measurements can be described using the von Neumann description of a quantum measurement at time t_0 of a time independent observable A that describes a quantum system in an initial fixed pre-selected state $|\psi_i\rangle = \sum_J c_j |a_j\rangle$ at t_0, where the set J indexes the eigenstates $|a_j\rangle$ of the operator $\hat{A}$. In this description, the interaction between the measurement apparatus—i.e., the pointer—and the quantum system is described by the von Neumann interaction operator $\hat{V}$ given by:

$$\hat{V} = e^{-\frac{i}{\hbar}\int \hat{H}dt} = e^{-\frac{i}{\hbar}\gamma\hat{A}\hat{p}},$$

where $\gamma = \int \gamma\delta(t - t_0)dt$ defines the strength of the measurement's impulsive interaction at t_0 and $\hat{p}$ is the momentum operator for the pointer of the measurement apparatus, which is in the initial state $|\phi\rangle$. Let $\hat{q}$ be the pointer's position operator that is conjugate to $\hat{p}$ and assume that $\phi(q) \equiv \langle q|\phi\rangle$ is real valued. When the observable A to be measured is a projector—as was the case in the 2010 experiment—the interaction operator is given exactly by [18]:

$$\hat{V} = \hat{1} - \hat{A} + \hat{A}\hat{T},$$

where $\hat{T} = e^{-\frac{i}{\hbar}\gamma\hat{p}}$ is the spatial translation operator defined by the action $\langle q|\hat{T}|\phi\rangle = \phi(q - \gamma)$.

Prior to the measurement of projector $\hat{A}$, the pre-selected system and the pointer are in the tensor product state $|\psi_i\rangle|\phi\rangle$. Immediately after the measurement, the combined system is in the PSO pointer state given—for arbitrary interaction strength—exactly by:

$$|\Phi\rangle = \hat{V}|\psi_i\rangle|\phi\rangle = (\hat{1} - \hat{A} + \hat{A}\hat{T})|\psi_i\rangle|\phi\rangle \tag{1}$$

and which yields:

$$\langle\Phi|\hat{q}|\Phi\rangle = \langle\phi|\hat{q}|\phi\rangle + \gamma\langle\psi_i|\hat{A}|\psi_i\rangle \tag{2}$$

and:

$$|\langle q|\Phi\rangle|^2 = (1 - \langle\psi_i|\hat{A}|\psi_i\rangle)|\langle q|\phi\rangle|^2 + \langle\psi_i|\hat{A}|\psi_i\rangle|\langle q|\hat{T}|\phi\rangle|^2 \tag{3}$$

as the exact mean PSO pointer position and pointer distribution profile, respectively. Note that since there are no interference cross terms in Equation (3), the profile $|\langle q|\Phi\rangle|^2$ does not exhibit interference.

If the state $|\psi_f\rangle$, $\langle\psi_f|\psi_i\rangle \neq 0$, is post-selected at t_0, the resulting expression for the PPS pointer state is given—regardless of interaction strength—exactly by:

$$|\Psi\rangle = \langle\psi_f|\Phi\rangle = \frac{e^{i\chi}}{N}[(1 - A_w)\hat{1} + A_w\hat{T}]|\phi\rangle, \tag{4}$$

where:

$$A_w \equiv \frac{\langle \psi_f | \hat{A} | \psi_i \rangle}{\langle \psi_f | \psi_i \rangle} \equiv (A^1)_w$$

is the complex valued *weak value* of A (e.g., [2,3]); χ is the Pancharatnam phase defined by:

$$e^{i\chi} = \frac{\langle \psi_f | \psi_i \rangle}{|\langle \psi_f | \psi_i \rangle|};$$

and:

$$N = \sqrt{a + J(\hat{1})}$$

is the normalization factor. Here:

$$a = 1 - 2Re A_w + 2|A_w|^2$$

and for any Hermitean $\hat{x}$:

$$J(\hat{x}) = A_w(1 - A_w^*)\langle \phi | \hat{x}\hat{T} | \phi \rangle + A_w^*(1 - A_w)\langle \phi | \hat{T}^\dagger \hat{x} | \phi \rangle.$$

The associated mean PPS pointer position and pointer distribution profile are given exactly by:

$$\langle \Psi | \hat{q} | \Psi \rangle = \frac{1}{N^2}[a\langle \phi | \hat{q} | \phi \rangle + J(\hat{q}) + \gamma |A_w|^2] \tag{5}$$

and:

$$|\langle q | \Psi \rangle|^2 = \left(\frac{1}{N^2}\right) \left\{ \begin{array}{l} |1 - A_w|^2 |\langle q | \phi \rangle|^2 + |A_w|^2 |\langle q | \hat{T} | \phi \rangle|^2 \\ +2Re[A_w(1 - A_w^*)\langle q | \phi \rangle^* \langle q | \hat{T} | \phi \rangle] \end{array} \right\} \tag{6}$$

Note that since Equation (6) contains interference cross terms, the profile $|\langle q | \Psi \rangle|^2$ exhibits interference. Of course, the PPS states are selected at times $t_i < t_0 < t_f$ and must be evolved forward and backward in time, respectively, to the measurement time t_0. It is important to note here that such PPS systems imply retro-causality, since the weak value of A is measured and $|\psi_f\rangle$ is the post-selected state at measurement time $t_0 < t_f$.

A weak measurement of A occurs when the interaction strength γ is sufficiently small so that the system is essentially undisturbed by the measurement and the pointer's position uncertainty Δq is much larger than the separation between $\hat{A}$'s eigenvalues (if a PPS system undergoes a weak measurement of A, then the resulting value of A is its weak value A_w). In order for measurements to qualify as weak measurements, the associated momentum uncertainty of the pointer must simultaneously satisfy the following two formal weakness conditions which define the extreme upper bound γ_w for the weak measurement regime, e.g., [14,19]:

$$\Delta p \ll \frac{\hbar}{\gamma}|A_w|^{-1}$$

and

$$\Delta p \ll \min_{(n=2,3,\cdots)} \frac{\hbar}{\gamma} \left| \frac{A_w}{(A^n)_w} \right|^{1/(n-1)}.$$

A measurement performed in accordance with these inequalities such that $\gamma \ll \gamma_w$ is a weak measurement whereas a measurement performed with a sufficiently large $\gamma \gg \gamma_w$ is a strong measurement. A measurement which is neither weak nor strong is a transition measurement, i.e., it is a measurement performed in the transition region between a weak measurement and a strong measurement. Although the measurements in the 2010 experiment were made—not only in the weak measurement regime—but also in the transition and strong measurement regions, Equations (1) and (4) still apply over this range of interaction strengths since these expressions are valid regardless of the interaction strength.

3. Capacity and Entropy Models for Retro-Causal Channels Present in PPS and PSO Mixtures

Suppose a projector measurement of fixed interaction strength of an ensemble of quantum systems produces independent measurement pointer distributions for both PPS and PSO systems (i.e., there is no phase relationship between the PPS and PSO states) such that the associated distribution profile $|\langle q|\Lambda\rangle|^2$ for the measurement pointer can be modelled as the statistical mixture

$$|\langle q|\Lambda\rangle|^2 = \alpha|\langle q|\Psi\rangle|^2 + \beta|\langle q|\Phi\rangle|^2. \tag{7}$$

Here $|\langle q|\Psi\rangle|^2$ and $|\langle q|\Phi\rangle|^2$ are the normalized distributions for the PPS and PSO measurement pointers given by Equations (6) and (3), respectively, and $0 \leq \alpha \leq 1$ ($\beta = 1 - \alpha$) is the fraction of PPS (PSO) systems produced by the measurement. Clearly, if $\alpha = 1$ ($\beta = 0$), the measured ensemble is comprised only of PPS systems, whereas if $\beta = 1$ ($\alpha = 0$), the ensemble is comprised entirely of PSO systems.

The model also assumes that: (i) only the measured presence of weak values in PPS systems induces retro-causal channels which permit the backward in time evolution of post-selected states from t_f to t_0; and (ii) since the weak value measurement of PPS systems implies retro-causality, the presence of such PPS systems in a mixture indicates the presence of a retro-causal channel during a measurement process. Based upon these assumptions and Equation (7), the capacity C for the retro-causal channel in a statistical mixture of PPS and PSO systems can be defined as the fraction of the mixture that is comprised of PPS systems, i.e.,

$$C \equiv \alpha.$$

When the fraction α is unknown, then C can be determined from knowledge of the associated mean pointer positions. To see this, observe that Equation (7) can be used to relate the mean pointer position for a mixture to the mean pointer positions for the PPS and PSO constituents of the mixture. In particular:

$$\int q|\langle q|\Lambda\rangle|^2 dq = \alpha \int q|\langle q|\Psi\rangle|^2 dq + \beta \int q|\langle q|\Phi\rangle|^2 dq.$$

After identifying each integral in the last equation with the appropriate mean pointer position and setting $\alpha = C$ and $\beta = 1 - C$, the expression:

$$\langle \Lambda|\hat{q}|\Lambda\rangle = \alpha\langle \Psi|\hat{q}|\Psi\rangle + \beta\langle \Phi|\hat{q}|\Phi\rangle = \langle \Phi|\hat{q}|\Phi\rangle + C(\langle \Psi|\hat{q}|\Psi\rangle - \langle \Phi|\hat{q}|\Phi\rangle)$$

is obtained which can be readily solved for C to yield:

$$C = \frac{\langle \Lambda|\hat{q}|\Lambda\rangle - \langle \Phi|\hat{q}|\Phi\rangle}{\langle \Psi|\hat{q}|\Psi\rangle - \langle \Phi|\hat{q}|\Phi\rangle}, \quad \langle \Psi|\hat{q}|\Psi\rangle \neq \langle \Phi|\hat{q}|\Phi\rangle. \tag{8}$$

Here $\langle \Lambda|\hat{q}|\Lambda\rangle$, $\langle \Phi|\hat{q}|\Phi\rangle$, and $\langle \Psi|\hat{q}|\Psi\rangle$ are the mean pointer positions for the mixture, the PSO systems in the mixture (Equation (2)), and the PPS systems in the mixture (Equation (5)), respectively. It is important to note that Equation (8) cannot be used to determine C when the associated PPS and PSO pointer positions are equal (since C is undefined) or when the pointer positions are such that $C < 0$ (since $0 \leq C \leq 1$). Also, observe that Equation (8) has the following requisite properties: (i) if $C = 1$, then $\langle \Lambda|\hat{q}|\Lambda\rangle = \langle \Psi|\hat{q}|\Psi\rangle$; and (ii) if $C = 0$, then $\langle \Lambda|\hat{q}|\Lambda\rangle = \langle \Phi|\hat{q}|\Phi\rangle$.

In order to associate an entropy with such a retro-causal channel it is further assumed that—for a fixed interaction strength and a fixed C—a measured system can be treated as a random variable in a Bernoulli process such that after the measurement it is either a PPS system with probability C or a PSO system with probability $1 - C$. The classical entropy H of the channel (in Shannons) is then defined here as the binary entropy function for the Bernoulli process given by:

$$H = -C \log_2 C - (1 - C) \log_2 (1 - C). \tag{9}$$

Relying upon the standard interpretation of a binary entropy function, $0 \leq H \leq 1$ can be viewed as a measure of the uncertainty associated with a measurement outcome: when $C = 1$ (0), then it is certain that the measurement produces a PPS (PSO) system and $H = 0$; maximum uncertainty is achieved when $C = \frac{1}{2}$ in which case $H = 1$. It is easy to see from Equations (7) and (8) that—as anticipated—maximum uncertainty is achieved when:

$$|\langle q|\Lambda\rangle|^2 = \frac{1}{2}(|\langle q|\Psi\rangle|^2 + |\langle q|\Phi\rangle|^2)$$

or when:

$$\langle \Lambda|\hat{q}|\Lambda\rangle = \frac{1}{2}(\langle \Psi|\hat{q}|\Psi\rangle + \langle \Phi|\hat{q}|\Phi\rangle)$$

4. The Retro-Causal Channel in the 2010 Twin Mach-Zehnder Interferometer Experiment

Now consider the 2010 twin Mach-Zehnder experiment mentioned above [14,16]. In that experiment, the mean pointer position measurements observed at the output port of the third beamsplitter were essentially derived from the overlapping at the second beamsplitter BS2 of the two beams traversing the arms in the interferometer between the first and second beamsplitters. When $\gamma = 0$, i.e., there is no measurement—the two beams completely overlap at BS2. However, as γ increases the beam overlap at BS2 decreases and the non-overlap region of the beams on BS2 increases.

To apply the channel capacity model to the experimental data note that—from an operational perspective—interference only occurs in the overlap region (at BS2) and that—from a theoretical perspective—only the PPS distribution given by Equation (6) exhibits interference. Pursuant to assumptions (i) and (ii) in the last section, the beam overlap region corresponds to the retro-causal channel in the apparatus. By similar reasoning, since interference does not occur in the non-overlap region and only the PSO distribution given by Equation (3) exhibits no interference, then the non-overlap region does not correspond to a retro-causal channel (operationally, the second MZI in the apparatus effectively responds to photons in the non-overlap region as though they never traversed the first MZI and enter as such the input ports of the second MZI as PSO systems). Consequently, the measured pointer distribution can be modelled as a mixture of independent PSO and PPS pointer distributions. It follows that as the beam overlap decreases, $|\langle q|\Phi\rangle|^2$ becomes increasingly dominant in the mixture and $\langle \Lambda|\hat{q}|\Lambda\rangle$ is increasingly dominated by $\langle \Phi|\hat{q}|\Phi\rangle$.

Here such measurements are viewed from a simplified perspective as a classical Bernoulli process which "sorts" measured systems into PSO and PPS "bins": measured systems which do not "intercept" post-selected states at the time of measurement go into the PSO bin, whereas those that do go into the PPS bin. As γ increases, the number of systems occupying the PPS bin (i.e., the capacity C) decreases.

Although the channel capacity is not directly measured in this experiment, it can be indirectly estimated using Equation (8) and the measurement pointer data presented in Figure 1 in [18] for the case $A_w = 1$. In that figure the vertical axis corresponds to pointer position in μm referenced to $\langle \phi|\hat{q}|\phi\rangle = 0$ and the x values along the horizontal axis correspond to the interaction strengths $\gamma = -1.5x$ μm. The lower curve in the figure labeled "statistical mixture $A_w = 1$" corresponds to the pointer positions $\langle \Lambda|\hat{q}|\Lambda\rangle$, the line labeled "PSO theoretical $\langle A \rangle =$" corresponds to $\langle \Phi|\hat{q}|\Phi\rangle$, and the line labeled "PPS theoretical no 'collapse' $A_w = 1$" corresponds to $\langle \Psi|\hat{q}|\Psi\rangle$. Referring to the figure, if $x = 300$ μm (i.e., $\gamma = -450$ μm and the measurement is a transition measurement), then $\langle \Lambda|\hat{q}|\Lambda\rangle \cong -325$ μm, $\langle \Phi|\hat{q}|\Phi\rangle \cong -225$ μm, and $\langle \Psi|\hat{q}|\Psi\rangle \cong -450$ μm. Substituting these values into Equation (8) yields $C \cong 0.444$ as the empirical estimate for the capacity of the associated retro-causal channel and corresponds to the fraction of PPS systems in the mixture when $\gamma = -450$ μm. The classical entropy of the channel can also be estimated using this value for the capacity in Equation (9) to obtain $H \cong -0.444 \log_2 (0.444) - 0.556 \log_2 (0.556) = 0.991$. Thus, when $\gamma = -450$ μm the uncertainty is nearly maximum as to whether the outcome of a measurement will be a PSO system or a PPS system.

Also, observe from Figure 1 that as x (i.e., γ) increases and the measurement becomes stronger, the pointer position $\langle \Lambda | \hat{q} | \Lambda \rangle$ for the mixture converges towards the pointer position $\langle \Phi | \hat{q} | \Phi \rangle$ for PSO systems, whereas the pointer position $\langle \Psi | \hat{q} | \Psi \rangle$ for PPS systems diverges from $\langle \Phi | \hat{q} | \Phi \rangle$. Application of these trends to Equations (8) and (9) shows the expected behavior that as the interaction strength increases and approaches that of a strong measurement, both the channel capacity and entropy approach zero. Conversely, as x approaches 0, $\langle \Lambda | \hat{q} | \Lambda \rangle$ converges towards $\langle \Psi | \hat{q} | \Psi \rangle$ while both $\langle \Psi | \hat{q} | \Psi \rangle$ and $\langle \Phi | \hat{q} | \Phi \rangle$ approach one another at $x = 0$. Using these trends in Equations (8) and (9) again shows the expected behavior, that as the measurement becomes weak, the capacity approaches unity and the entropy vanishes.

5. Discussion

Although the models presented here are simple, they provide an intuitive description of the capacity and entropy of the apparent retro-causal channel in the 2010 experimental data. This includes the perhaps unexpected possibility that non-vanishing retro-causal channels persist in weak value-measured PPS ensembles even when the measurements are strong (e.g., if the two apparently non-overlapping beams resulting from a strong measurement have Gaussian distributions—as was the case for the 2010 experiment—the associated capacity theoretically approaches zero asymptotically as $\gamma \to \infty$ since the wings of these distributions still overlap).

Regardless of the fact that the model assumes the use of a projector measurement (because the 2010 experiment involved projector measurements and the presence (absence) of interference in the associated pointer theories provide the basis for assigning PPS (PSO) systems to the overlap (non-overlap) region), the capacity model should be applicable for non-projector measurements of ensembles which have pointer probability distribution functions that can be reasonably represented as a statistical mixture of a weak value measured PPS pointer distribution function and a PSO pointer distribution function. The model is not valid for situations which do not possess an associated weak value measured PPS pointer distribution because assumptions (i) and (ii) in Section 3 are violated. As implied in Section 3, the utility of the model is also limited by the fact that Equation (8) can only be used to estimate the capacity when values for the mean pointer distributions for the mixture and the PPS and PSO components are known.

Before closing, it is noted that although the pointer distributions associated with the capacity model are quantum mechanical, the entropy H used here is effectively the classical information theoretic Shannon entropy and was selected for its simple adequate description of the measurement process as a series of Bernoulli trials. Although the Shannon entropy has a natural extension to quantum systems via the von Neumann entropy $S = -Tr\{\hat{\rho} \log \hat{\rho}\}$, where $\hat{\rho}$ is the statistical operator for the system, H was chosen for use here instead of its quantum mechanical counterpart to avoid introducing unnecessary complexity into describing the measurement process, e.g., [20].

Author Contributions: S.S. designed and performed the experiment; A.P. and S.S. analyzed the data; A.P. developed the models and wrote the paper.

Funding: This research was funded by a grant from the Naval Surface Warfare Center Dahlgren Division's In-house Laboratory Independent Research Program. This program also provided funds to cover the associated open access publication costs.

Acknowledgments: The authors thank J. E. Gray for helpful suggestions concerning the preparation of this manuscript.

Conflicts of Interest: The authors declare no conflict of interest.

References

1. Aharonov, Y.; Bergmann, P.; Lebowitz, J. Time symmetry in the quantum process of measurement. *Phys. Rev.* **1964**, *134*, B1410–B1416. [CrossRef]
2. Aharonov, Y.; Albert, D.; Vaidman, L. How the result of a measurement of a component of the spin of a spin-$\frac{1}{2}$ particle can turn out to be 100. *Phys. Rev. Lett.* **1988**, *60*, 1351–1354. [CrossRef] [PubMed]
3. Aharonov, Y.; Vaidman, L. Properties of a quantum system during the time interval between two measurements. *Phys. Rev. A* **1990**, *41*, 11–20. [CrossRef] [PubMed]
4. Aharonov, Y.; Popescu, S.; Tollaksen, J. A time-symmetric formulation of quantum mechanics. *Phys. Today* **2010**, *63*, 27–32. [CrossRef]
5. Nauenberg, M.; Hobson, A.; Mukamel, S.; Griffiths, R.; Aharonov, Y.; Popescu, S.; Tollaksen, J. Time-symmetric quantum mechanics questioned and defended. *Phys. Today* **2011**, *64*, 62–63. [CrossRef]
6. Price, H. Does Time-Symmetry Imply Retrocausality? How the Quantum World Says "Maybe". *Stud. Hist. Philos. Sci. Part B Stud. Hist. Philos. Mod. Phys.* **2012**, *43*, 75–83. [CrossRef]
7. Leifer, M.; Pusey, M. Is a time symmetric interpretation of quantum theory possible without retrocausality? *Proc. R. Soc. A* **2017**, *473*, 20160607. [CrossRef] [PubMed]
8. Ritchie, N.; Story, J.; Hulet, R. Realization of a measurement of a 'weak value'. *Phys. Rev. Lett.* **1991**, *66*, 1107–1110. [CrossRef] [PubMed]
9. Parks, A.; Cullin, D.; Stoudt, D. Observation and measurement of an optical Aharonov-Albert-Vaidman effect. *Proc. R. Soc. A* **1998**, *454*, 2997–3008. [CrossRef]
10. Resch, K.; Lundeen, J.; Steinberg, A. Experimental realization of the quantum box problem. *Phys. Lett. A* **2004**, 125–131. [CrossRef]
11. Wang, Q.; Sun, F.; Zhang, Y.; Li, J.; Huang, Y.; Guo, G. Experimental demonstration of a method to realize weak measurement of the arrival time of a single photon. *Phys. Rev. A* **2006**, *73*, 02384. [CrossRef]
12. Hosten, O.; Kwiat, P. Observation of the spin Hall effect in light via weak measurements. *Science* **2008**, 787–790. [CrossRef] [PubMed]
13. Tollaksen, J.; Aharonov, Y.; Casher, A.; Kaufherr, T.; Nussinov, S. Quantum interference experiments, modular variables and weak measurements. *New J. Phys.* **2010**, *12*, 01302. [CrossRef]
14. Spence, S.; Parks, A. Experimental Evidence for a Dynamical Non-Locality Induced Effect in Quantum Interference Using Weak Values. *Found. Phys.* **2012**, *42*, 803–815. [CrossRef]
15. Spence, S.; Parks, A.; Niemi, D. Methods used to observe a dynamical nonlocality effect in a twin Mach-Zehnder interferometer. *Appl. Opt.* **2012**, *51*, 7853–7857. [CrossRef] [PubMed]
16. Spence, S.; Parks, A. Experimental evidence for retro-causation in quantum mechanics using weak values. *Quantum Stud. Math. Found.* **2017**, *4*, 1–6. [CrossRef]
17. Parks, A.; Spence, S. An observed manifestation of a parity symmetry in the backward in time evolved post-selected state of a twin Mach-Zehnder interferometer. *Quantum Stud. Math. Found.* **2017**, *4*, 315–322. [CrossRef]
18. Parks, A.; Spence, S.; Gray, J. Exact pointer theories for von Neumann projector measurements of pre- and postselected and preselected-only quantum systems: Statistical mixtures and weak value persistence. *Proc. R. Soc. A* **2014**, *470*, 20130651. [CrossRef]
19. Gray, J.; Parks, A. A note on the "Aharonov-Vaidman operator action representation theorem". *Quantum Stud. Math. Found.* **2018**, *5*, 213–217. [CrossRef]
20. Peres, A. *Quantum Theory: Concepts and Methods*; Kluwer Academic Publishers: Dordrecht, The Netherlands, 1995; pp. 280–281. ISBN 0-7923-3632-1.

entropy

MDPI

Article
Non-Classical Correlations in n-Cycle Setting

Kishor Bharti [1], Maharshi Ray [1] and Leong-Chuan Kwek [1,2,3],*

[1]   Centre for Quantum Technologies, National University of Singapore, 3 Science Drive 2, Singapore 117543, Singapore; e0016779@u.nus.edu (K.B.); maharshi91@gmail.com (M.R.)
[2]   MajuLab, CNRS-UNS-NUS-NTU International Joint Research Unit, Singapore UMI 3654, Singapore
[3]   National Institute of Education, Nanyang Technological University, Singapore 637616, Singapore
*   Correspondence: cqtklc@gmail.com

Received: 14 December 2018; Accepted: 30 January 2019; Published: 1 February 2019

Abstract: Quantum communication and quantum computation form the two crucial facets of quantum information theory. While entanglement and its manifestation as Bell non-locality have been proved to be vital for communication tasks, contextuality (a generalisation of Bell non-locality) has shown to be the crucial resource behind various models of quantum computation. The practical and fundamental aspects of these non-classical resources are still poorly understood despite decades of research. We explore non-classical correlations exhibited by some of these quantum as well as super-quantum resources in the n-cycle setting. In particular, we focus on correlations manifested by Kochen–Specker–Klyachko box (KS box), scenarios involving n-cycle non-contextuality inequalities and Popescu–Rohlrich boxes (PR box). We provide the criteria for optimal classical simulation of a KS box of arbitrary n dimension. The non-contextuality inequalities are analysed for n-cycle setting, and the condition for the quantum violation for odd as well as even n-cycle is discussed. We offer a simple extension of even cycle non-contextuality inequalities to the phase space case. Furthermore, we simulate a generalised PR box using KS box and provide some interesting insights. Towards the end, we discuss a few possible interesting open problems for future research. Our work connects generalised PR boxes, arbitrary dimensional KS boxes, and n-cycle non-contextuality inequalities and thus provides the pathway for the study of these contextual and nonlocal resources at their junction.

Keywords: KS Box; PR Box; Non-contextuality inequality

1. Introduction

The quantum mechanical description of nature is incompatible with any local hidden variable theory and consequently is said to exhibit Bell non-locality [1]. This counter-intuitive phenomenon finds applications in various quantum information processing tasks such as randomness certification [2], self-testing [3–6] and distributed-computing [7]. The Bell non-locality can be thought of as a particular case of another under-appreciated phenomenon, referred to as contextuality [8–10]. Recently, contextuality has been shown to be useful for quantum cryptography [11,12], self-testing [13] and various models of quantum computing [14,15]. These non-classical correlations are not only present in quantum theory, but post-quantum theories as well [9,16]. It is still not clear if the quantum theory is the only physical theory despite decades of research which makes it pertinent to understand these resources for not only quantum theory but also post-quantum theories, for fundamental as well as practical manifestations [9,17]. In this light, we study some of the relatively less explored nonlocal and contextual resources and discuss possible inter-connections among themselves.

Our focus revolves around the correlations manifested by three different objects from quantum and post-quantum theories with underlying structure governed by the n-cycle graph. In particular, we explore the correlations manifested by Kochen–Specker–Klyachko box (KS box), Popescu–Rohlrich boxes (PR box) and scenarios involving n-cycle non-contextuality inequalities. The KS box was first introduced by Bub et al. in 2009 [18] who analysed it for a five-dimensional case. The box has a tunable parameter (denoted by p), which determines the nature of box namely classical, quantum and post-quantum. Bub et al. showed that it is impossible to simulate the KS box statistics for $p = \frac{1}{3}$ using any classical strategy. For the aforementioned value of p, the best classical strategy has a success probability of approximately 0.94667 [18] . The authors showed that the KS box is sturdy enough to simulate the famous PR box—the most nonlocal no-signalling box for the simplest Bell non-locality scenario [16]. It is important to note that PR box has played a crucial role in the understanding of concepts from communication complexity and provide the primary test bed to check against the physical principles to single out quantum theory, which further demands a careful study of these no-signalling nonlocal boxes [17,19,20]. For $p = \frac{1}{2}$, KS box efficiently simulates the PR box [18].

Now, we turn to the last object of our study. The Bell nonlocal nature of theories can be witnessed via the violation of certain inequalities, referred to as Bell inequalities and non-contextuality inequalities in the general case of contextuality [9]. In their seminal paper, Cabello, Severini and Winter showed that certain graph-theoretic numbers give the bounds on these inequalities for classical, quantum and more general theories, namely independence number, Lovász theta number and fractional packing number, respectively [8]. Using the tools from the aforementioned work, Araújo et al. [21] provided the construction for the maximal violation of the odd cycle generalisation of the well-known Klyachko–Can–Binicioğlu–Shumovsky (KCBS) inequality for qutrits [22,23] and even cycle generalisation of Clauser–Horne–Shimony–Holt (CHSH) inequality for two-qubits [24]. Note that the even cycle generalisation of CHSH inequality is similar to Braunstein–Caves inequalities [25], which have been heavily investigated in the literature. For the four cycle case, a simple extension of these even cycle non-contextuality inequalities to the phase space case was provided by Arora et al. [26].

It is important to observe the connections among KS box, non-contextuality inequalities and PR boxes.

1. The KS boxes were motivated by KCBS non-contextuality inequality [18].
2. The KCBS non-contextuality inequality belongs to the same family of inequalities as the CHSH inequality [9].
3. The maximum value of CHSH inequality for no-signalling theories is provided by PR boxes [16].
4. The PR box can be simulated by KS box for $p = \frac{1}{2}$ [18].

In this paper, we further explore the interconnections among the generalised versions of the aforementioned objects, namely generalised PR boxes, arbitrary dimensional KS boxes and n-cycle non-contextuality inequalities. Our work provides the pathway for the study of these generalised contextual and nonlocal resources at their junction.

Paper Structure

In Section 2, we start with studying KS box for the n-dimensional case and provide the optimal classical strategy as well as corresponding success probability for simulating the box using classical resources. Our results provide the minimum gap between the optimal classical strategy and the KS box based strategies for arbitrary p and n. We observe that the optimal success probability for classical simulation decreases monotonically with the dimension of the KS box.

In Section 3, we study the n-cycle contextuality scenario and the corresponding non-contextuality inequalities. We explore the odd cycle generalisation of the well-known Klyachko–Can–Binicioğlu–Shumovsky (KCBS) inequality [22,23] and even cycle generalisation of Clauser–Horne–Shimony–Holt (CHSH) inequality [24]. Following the construction provided by Araújo et al. [21], we discuss the necessary and sufficient condition for the violation of the generalised KCBS inequality in Section 3.4 and necessary condition for the violation of even-cycle generalisation of CHSH inequality in Section 3.4. Furthermore, we provide a simple phase space extension of even cycle generalisation of CHSH inequality by harnessing the techniques provided by Arora et al. [26].

Within no-signalling theories, the maximum violation of CHSH inequality is obtained by Popescu–Rohlrich box, also known as PR box [16]. The PR box and its analogue for even-cycle generalisation of CHSH inequality are the contents of Section 4. In their seminal work [18], Bub et al. studied the simulation of a PR box using KS box. We extend the idea to arbitrary dimensional KS box and PR box. We study the joint probability distribution for the KS box and find the criteria for the violation of even-cycle generalisation of CHSH inequality. Given the even cycle generalisation of CHSH inequality, we provide the bound on p (tunable parameter) for the KS box required to saturate classical, quantum and no-signalling bounds.

Finally, we conclude in Section 5. We discuss the implications of our study and some interesting open problems for future work.

2. Simulating KS Box

A Kochen–Specker–Klyachko box or KS box is a bipartite no-signalling box with two inputs and two outputs (depicted in Figure 1). No-signalling means that the inputs of one sub-part of the box are independent of the output of the complementary part. The outputs are always binary; however, the inputs depend on the dimensionality of the box.

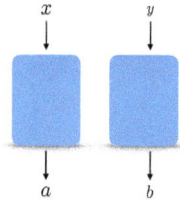

Figure 1. KS box is a bipartite no-signalling box. The value of a does not depend on y and similarly b does not depend on x. The box exhibits nonlocal correlations.

Formally speaking, no-signalling enforces the following constraints:

$$\sum_b P(a,b|x,y) = \sum_b P(a,b|x,y'),\tag{1}$$

$$\sum_a P(a,b|x,y) = \sum_a P(a,b|x',y),\tag{2}$$

where $P(a,b|x,y)$ denotes the probability of getting a and b, when x,y are the input. One can define the box formally as following.

Definition 1. *An N-dimensional Kochen–Specker–Klyachko box or KS box, defined in [18] is a no-signalling resource with two inputs, $x,y \in \{1,2,\cdots,N\}$ and two outputs $a,b \in \{0,1\}$, which satisfies the following constraints:*

1. *$a = b$ if $x = y$, and*
2. *$a.b = 0$ if $x \neq y$.*

A KS box with marginal probability p for the output "1" is referred to as KS_p box. For example, the fraction of "1"s in a $KS_{\frac{1}{5}}$ box is $\frac{1}{5}$. We refer to the KS box condition corresponding to $a.b = 0$ for unequal inputs as $\perp$. Given two parties, e.g., Alice and Bob, who are space-like separated, it is not possible to simulate the KS box statistics with full accuracy for arbitrary p using classical resources only (for example, some shared randomness) [18] . We want to find the probability of successful simulation of KS box statistics for various strategies. This is an important question because any classical strategy will only produce the best Bell-local statistics and thus the amount by which it fails to simulate a Bell-nonlocal resource such as KS can can be used to quantify the Bell non-locality of KS box.

To capture the essence of classical strategies, we use the language from graph theory. Consider an N-gon with a 0/1 assignment to its vertices. A 0/1 assignment with M "1s" for a given N-gon corresponding to an N-dimensional KS box is referred to as a chart of degree M, in short C_M. For example, chart C_1 for a five-dimensional KS box will assign "1" to one of the vertices and "0" to the rest. Please refer to Figure 2, for a pictorial understanding. To simulate the statistics corresponding to KS box, the spatially separated parties (Alice and Bob) will use their pre-shared strategy. The possible strategies can be captured using the charts discussed before. No-communication is allowed between Alice and Bob once the simulation starts. The only classical resource they share is the access to such charts and some shared randomness to decide which chart to use. The shared randomness determines the fraction of times a particular chart can be used in a strategy. For example, suppose they agree to simulate $P = \frac{1}{3}$ using charts C_1 and C_2. Then, they must use chart C_1 with probability $\frac{1}{3}$ and chart C_2 with probability $\frac{2}{3}$. This can be achieved by using a biased coin which gives head with probability $\frac{2}{3}$ and tail with probability $\frac{1}{3}$. Using chart C_0 and C_1 will always satisfy the $\perp$ condition. All other charts will violate the $\perp$ condition up to varying proportion.

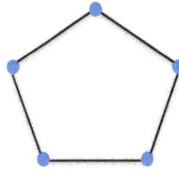

Figure 2. Chart C_2 for a five-dimensional KS box corresponds to two "1s" and three "0s". The red entries correspond to inputs and the outputs are in green. The above chart fails to simulate the KS box statistics when the inputs are 2 and 5.

Simulating a KS_p box essentially requires the satisfaction of the $\perp$ conditions along with the marginal condition. The use of charts already guarantees equal outputs for same inputs.

Proposition 1. *Given the chart C_M, the probability of successful simulation of the $\perp$ condition is given by*

$$P_\perp(C_M) = \frac{N^2 - M^2 + M}{N^2}. \tag{3}$$

Proof. For an N-dimensional KS box, the total number of possible input pairs for Alice and Bob are N^2. If they use the chart C_M to simulate the KS box, then the probability of failure corresponds to the probability of choosing different inputs with output 1. The number of such edges (with ordering) whose vertices correspond to output 1 is $M(M-1)$. Thus, the probability of successful simulation is

$$1 - \frac{M(M-1)}{N^2} = \frac{N^2 - M^2 + M}{N^2}. \tag{4}$$

This completes the proof. □

For $p \le \frac{1}{N}$, Alice and Bob can use chart C_0 and C_1 to simulate the KS box. However, we observe that, to satisfy the marginal constraints for $p > \frac{1}{N}$, one needs to use charts of higher degree, which in turn violates the $\perp$ conditions. Therefore, perfect classical simulation of the KS_p box only exists for $p \leq \frac{1}{N}$. We now fix a $p \leq 0.5$ and compute the optimal classical simulation probability of the KS_p box. Now, we present our result concerning the optimal probability of successful simulation for an N-dimensional KS_p box for arbitrary p.

Theorem 1. *For a given $p \leq 0.5$, the charts C_{M-1} and C_M (only chart C_M in case Np is an integer) are optimal for simulating N-dimensional KS_p box, where $M = \lceil Np \rceil$ (ceiling integral value) and the optimal probability of simulation is given by*

$$P_{optimal}(M, N, p) = 1 - \frac{(2Np - M)(M - 1)}{N^2}.$$

Proof. Assume that Alice and Bob play the charts C_i with probability p_i, for $i \in \mathbb{Z}^{\geq}$, i.e., the set of non-negative integers. For a given probability distribution $\{p_i\}$ over charts, the probability of successful simulation of KS_p box is given by $\sum_i p_i P_\perp(C_i)$. Hence, the optimal simulation probability is given by the following linear program:

$$\max_{\{p_i\}} \sum_i p_i P_\perp(C_i) \qquad \text{(success probability)}$$

$$\text{s.t} \sum_i p_i i = Np \qquad \text{(mean condition)}$$

$$\sum_i p_i = 1, p_i \geq 0 \, \forall i \qquad \text{(valid probability)}$$

Now, observe that the objective function is

$$\sum_i p_i P_\perp(C_i) = \frac{1}{N^2} \sum_i p_i \left(N^2 - i^2 + i\right)$$

$$= 1 + \frac{p}{N} - \frac{1}{N^2} \sum_i p_i i^2,$$

where in the second equality we used the mean condition along with the valid probability condition. Hence, maximising the objective function corresponds to minimising the variance term with respect to the probability distribution $\{p_i\}$. The optimisation problem of minimising the variance of a random variable defined on a set of non-negative integral points, over all possible probability distributions, for a fixed given mean, has support size at most two. A simple proof for this is given in Appendix A Proposition A1. Specifically, if the mean (Np) is an integer (e.g., $= M$), the least variance solution will be $p_M = 1$ and $p_i = 0, \forall i \neq M$. For the case when the mean is not an integer, the least variance solution corresponds to a support containing $M - 1$ and M, with $M = \lceil Np \rceil$, which follows from simple convexity arguments. With this support, we can compute p_{M-1} and p_M using the mean condition, which evaluates to $p_{M-1} = M - Np$ and $p_M = Np - M + 1$. Plugging this into the success probability function gives us the optimal simulation probability of the KS_p box

$$P_{optimal}(M, N, p) = \frac{1}{N^2}\left(2Np - 2NpM + N^2 + M^2 - M\right) = 1 - \frac{(2Np - M)(M - 1)}{N^2}$$

This completes the proof. $\square$

Let us have a look at the simulation efficiency in a bit detail. Numerical evidence (refer to Figure 3) suggests that the simulation efficiency decreases with the dimension of the KS box. Please refer to Table 1 for the specific case of $p = 0.4$.

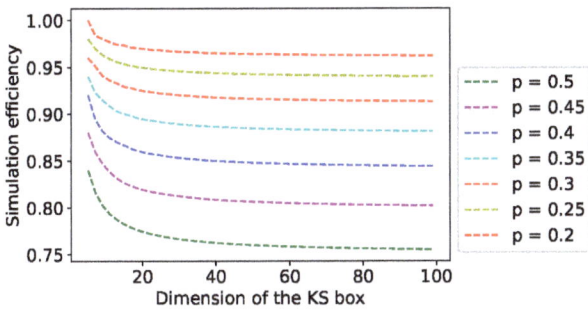

Figure 3. The simulation efficiency has been plotted here as a function of the dimension of the KS box for various marginal probabilities, p. It can be seen that the simulation efficiency decreases with dimension for a particular p.

For a particular value of p, the nonlocal nature of KS box increases with dimension of the box and hence the simulation efficiency for the optimal classical strategy decreases. Moreover, for a KS box with fix dimension, its nonlocal nature increases with increase in p.

Table 1. The simulation efficiency decreases with the dimension of the KS box.

Dimension	Marginal Probability	Simulation Efficiency
5	0.4	0.92
7	0.4	0.893878
9	0.4	0.881481
11	0.4	0.87438
13	0.4	0.869822
15	0.4	0.866667
17	0.4	0.862976

Having studied the KS box, we move next to the n-cycle non-contextuality inequalities.

3. Analysing n-Cycle Non-Contextuality Inequalities

Before we analyse the n-cycle generalisation of KCBS and CHSH inequalities, we would like to discuss the prior art briefly required to understand our work.

3.1. KCBS Inequality

The observables in quantum mechanics are represented as Hermitian matrices. Unlike real or complex numbers, matrices do not commute in general. More importantly, it is possible to have three observables A, B and C such that $[A, B] = 0$, $[A, C] = 0$ but $[B, C] \neq 0$. The maximal set of commuting observables defines a context. In the previous example, the observable A lies in two contexts defined by the sets $\{A, B\}$ and $\{A, C\}$. Since the observables in a context commute among themselves, they can be measured simultaneously. Given a theory, if the value of an observable in the experiment depends on the context in which it has been measured, the theory is called contextual, otherwise non-contextual. Quantum mechanics is a contextual theory [10]. The experimental tests which can be used to probe the contextual nature of a theory are referred to as contextuality tests. These tests can be often written in terms of an algebraic inequality whose violation witnesses contextuality of the underlying theory. Klyachko–Can–Binicioğlu–Shumovsky (KCBS) inequality is one of the extensively studied state-dependent non-contextuality inequality [8,22,23]. The violation of KCBS inequality by any probabilistic theory rules out its possible completion by a non-contextual hidden variable model. To understand this and the KCBS inequality, let us have a look at the following algebraic quantity (related to KCBS inequality):

$$K = A_1 A_2 + A_2 A_3 + A_3 A_4 + A_4 A_5 + A_5 A_1, \tag{5}$$

where all the A_i can be either $+1$ or -1. Now, in any theory, where the values of all the A_i are predetermined (such is the case in a non-contextual hidden variable theory i.e., the probability theories which can have a non-contextual completion), the average value of Equation (5) is lower bounded by -3. Formally, the KCBS non-contextuality inequality is given by

$$\langle K \rangle \geq -3, \tag{6}$$

where $\langle K \rangle$ refers to expectation value of K. Now, it is possible to have a theory which violates the bound in Equation (6). For example, quantum theory achieves up-to $5 - 4\sqrt{5}$, which is approximately -3.94427 and hence less than -3. This proves that quantum mechanics is a contextual theory [22]. The measurement setting and the state corresponding to optimal violation of KCBS inequality withIn quantum theory is given by

$$A_i = 2|v_i\rangle\langle v_i| - \mathbb{I}, \tag{7}$$

$$|v_i\rangle = \left(\sin(\theta) \cos\left(\frac{4\pi i}{5}\right), \sin(\theta) \sin\left(\frac{4\pi i}{5}\right), \cos(\theta) \right)^T, \tag{8}$$

$$|\psi\rangle = (0, 0, 1)^T, \tag{9}$$

where $\mathbb{I}$ refers to identity, $i \in 1, 2, 3, 4, 5$ and $\cos^2(\theta) = \frac{\cos\left(\frac{\pi}{5}\right)}{1 + \cos\left(\frac{\pi}{5}\right)}$. By doing a basis transformation, one can view the KCBS inequality as a state-dependent non-contextuality inequality with five dichotomic measurements with $0/1$ outcome. Explicitly,

$$P_i = \frac{1 + A_i}{2},$$

transforms A_i with ± 1 outcome space to P_i with $0/1$ outcome space such that

- P_i and $P_i + 1$ are compatible and
- P_i and P_{i+1} are exclusive.

Here, addition is taken modulo 5 and exclusivity means that P_i and P_{i+1} cannot have outcome 1. This exclusivity corresponding to projective measurements and their outcomes can be captured using a graph, known as "exclusivity graph". The exclusivity graph approach to contextuality has been studied extensively in the literature and it is important to review the basics of this framework [8]. The nodes of an exclusivity graph correspond to event where an event is constituted by the combination of measurement and corresponding outcome. For example, $(a|i)$ is an event which corresponds to getting outcome "a" for measurement "i". Let us represent the probability of getting outcome "1" given the input was "i" as $P(1|i)$. The events follow exclusivity relation according to the exclusivity graph (a pentagon in the case of KCBS). The exclusivity relation induces following constraint:

$$P(1|i) + P(1|j) \leq 1, \tag{10}$$

$\forall i, j \in E$, where E corresponds to the edge set of the exclusivity graph. The KCBS inequality corresponds to sum of probabilities assigned to five events of the kind $(1|i)$ with exclusivity relation following a pentagon (refer to Figure 4).

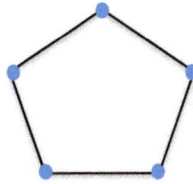

Figure 4. The exclusivity graph corresponding to the KCBS inequality is a pentagon. The inequality involves five events of type $(1|i)$ where $i \in \{1, 2, 3, 4, 5\}$. The bound on the inequality for non-contextual hidden variable theories is 2. Quantum theory achieves up to $\sqrt{5}$ and thus manifests the contextual nature of quantum theory.

Given a non-contextuality inequality, the upper bound for non-contextual hidden variable (NCHV) theories is given by independence number of the underlying exclusivity graph, denoted by $\alpha(G)$ [8]. The upper bound for quantum theories is given by Lovász theta number, represented as $\vartheta(G)$ [8]. Formally, in graph theoretic language, the KCBS inequality is given by

$$\sum_{i=1}^{5} P(1|i) \leq \alpha(C_5), \tag{11}$$

where C_5 represents pentagon and $\alpha(C_5)$ is equal to 2. The quantum bound corresponds to $\vartheta(C_5)$ and is equal to $\sqrt{5}$. Since $\vartheta(C_5) > \alpha(C_5)$, it witnesses the contextual nature of quantum theory [8,22].

3.2. CHSH Inequality

CHSH inequality is a special case of non-contextuality inequality where the context is provided by space-like separation of parties involved, e.g., Alice and Bob. The scenario corresponding to inequality corresponds to four measurements, two for each party. Each of Alice's measurements are compatible with Bob's measurements and vice versa. Suppose Alice's measurements are given by A_1, A_2 and Bob's measurements are given by B_1, B_2. The outcomes corresponding to the measurements are either $+1$ or -1. The CHSH inequality is given by

$$\langle C_4 \rangle = \langle A_1 B_1 \rangle + \langle A_1 B_2 \rangle + \langle A_2 B_1 \rangle - \langle A_2 B_2 \rangle \leq 2. \tag{12}$$

The local hidden variable theories respect the bound in (12), however quantum theory achieves up-to $2\sqrt{2}$ with appropriate measurement settings and state [24]. These optimal measurement settings and state corresponding to maximal quantum violation are given by,

$$A_1 = Z \otimes \mathbb{I}, \quad A_2 = X \otimes \mathbb{I}, \tag{13}$$

$$B_1 = \mathbb{I} \otimes \frac{-Z - X}{\sqrt{2}}, \quad B_2 = \mathbb{I} \otimes \frac{Z - X}{\sqrt{2}}, \tag{14}$$

$$|\psi\rangle = \frac{|01\rangle - |10\rangle}{\sqrt{2}}, \tag{15}$$

where X and Z are Pauli matrices, $\mathbb{I}$ is identity and $|\psi\rangle$ is a Bell state.

3.3. Analysing the Generalised KCBS Inequality

The inequality in Equation (11) has been further extended to general odd cycle, which is

$$\sum_{i=1}^{n} P(1|i) \leq \frac{n-1}{2}. \tag{16}$$

The odd cycle generalisation of KCBS inequality has been studied extensively in literature [13,21,23,27]. Surprisingly, $\frac{n-1}{2}$ corresponds to independence number of the graph for odd cycle case [8,23,28]. The maximum quantum violation for generalised KCBS inequality corresponds to Lovász theta number (denoted by $\vartheta(G)$), which is $\frac{n\cos\left(\frac{\pi}{n}\right)}{1+\cos\left(\frac{\pi}{n}\right)}$.

We represent the density matrices in the standard basis $\{|i\rangle\}$ with matrix elements given by $\rho_{ij} = \langle i|\rho|j\rangle$. For the odd n-cycle generalisation of KCBS inequality, the projectors corresponding to the optimal quantum violation are given by

$$\Pi_j = |\psi_j\rangle\langle\psi_j|$$

where

$$|\psi_j\rangle = \left(\sin(\theta)\cos\left(\frac{j\pi(n-1)}{n}\right), \sin(\theta)\sin\left(\frac{j\pi(n-1)}{n}\right), \cos(\theta)\right)^T$$

and $\cos^2(\theta) = \frac{\cos\left(\frac{\pi}{n}\right)}{1+\cos\left(\frac{\pi}{n}\right)}$. Now, we present the condition under which a qutrit will violate the generalised KCBS inequality for the above measurement settings.

Proposition 2. *A qutrit violates the odd n-cycle generalisation of KCBS non-contextuality inequality if and only if $\rho_{33} \geq \left(\frac{\cos\left(\frac{\pi}{n}\right)(n-1)-1}{n\left(2\cos\left(\frac{\pi}{n}\right)-1\right)}\right)$.*

Proof. The generalised KCBS operator for the odd n-cycle scenario can be defined as

$$K_n = \sum_{j=1}^{n} \Pi_j.$$

Adding all the projectors (Π_js), we get

$$K_n = \sum_{i=1}^{3} k_i |\phi_i\rangle\langle\phi_i|$$

where

$$|\phi_1\rangle = \begin{pmatrix} 1 \\ 0 \\ 0 \end{pmatrix}, |\phi_2\rangle = \begin{pmatrix} 0 \\ 1 \\ 0 \end{pmatrix}, |\phi_3\rangle = \begin{pmatrix} 0 \\ 0 \\ 1 \end{pmatrix}$$

and

$$k_1 = \frac{1}{1+\cos\left(\frac{\pi}{n}\right)} \sum_{j=1}^{n} \cos^2\left(\frac{j\pi(n-1)}{n}\right)$$

$$k_2 = \frac{1}{1+\cos\left(\frac{\pi}{n}\right)} \sum_{j=1}^{n} \sin^2\left(\frac{j\pi(n-1)}{n}\right)$$

$$k_3 = n\cos^2(\theta) = \frac{n\cos\left(\frac{\pi}{n}\right)}{1+\cos\left(\frac{\pi}{n}\right)}.$$

Since $\sum_j \cos^2\left(\frac{j\pi(n-1)}{n}\right) = \sum_j \sin^2\left(\frac{j\pi(n-1)}{n}\right) = \frac{n}{2}$, we get

$$k_1 = k_2 = \frac{n}{2\left(1 + \cos\left(\frac{\pi}{n}\right)\right)}$$

$$k_3 = \frac{n\cos\left(\frac{\pi}{n}\right)}{1 + \cos\left(\frac{\pi}{n}\right)}$$

The odd n-cycle non-contextuality inequality is written as

$$\langle K_n \rangle \leq \frac{n-1}{2},$$

where $\langle K_n \rangle$ corresponds to the expectation value of the generalised KCBS operator with respect to the underlying preparation. In terms of quantum expectation, the inequality is given by

$$\text{Tr}\left(K_n \rho\right) \leq \frac{n-1}{2}.$$

Note that the generalised KCBS operator is diagonal in standard basis and leads to the following simplification:

$$\frac{n}{2\left(1 + \cos\left(\frac{\pi}{n}\right)\right)}[\rho_{11} + \rho_{22}] + \frac{n\cos\left(\frac{\pi}{n}\right)}{1 + \cos\left(\frac{\pi}{n}\right)}[\rho_{33}] \leq \frac{n-1}{2}.$$

Since the trace of a density matrix is always 1, the condition for the violation of odd n-cycle non-contextuality inequality becomes;

$$\rho_{33} > \frac{(n-1)\left(1 + \cos\left(\frac{\pi}{n}\right)\right) - n}{n\left(2\cos\left(\frac{\pi}{n}\right) - 1\right)}.$$

Simplifying the above expression, we get

$$\rho_{33} > \frac{\cos\left(\frac{\pi}{n}\right)(n-1) - 1}{n\left(2\cos\left(\frac{\pi}{n}\right) - 1\right)}. \tag{17}$$

This completes the proof. □

Remark 1. *We can see that the set of quantum states for qutrits, which can violate odd n-cycle non-contextuality inequality, shrinks as we increase n (See Figure 5). In the infinite n scenario, the only qutrit which violates the inequality is the pure state* $|\psi\rangle = (0,0,1)^T$!

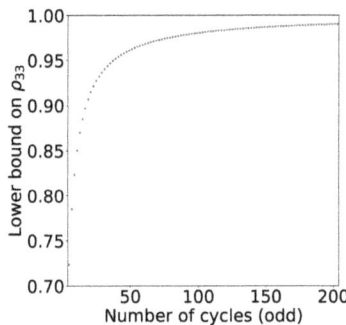

Figure 5. The condition for the quantum violation of the odd n-cycle generalisation of KCBS inequality is computed. Lower bound on ρ_{33} for odd n-cycle graph has been plotted as a function of n. The set of states which can violate the KCBS inequality corresponding to optimal measurement setting shrinks as we increase n.

3.4. Analysing Chained Bell Inequalities

The n-cycle generalisation of CHSH inequality is referred to as chained Bell inequality [21,25]. The even n-cycle scenario has n measurements i.e., $\{X_1, X_2, \cdots, X_n\}$. All of these are dichotomic measurements with possible outcomes ±1. The chained Bell inequality of cycle n is given by

$$\sum_{j=1}^{n-1} \langle X_j X_{j+1} \rangle - \langle X_n X_1 \rangle \leq n - 2. \tag{18}$$

The optimal construction [21] for violation of this inequality corresponds to $X_j = \widetilde{X_j} \otimes \mathbb{I}$ for even j and $X_j = \mathbb{I} \otimes \widetilde{X_j}$ for odd j, where

$$\widetilde{X_j} = \cos\left(\frac{j\pi}{n}\right)\sigma_x + \sin\left(\frac{j\pi}{n}\right)\sigma_z. \tag{19}$$

We now provide the necessary condition for the quantum violation of a chained Bell inequality corresponding to optimal quantum measurement settings.

Proposition 3. *For a given two qubit state, the necessary condition for the quantum violation of chained Bell inequality of cycle n is given by the difference of its extremal eigenvalues i.e.,*

$$\lambda_1 - \lambda_4 > \frac{n-2}{n}. \tag{20}$$

Proof. For even j,

$$X_j X_{j+1} = \widetilde{X_j} \otimes \widetilde{X_{j+1}}$$

$$= \left[\cos\left(\frac{j\pi}{n}\right)\sigma_x + \sin\left(\frac{j\pi}{n}\right)\sigma_z\right] \otimes \left[\cos\left(\frac{(j+1)\pi}{n}\right)\sigma_x + \sin\left(\frac{(j+1)\pi}{n}\right)\sigma_z\right]. \tag{21}$$

Similarly for odd j,

$$X_j X_{j+1} = \left[\cos\left(\frac{(j+1)\pi}{n}\right)\sigma_x + \sin\left(\frac{(j+1)\pi}{n}\right)\sigma_z\right] \otimes \left[\cos\left(\frac{j\pi}{n}\right)\sigma_x + \sin\left(\frac{j\pi}{n}\right)\sigma_z\right]. \tag{22}$$

Further,

$$X_n X_1 = \widetilde{X_n} \otimes \widetilde{X_1}$$

$$= -\cos\left(\frac{\pi}{n}\right)\sigma_x \otimes \sigma_x - \sin\left(\frac{\pi}{n}\right)\sigma_x \otimes \sigma_z. \tag{23}$$

Using Equations (21)–(23) and basic arithmetics, the n-cycle chained Bell inequality for quantum systems transforms as

$$\frac{n}{2}\cos\left(\frac{\pi}{n}\right)\langle\sigma_x \otimes \sigma_x\rangle + \frac{n}{2}\cos\left(\frac{\pi}{n}\right)\langle\sigma_z \otimes \sigma_z\rangle + \frac{n}{2}\sin\left(\frac{\pi}{n}\right)\langle\sigma_x \otimes \sigma_z\rangle - \frac{n}{2}\sin\left(\frac{\pi}{n}\right)\langle\sigma_z \otimes \sigma_x\rangle \leq n - 2,$$

which further simplifies to

$$\cos\left(\frac{\pi}{n}\right)[\langle\sigma_x \otimes \sigma_x\rangle + \langle\sigma_z \otimes \sigma_z\rangle] + \sin\left(\frac{\pi}{n}\right)[\langle\sigma_x \otimes \sigma_z\rangle - \langle\sigma_z \otimes \sigma_x\rangle] \leq \frac{2(n-2)}{n}.$$

For a two qubit density matrix ρ, this translates into

$$\text{Tr}\,(O_n\rho) \leq \frac{2\,(n-2)}{n}, \tag{24}$$

where $O_n = \cos\left(\frac{\pi}{n}\right)[\sigma_x \otimes \sigma_x + \sigma_z \otimes \sigma_z] + \sin\left(\frac{\pi}{n}\right)[\sigma_x \otimes \sigma_z - \sigma_z \otimes \sigma_x]$.

The condition for violation of n-cycle chained Bell inequality becomes

$$\text{Tr}\,(O_n\rho) > \frac{2\,(n-2)}{n} \tag{25}$$

The eigenvalues of O_n are $2, 0, 0, -2$. Suppose the eigenvalues of ρ are $\lambda_1 \geq \lambda_2 \geq \lambda_3 \geq \lambda_4$, then

$$\text{Tr}\,(O_n\rho) \leq 2\,(\lambda_1 - \lambda_4). \tag{26}$$

Using Equations (26) and (25), the necessary condition for the violation of n-cycle chained Bell inequality turns out to be

$$\lambda_1 - \lambda_4 > \frac{n-2}{n}. \tag{27}$$

This completes the proof. $\square$

The set of quantum states form a convex set. Since the non-contextuality inequality in Equation (18) is a linear inequality, its maximum over quantum sets is attained at the extreme points i.e., for pure states. Mixed-ness may lead to non-violation of the aforementioned linear inequality. In this light, it is necessary to study the upper bound on λ_4 (if any).

Since the system under consideration is a two qubit density matrix. We have the following constraints on the eigenvalues:

$$0 \leq \lambda_i \leq 1 \ \forall i \in \{1,2,3,4\} \tag{28}$$

and

$$\sum_{i=1}^{4} \lambda_i = 1. \tag{29}$$

The constraints in Equations (28) and (29) imply that

$$\lambda_1 + \lambda_4 \leq 1. \tag{30}$$

Using Equations (20) and (30), we get

$$\lambda_4 \leq \frac{1}{n}. \tag{31}$$

The Equation (31) provides an upper bound on λ_4.

Remark 2. *It is easy to see that set of two-qubit quantum states that can violate chained Bell inequality shrinks as we increase n (See Figure 6). In the infinite n scenario, the only two qubit state that violates the inequality is a Bell state!*

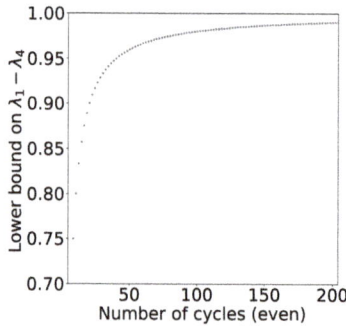

Figure 6. Here, we plot the lower bound on the difference of extremal eigenvalues of a two qubit density matrix as a function of even values of n. The set of two qubit quantum states, which could potentially violate chained Bell inequality (as our is necessary and not sufficient), shrinks as we increase n. In the infinite n scenario, the only two qubit state that might violate the inequality is Bell state!

The even cycle non-contextuality inequalities can be extended to phase space case quite easily following the work of Arora et al. [26], where the authors provided the phase space extension for $n = 4$. We have already discussed the construction corresponding to the maximal violation of inequality in Equation (18). The inequality is maximally violated by $\left(0, 1/\sqrt{2}, -1/\sqrt{2}, 0\right)^T$ and the maximum violation is $n \cos\left(\pi/n\right)$. We define the following non-contextuality operator in this regard,

$$C_n = \sum_{j=1}^{n-1} X_j X_{j+1} - X_n X_1. \tag{32}$$

We know that

$$\exp\left(\imath \theta n.\sigma\right) \sigma \exp\left(-\imath \theta n.\sigma\right) = \sigma \cos\left(2\theta\right) + n \times \sigma \sin\left(2\theta\right) + n \, n.\sigma \left(1 - \cos\left(2\theta\right)\right) \tag{33}$$

For $\sigma = \sigma_x \hat{x}$ and $n = \hat{z}$,

$$\exp\left(\imath \theta \sigma_z\right) \sigma_x \exp\left(-\imath \theta \sigma_z\right) = \sigma_x \cos\left(2\theta\right) + \sigma_y \sin\left(2\theta\right) \tag{34}$$

Let us look back at the operator in Equation (19) more closely. This can be thought of as σ_x rotated around z-axis with angle $\left(\frac{j\pi}{2n}\right)$. To get the phase space representation, let us start with the quantum mechanical translational operator $\exp\left(\frac{-\imath pL}{\hbar}\right)$, which translates a particle by distance L. This operator is not Hermitian and hence we introduce the following symmetric combination to make it Hermitian,

$$\mathcal{X}\left(0\right) \equiv \frac{e^{-\imath pL/\hbar} + e^{\imath pL/\hbar}}{2} = \cos\left(\frac{pL}{\hbar}\right). \tag{35}$$

Let $\mathcal{U}(\phi) = \exp\left(\frac{\imath \mathcal{Z}\phi}{2}\right)$ where $\mathcal{Z} = \text{sgn}\left(\sin\left(\frac{q\pi}{L}\right)\right)$. One can easily see that $\mathcal{X}(\phi) \equiv \mathcal{U}^\dagger\left(\phi\right) \mathcal{X}\left(0\right) \mathcal{U}^\dagger\left(\phi\right)$ and $\widetilde{X}_j = \mathcal{X}\left(\frac{j\pi}{n}\right)$.

Let $\phi\left(q\right) = \langle q|\phi\rangle$ be the localised quantum state symmetric about $q = \frac{1}{2}$, for some length scale L. and $\phi_n\left(q\right) \equiv \phi\left(q - nL\right)$. Using this construction, the following states are defined:

$$|\psi_0\rangle \equiv \frac{1}{\sqrt{M}} \sum_{n=-\frac{M}{2}}^{n=\frac{M-1}{2}} |\phi_{2n+1}\rangle \tag{36}$$

$$|\psi_1\rangle \equiv \frac{1}{\sqrt{M}} \sum_{n=-\frac{M}{2}}^{n=\frac{M-1}{2}} |\phi_{2n}\rangle \tag{37}$$

Let $|\psi_+\rangle \equiv \frac{|\psi_0\rangle + |\psi_1\rangle}{\sqrt{2}}$ and $|\psi_-\rangle \equiv \frac{|\psi_0\rangle - |\psi_1\rangle}{\sqrt{2}}$. Interestingly, for $N = 2M$,

$$\langle\psi_+|\mathcal{X}|\psi_+\rangle = \left(\frac{N-1}{N}\right), \tag{38}$$

and

$$\langle\psi_-|\mathcal{X}|\psi_-\rangle = -\left(\frac{N-1}{N}\right). \tag{39}$$

The appropriate entangled state which shows the violation is

$$|\psi\rangle \equiv \frac{|\psi_+\rangle_1|\psi_-\rangle_2 - |\psi_-\rangle_1|\psi_+\rangle_2}{\sqrt{2}}. \tag{40}$$

The quantum violation for the state in Equation (40) corresponds to the maximum quantum violation i.e., $n\cos\left(\frac{\pi}{n}\right)$ for large N. The experimental implementation of the phase space extension is quite simple and follows directly from the work of Arora et al. [26].

4. Simulating PR Box

The KS box is a powerful resource which can be used to efficiently simulate the most non-local no-signalling box, i.e., PR box [16]. The PR box has initially been defined as the box which allows maximum violation of the CHSH inequality in no-signalling theories. One can generalise the notion of PR box corresponding to chained Bell inequalities.

Definition 2. *A PR box is a no-signalling resource with input pair x, y and corresponding output pair a, b where each of these variables takes their values from the set $\{0,1\}$. The statistics of the PR box follows the following relation:*

$$xy = a \oplus b, \tag{41}$$

which means that the outputs are different if and only if the inputs are $x = y = 1$, otherwise the outputs are same. The PR box can be generalised for input pair $(x, y) \in \{1, 2, \cdots, n\}^2$ and output from the set $\{0,1\}$ such that outputs are same when inputs are anything except $\{1,1\}$. When inputs are $\{1,1\}$, the outputs must be different.

Now, suppose Alice and Bob are equipped with an arbitrary dimensional KS box. Table 2 gives the joint probabilities for an n-dimensional KS_p box.

KS box is more powerful than PR box and can be used to simulate the same [18]. We ask whether Alice and Bob can simulate a generalised PR box (as defined before) using KS_p box. The answer is in the affirmative, and we provide a simple strategy to do so.

Proposition 4. *A PR box of dimension (number of inputs for each party) n can be simulated efficiently using a KS box of dimension $2n - 1$ with marginal value of $p = \frac{1}{2}$.*

Proof. To prove our claim, we provide the following strategy: Alice relabels her inputs for PR box as follows:

$$1 \to 1, 2 \to 2, 3 \to 4, 4 \to 6 \cdots, n \to 2n - 2.$$

Similarly, Bob relabels his inputs as follows:

$$1 \to 1, 2 \to 3, 3 \to 5, 4 \to 7 \cdots, n \to 2n - 1.$$

The relabelled inputs are used as fresh input for the $KS_{\frac{1}{2}}$ box. Alice outputs what she gets as output from the $KS_{\frac{1}{2}}$. Bob flips his output from $KS_{\frac{1}{2}}$ box in every round and outputs the resultant value. This strategy simulates the statistics corresponding to generalised PR box. $\square$

Table 2. The table displays the joint probabilities for an n-dimensional KS_p box. Note that each of the blocks along the diagonal are same and similarly all the off diagonal blocks are same. Within a block, the top left element is the probability of getting $(0,0)$, top right signifies the probability of getting $(0,1)$, bottom left indicates the corresponding value for $(1,0)$ and, the probability for $(1,1)$ is indicated by the bottom right entry.

x / y	1		2		...	n	
1	$1-p$	0	$1-2p$	p	$\ddots$	$1-2p$	p
	0	p	p	0		p	0
2	$1-2p$	p	$1-p$	0	$\ddots$	$1-2p$	p
	p	0	0	p		p	0
$\vdots$	$\ddots$		$\ddots$		$\ddots$	$\ddots$	
n	$1-2p$	p	$1-2p$	p	$\ddots$	$1-p$	0
	p	0	p	0		0	p

Given the even cycle generalisation of CHSH inequality, the marginal probabilities p in the KS_p required to saturate classical bound, quantum bound and no-signalling bound are given by

$$p_c \leq \frac{n-2}{2(n-1)},$$

$$p_q \leq \frac{n \left(\cos \left(\frac{\pi}{n} \right) + 1 \right) - 2}{4(n-1)}$$

and

$$p_{NS} \leq \frac{1}{2} \tag{42}$$

respectively.

Remark 3. *For a large value of n, all the above probability expressions tend to one half. However, the quantum probability approaches the PR box limit of $\frac{1}{2}$ significantly faster than the classical probability. For large n, all these probabilities approach $\frac{1}{2}$ (see Figure 7).*

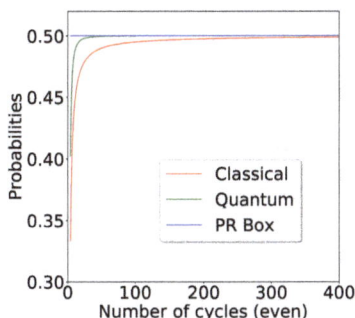

Figure 7. We look at KS box probabilities in various regimes. Note that the quantum probability approaches the PR box limit faster than classical probability as we increase the number of cycles.

5. Conclusions

We studied arbitrary dimensional KS box, generalised PR box and n-cycle non-contextuality inequalities in this work. We provided the optimal classical strategy and the corresponding success probability for classically simulating the KS box. For future work, it is worth exploring the optimal quantum strategy for this purpose. We provided the sufficient condition for the violation of the generalised KCBS inequality and necessary condition for the violation of even-cycle generalisation of CHSH inequality. We also discussed the phase space extension of even-cycle generalisation of CHSH inequality. We leave the phase space extension of KCBS and generalised KCBS inequality for future work. We also studied the strategy for simulating a generalised PR box using KS box. It is also interesting to explore further how the generalised PR box, arbitrary dimensional KS box and n-cycle non-contextuality inequalities are related to each other and their implications.

Our work helps quantify the Bell non-locality of KS box in terms of impossibility of classical simulation for general n-dimensional case. We also provided the sufficient condition for violation of odd n-cycle non-contextuality inequalities. Since contextuality is the chief resource behind various models of quantum computation, our result can help select the resources required to get necessary quantum speed-up. Moreover, our phase space extension of chained Bell inequalities make them suitable for experimental purposes and also harness the underlying Bell non-locality for various quantum communication tasks such as secure key distribution for example.

Author Contributions: Conceptualization, K.B., M.R. and L.-C.K.; software, K.B.; formal analysis, K.B.; investigation, L.-C.K.; writing—original draft preparation, K.B.; writing—review and editing, K.B., M.R. and L.-C.K.; and supervision, L.-C.K.

Funding: This research was funded by the National Research Foundation of Singapore and the Ministry of Education of Singapore.

Acknowledgments: We thank Atul Singh Arora, Naresh Boddu and Bangaliya for useful discussions. K.B. and M.R. acknowledge the CQT Graduate Scholarship.

Conflicts of Interest: The authors declare no conflict of interest.

Appendix A

Proposition A1. *The optimisation problem of minimising the variance of a random variable defined on a set of non-negative integral points, over all possible probability distributions, for a fixed given mean, say = m has support size at most two.*

Proof. We make use of the Karush–Kuhn–Tucker (KKT) conditions for deriving necessary optimality conditions. KKT conditions ensure that the gradient of the objective function is perpendicular to the constrained set and the constraints are satisfied. Note that these KKT conditions are just extensions of the Lagrange multiplier method, where now we also have inequality constraints. The Lagrangian for the above optimisation problem is given by

$$L(p, \lambda, \mu, v) = \sum_i i^2 p_i - \lambda \left(\sum_i p_i - 1 \right) + \mu \left(\sum_i i p_i - m \right) - \sum_i v_i p_i,$$

where λ and $v_i \geq 0$ for all i, are the KKT multiplier corresponding to the valid probability constraint, and μ is the KKT multiplier corresponding to the fixed mean condition. A necessary condition for optimality is derived by taking the partial derivative of the Lagrangian with respect to p_i and setting it to zero, thus getting $i^2 - \lambda + \mu i = v_i$. Another necessary KKT condition for optimality is the complementary slackness, which implies $v_i p_i = 0$, for all i. These two conditions give us $p_i (i^2 - \lambda + \mu i) = 0$, for all i. The term inside the brackets is a quadratic expression in terms of i, implying that it can be equal to zero for at most two distinct values of i. This, in turn, tells us that $p_i = 0$, for at least all but two i, or, in other words, p has support size at most 2. □

References

1. Bell, J.S. On the Einstein Podolsky Rosen paradox. *Phys. Phys. Fizika* **1964**, *1*, 195–200. [CrossRef]
2. Pironio, S.; Acín, A.; Massar, S.; de La Giroday, A.B.; Matsukevich, D.N.; Maunz, P.; Olmschenk, S.; Hayes, D.; Luo, L.; Manning, T.A.; et al. Random numbers certified by Bell's theorem. *Nature* **2010**, *464*, 1021. [CrossRef]
3. Popescu, S.; Rohrlich, D. Generic quantum nonlocality. *Phys. Lett. A* **1992**, *166*, 293–297. [CrossRef]
4. Mayers, D.; Yao, A. Self testing quantum apparatus. *arXiv* **2003**, arXiv:quant-ph/0307205.
5. Summers, S.J.; Werner, R. Bell's inequalities and quantum field theory. I. General setting. *J. Math. Phys.* **1987**, *28*, 2440–2447. [CrossRef]
6. Tsirel'son, B.S. Quantum analogues of the Bell inequalities. The case of two spatially separated domains. *J. Soviet Math.* **1987**, *36*, 557–570. [CrossRef]
7. Cleve, R.; Buhrman, H. Substituting quantum entanglement for communication. *Phys. Rev. A* **1997**, *56*, 1201. [CrossRef]
8. Cabello, A.; Severini, S.; Winter, A. Graph-theoretic approach to quantum correlations. *Phys. Rev. Lett.* **2014**, *112*, 040401. [CrossRef]
9. Amaral, B.; Cunha, M.T. *On Graph Approaches to Contextuality and Their Role in Quantum Theory*; Springer: Berlin, Germany, 2018.
10. Kochen, S.; Specker, E.P. The problem of hidden variables in quantum mechanics. In *The Logico-Algebraic Approach to Quantum Mechanics*; Springer: Berlin, Germany, 1975; pp. 293–328.
11. Singh, J.; Bharti, K.; Arvind. Quantum key distribution protocol based on contextuality monogamy. *Phys. Rev. A* **2017**, *95*, 062333. [CrossRef]
12. Cabello, A.; D'Ambrosio, V.; Nagali, E.; Sciarrino, F. Hybrid ququart-encoded quantum cryptography protected by Kochen-Specker contextuality. *Phys. Rev. A* **2011**, *84*, 030302. [CrossRef]
13. Bharti, K.; Ray, M.; Varvitsiotis, A.; Warsi, N.A.; Cabello, A.; Kwek, L.C. Robust self-testing of quantum systems via noncontextuality inequalities. *arXiv* **2018**, arXiv:1812.07265.
14. Raussendorf, R. Contextuality in measurement-based quantum computation. *Phys. Rev. A* **2013**, *88*, 022322. [CrossRef]
15. Howard, M.; Wallman, J.; Veitch, V.; Emerson, J. Contextuality supplies the 'magic' for quantum computation. *Nature* **2014**, *510*. [CrossRef] [PubMed]
16. Popescu, S.; Rohrlich, D. Quantum nonlocality as an axiom. *Found. Phys.* **1994**, *24*, 379–385. [CrossRef]
17. Pawłowski, M.; Paterek, T.; Kaszlikowski, D.; Scarani, V.; Winter, A.; Żukowski, M. Information causality as a physical principle. *Nature* **2009**, *461*, 1101. [CrossRef] [PubMed]
18. Bub, J.; Stairs, A. Contextuality and nonlocality in 'no signaling'theories. *Found. Phys.* **2009**, *39*, 690–711. [CrossRef]
19. Popescu, S. Nonlocality beyond quantum mechanics. *Nat. Phys.* **2014**, *10*, 264.
20. van Dam, W. Implausible consequences of superstrong nonlocality. *Nat. Comput.* **2013**, *12*, 9–12.
21. Araújo, M.; Quintino, M.T.; Budroni, C.; Cunha, M.T.; Cabello, A. All noncontextuality inequalities for then-cycle scenario. *Phys. Rev. A* **2013**, *88*. [CrossRef]

Entropy **2019**, *21*, 134

22. Klyachko, A.A.; Can, M.A.; Binicioğlu, S.; Shumovsky, A.S. Simple Test for Hidden Variables in Spin-1 Systems. *Phys. Rev. Lett.* **2008**, *101*, 020403. [CrossRef]
23. Liang, Y.C.; Spekkens, R.W.; Wiseman, H.M. Specker's parable of the overprotective seer: A road to contextuality, nonlocality and complementarity. *Phys. Rep.* **2011**, *506*, 1–39. [CrossRef]
24. Clauser, J.F.; Horne, M.A.; Shimony, A.; Holt, R.A. Proposed Experiment to Test Local Hidden-Variable Theories. *Phys. Rev. Lett.* **1969**, *23*, 880–884. [CrossRef]
25. Braunstein, S.L.; Caves, C.M. Wringing out better Bell inequalities. *Ann. Phys.* **1990**, *202*, 22–56. [CrossRef]
26. Arora, A.S.; Asadian, A. Proposal for a macroscopic test of local realism with phase-space measurements. *Phys. Rev. A* **2015**, *92*, 062107. [CrossRef]
27. Bharti, K.; Arora, A.S.; Kwek, L.C.; Roland, J. A simple proof of uniqueness of the KCBS inequality. *arXiv* **2018**, arXiv:1811.05294.
28. Knuth, D.E. The sandwich theorem. *Electron. J. Comb.* **1994**, *1*, 1.

entropy

MDPI

Article

On the Significance of the Quantum Mechanical Covariance Matrix

Avishy Carmi [1,*] and Eliahu Cohen [2]

[1]   Center for Quantum Information Science and Technology & Faculty of Engineering Sciences,
    Ben-Gurion University of the Negev, Beersheba 8410501, Israel
[2]   Physics Department, Centre for Research in Photonics, University of Ottawa, Advanced Research Complex,
    25 Templeton, Ottawa, K1N 6N5, Canada; eli17c@gmail.com
*   Correspondence: avcarmi@bgu.ac.il

Received: 30 May 2018; Accepted: 26 June 2018; Published: 28 June 2018

Abstract: The characterization of quantum correlations, being stronger than classical, yet weaker than those appearing in non-signaling models, still poses many riddles. In this work, we show that the extent of binary correlations in a general class of nonlocal theories can be characterized by the existence of a certain covariance matrix. The set of quantum realizable two-point correlators in the bipartite case then arises from a subtle restriction on the structure of this general covariance matrix. We also identify a class of theories whose covariance has neither a quantum nor an "almost quantum" origin, but which nevertheless produce the accessible two-point quantum mechanical correlators. Our approach leads to richer Bell-type inequalities in which the extent of nonlocality is intimately related to a non-additive entropic measure. In particular, it suggests that the Tsallis entropy with parameter $q = 1/2$ is a natural operational measure of non-classicality. Moreover, when generalizing this covariance matrix, we find novel characterizations of the quantum mechanical set of correlators in multipartite scenarios. All these predictions might be experimentally validated when adding weak measurements to the conventional Bell test (without adding postselection).

Keywords: quantum correlations; quantum bounds; nonlocality; tsallis entropy

1. Introduction

The extent of nonlocality is commonly determined by a set of correlations. In the simplest bipartite scenario, the four two-point correlators c_1, c_2, c_3 and c_4, corresponding to the four pairs of possible outcomes of Alice and Bob, may render the theory classical, quantum, or stronger-than-quantum. In this paper, we tell the richer story provided by a certain covariance matrix presented in the next section. This matrix, which may be defined in any statistical theory, implies a bound on two-point correlators analogous to that of quantum mechanics. We thus prove that all potential theories having a covariance structure similar to that of quantum mechanics have a similar set of realizable correlators. Interestingly, this is yet less than the structure imposed by quantum mechanics and theories having almost quantum correlations [1]. These results cast light on the origin of quantum correlations; they suggest that other hypothetical theories might exist whose correlations are indistinguishable from both quantum and almost quantum correlations. In this sense, our work can be seen as part of the efforts (see, e.g., [2–9]) to achieve better qualitative and quantitative understanding of quantum nonlocality.

This paper has two main parts. The first is general and does not rely on the quantum-mechanical formalism to characterize nonlocality. The second, which builds on these general results, assumes a quantum structure to derive new bounds on bipartite and tripartite two-point correlators.

Among the preceding papers in this area, there are mainly two other works where covariance and second moment matrices, different from the ones considered here, are used for characterizing quantum mechanical correlations and probability distributions: the NPA test [3], which significantly extends the approach previously employed in [10]. We note the following primary difference between these works and the paper at hand. While the positive semi-definiteness property plays a role in both, the particular covariance here leads to the identification of fundamental relations between the entries in this matrix. These relations alone are shown to govern the set of realizable binary bipartite correlators not only in quantum mechanics but in any nonlocal theory, and to imply new tighter bounds on this set.

2. Covariance-Based Certificate of Nonlocality

We restrict ourselves for the moment to the Bell–CHSH [11,12] setup where two experimenters perform measurements with one of their measurement devices. Alice measures using either her device 0 or device 1, and similarly Bob measures using either his device 0 or device 1. Both Alice's outcome a_i when she measured using device i and Bob's outcome b_j when he measured using device j may either be 1 or -1. We consider the products $x_{1+i+2j} = a_i b_j$ in different experiments where Alice and Bob used the pair of devices i, j. In a local hidden variables theory, the Bell–CHSH inequality, $|E[x_1] + E[x_2] + E[x_3] - E[x_4]| \leq 2$, holds [12].

Suppose now there exists a covariance matrix underlying the products $x_1, \ldots, x_4$. This 4×4 matrix is defined as

$$C \stackrel{\text{def}}{=} \mathcal{M} - VV^T, \tag{1}$$

where $\mathcal{M}$ is a positive semi-definite second moment matrix whose diagonal entries all equal 1, and $V^T = [c_1, \ldots, c_4]$ is the vector of two-point correlators. If the product x_i is a realization of the random variable X_i, then $\mathcal{M}_{ij} \stackrel{\text{def}}{=} E[X_i X_j]$ and $c_i \stackrel{\text{def}}{=} E[X_i]$, and, if it is associated with an operator X_i (as in quantum mechanics), then $\mathcal{M}_{ij} \stackrel{\text{def}}{=} \frac{1}{2}\langle\{X_i, X_j\}\rangle$ and $c_i \stackrel{\text{def}}{=} \langle X_i \rangle$, where $\{X_i, X_j\} \stackrel{\text{def}}{=} X_i X_j + X_j X_i$ is the anti-commutator. The covariance is by construction real, symmetric and positive semi-definite.

However, even without specifying how the covariance is evaluated, $C \succeq 0$ (which means hereinafter that C is positive semidefinite) may be understood as an algebraic constraint on the vector of correlators that allows a covariance matrix to be defined in the underlying theory. In particular,

$$VV^T \preceq \mathcal{M}, \tag{2}$$

which geometrically means that V is confined to the ellipsoid described by $\mathcal{M}$. For example, a theory having no constraints whatsoever on the correlators may have $\mathcal{M} = VV^T$. The PR-box is one such theory. It is worth noting that, in the language of [13], the left-hand side in Equation (2) is a Fisher information matrix associated with the vector V of correlators.

The constraint in Equation (2) leads to the following quantum-like characterization of realizable two-point correlators in any statistical theory. See Figure 1.

Theorem 1. *The correlators satisfy*

$$
\begin{aligned}
|c_1 c_2 - c_3 c_4 - \mathcal{M}_{12} + \mathcal{M}_{34}| &\leq \sigma_1 \sigma_2 + \sigma_3 \sigma_4 \\
|c_1 c_3 - c_2 c_4 - \mathcal{M}_{13} + \mathcal{M}_{24}| &\leq \sigma_1 \sigma_3 + \sigma_2 \sigma_4 \\
|c_2 c_3 - c_1 c_4 - \mathcal{M}_{23} + \mathcal{M}_{14}| &\leq \sigma_2 \sigma_3 + \sigma_1 \sigma_4,
\end{aligned}
\tag{3}
$$

where $\sigma_i^2 = 1 - c_i^2$.

Proof. The 4×4 matrix C can be partitioned into blocks as follows

$$C = \begin{bmatrix} D_{12} & N \\ N^T & D_{34,} \end{bmatrix} \tag{4}$$

where D_{12}, N and D_{34} are 2×2 matrices. Because $C \succeq 0$ so are $D_{12} \succeq 0$ and $D_{34} \succeq 0$. Therefore,

$$\det(D_{ij}) = \sigma_i^2 \sigma_j^2 - (\mathcal{M}_{ij} - c_i c_j)^2 \geq 0, \tag{5}$$

namely,

$$|\mathcal{M}_{ij} - c_i c_j| \leq \sigma_i \sigma_j, \tag{6}$$

for $i, j = 1, 2$ and $i, j = 3, 4$. This together with the triangle inequality imply

$$|c_1 c_2 - c_3 c_4 - \mathcal{M}_{12} + \mathcal{M}_{34}| \leq |\mathcal{M}_{12} - c_1 c_2| + |\mathcal{M}_{34} - c_3 c_4| \leq \sigma_1 \sigma_2 + \sigma_3 \sigma_4. \tag{7}$$

All other symmetries of this inequality in Equation (3) are obtained by swapping rows and the respective columns of $\mathcal{C}$. $\square$

The next corollary suggests that very little is needed to reproduce the set of quantum mechanical two-point binary correlators.

Corollary 1. *The correlators vector V is realizable in quantum mechanics if and only if Equation (2) holds for some positive semi-definite matrix $\mathcal{M}$ whose diagonal entries all equal 1, and for which one of the terms, $\mathcal{M}_{12} - \mathcal{M}_{34}, \ \mathcal{M}_{13} - \mathcal{M}_{24}, \ \mathcal{M}_{23} - \mathcal{M}_{14}$, vanishes. In such a case,*

$$\begin{aligned} |c_1 c_2 - c_3 c_4| &\leq \sigma_1 \sigma_2 + \sigma_3 \sigma_4 \\ |c_1 c_3 - c_2 c_4| &\leq \sigma_1 \sigma_3 + \sigma_2 \sigma_4 \\ |c_2 c_3 - c_1 c_4| &\leq \sigma_2 \sigma_3 + \sigma_1 \sigma_4. \end{aligned} \tag{8}$$

The condition in Equation (8), which from within quantum mechanics has been shown to be necessary and sufficient for quantum-realizable correlators independently by Tsirelson, Landau, and Masanes [10,14,15], is obtained here without assuming quantum mechanics, but rather from a subtle restriction on the structure of $\mathcal{M}$ in any statistical theory.

Proof. Suppose, for example, that $\mathcal{M}_{12} - \mathcal{M}_{34} = 0$, in which case the first inequality in Equation (3) coincides with the first inequality in Equation (8). All other symmetries of this inequality immediately follow for they are all equivalent (upon squaring, all these inequalities become identical: $2\sigma_1^2 \sigma_2^2 \sigma_3^2 \sigma_4^2 + 2c_1 c_2 c_3 c_4 + 2 - (c_1^2 + c_2^2 + c_3^2 + c_4^2) \geq 0$). $\square$

The Covariance in Quantum Mechanics

If all products can be factorized as $x_{1+i+2j} = a_i b_j$, where a_i and b_j are the local outcomes of Alice and Bob per their choices i and j (which actually amounts to the existence of local hidden variables), then Equation (3) reduces to the set of classical correlators [16]. The next theorem shows that when the products are associated with operators, a similar factorization leads to the set of quantum realizable two-point binary correlators. An important difference, then, between models of local hidden variables and quantum mechanics, is the non-commutativity of Alice's operators, as well as the non-commutativity of Bob's operators, which allows quantum mechanics to reach stronger correlations.

Theorem 2. *Let $X_1 \stackrel{\text{def}}{=} A_0 B_0$, $X_2 \stackrel{\text{def}}{=} A_1 B_0$, $X_3 \stackrel{\text{def}}{=} A_0 B_1$, and $X_4 \stackrel{\text{def}}{=} A_1 B_1$, where the commuting operators A_i and B_j are self-adjoint with ± 1 eigenvalues. Then, the correlations satisfy Equation (8).*

Proof. The entries, $\mathcal{M}_{12} = \langle X_1 X_2 + X_2 X_1 \rangle / 2 = \langle\{A_0, A_1\}\rangle / 2 = \langle X_3 X_4 + X_4 X_3 \rangle / 2 = \mathcal{M}_{34}$, and $\mathcal{M}_{13} = \langle X_1 X_3 + X_3 X_1 \rangle / 2 = \langle\{B_0, B_1\}\rangle / 2 = \langle X_2 X_4 + X_4 X_2 \rangle / 2 = \mathcal{M}_{24}$. By the preceding theorem this is all that is needed to produce the quantum set of realizable two-point binary correlators. $\square$

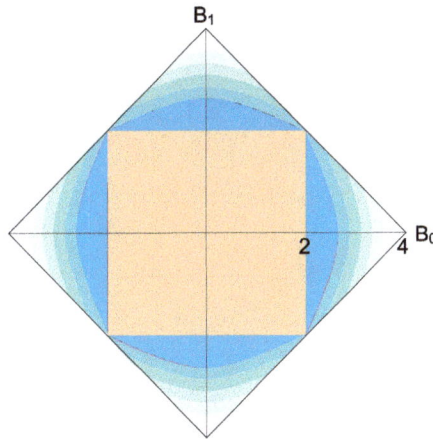

Figure 1. Quantum-like bounds on any statistical theory in Equation (3). The paler is the region, the larger is the difference $|\mathcal{M}_{12} - \mathcal{M}_{34}|$. The quantum bound on the two-point correlators, where this difference vanishes, is shown in dark blue. Classical correlators make the bounded square. In this figure, $\mathcal{B}_x \overset{\text{def}}{=} c_1 + c_2 + (-1)^x (c_3 - c_4)$ is a symmetry of the Bell–CHSH parameter.

This result naturally carries over to almost quantum correlations [1] where $A_i B_j |\psi\rangle = B_j A_i |\psi\rangle$ for some states, but not necessarily all of them. Thus, in quantum theory, as well as for almost quantum correlations the matrix $\mathcal{M}$ has both $\mathcal{M}_{12} - \mathcal{M}_{34}$ and $\mathcal{M}_{13} - \mathcal{M}_{24}$ vanish. Interestingly, due to the preceding theorem there may exist theories, where only one of these terms vanishes, which nevertheless produce the set of quantum mechanical two-point correlators.

The quantum covariance in which $\mathcal{M}_{12} - \mathcal{M}_{34} = 0$ and $\mathcal{M}_{13} - \mathcal{M}_{24} = 0$ will henceforth be denoted as $\mathcal{C}^Q$.

3. Nonlocality and Tsallis Entropy

In quantum theory and for almost quantum correlations, the extent of nonlocality may be quantified by a non-additive measure of entropy.

Theorem 3. *In quantum theory, as well as for almost quantum correlations*

$$|\mathcal{B}| \leq 2 + S(a, b) \tag{9}$$

where $\mathcal{B}$ is the Bell–CHSH parameter, and $S(a, b)$ is either $S(a)$ or $S(b)$, the smallest among them, where $S(a)$ and $S(b)$ are the Tsallis entropies [17] with parameter $1/2$ of a ± 1-valued random variables a and b whose means are, respectively, $\langle \{A_0, A_1\} \rangle / 2$ and $\langle \{B_0, B_1\} \rangle / 2$. The right hand side in this inequality takes values between the Bell limit, 2, and the Tsirelson's bound, $2\sqrt{2}$ (see Figure 2). The Bell bound is attained when one of the pairs, either A_0, A_1 or B_0, B_1, commute, and the Tsirelson's bound is attained when both anti-commute.

Proof. The covariance matrix in Equation (1) can be partitioned as

$$\mathcal{C}^Q = \begin{bmatrix} D_{12} & N \\ N^T & D_{34}, \end{bmatrix} \tag{10}$$

where D_{12}, N and D_{34} are 2×2 matrices. Because $\mathcal{C}^Q \succeq 0$ so are $D_{12} \succeq 0$ and $D_{34} \succeq 0$. Let $g \overset{\text{def}}{=} [1, \pm 1]$ and write

$$g D_{ij} g^T = 2(1 \pm \mathcal{M}_{ij}) - (\langle X_i \rangle \pm \langle X_j \rangle)^2 \geq 0, \tag{11}$$

namely,

$$|\langle X_i \rangle \pm \langle X_j \rangle| \leq \sqrt{2(1 \pm \mathcal{M}_{ij})}, \tag{12}$$

for $i, j = 1, 2$ and $i, j = 3, 4$. This together with the triangleinequality yield

$$|\mathcal{B}| \leq |\langle X_1 \rangle + \langle X_2 \rangle| + |\langle X_3 \rangle - \langle X_4 \rangle| \leq \sqrt{2(1 + d)} + \sqrt{2(1 - d)}. \tag{13}$$

where $d \stackrel{\text{def}}{=} \langle \{A_0, A_1\} \rangle / 2 = \mathcal{M}_{12} = \mathcal{M}_{34}$. Let y be a ± 1-valued random variable whose mean is d, i.e., $p(y = \pm 1) = (1 \pm d)/2$. The above relation can now be written as

$$|\mathcal{B}| \leq 2 + S(a), \tag{14}$$

where the Tsallis entropy of a is given by

$$S(a) \stackrel{\text{def}}{=} \frac{1}{q - 1} \left[1 - \sum_{i = \pm 1} p(a = i)^q. \right]. \tag{15}$$

with $q = 1/2$. Repeating all of the above calculations for $\tilde{\mathcal{C}}^Q$ instead of $\mathcal{C}^Q$, where $\tilde{\mathcal{C}}^Q$ is obtained by permuting the second and third columns of $\mathcal{C}^Q$ and then its second and third rows, the parameter $d \stackrel{\text{def}}{=} \langle \{B_0, B_1\} \rangle / 2 = \mathcal{M}_{13} = \mathcal{M}_{24}$. □

This may strengthen different approaches, e.g., [18], seeking for a natural relation between uncertainty and nonlocality.

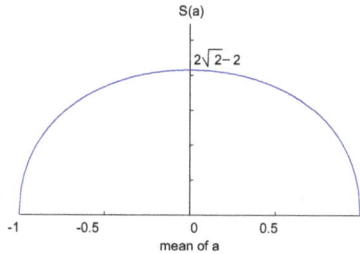

Figure 2. Tsallis entropy $S(a)$ quantifies the extent of nonlocality in the Bell–CHSH experiment.

Note also that quantum and almost quantum correlations will generally have different bounds in Equation (9), depending on the pairs A_0, A_1 and B_0, B_1.

4. Verification Using Weak Measurements

The above analysis extends the ordinary Bell–CHSH experiment by introducing $d = \langle \{A_0, A_1\} \rangle / 2$ or $d = \langle \{B_0, B_1\} \rangle / 2$, i.e., a pair of local operators at either Alice's side, Bob's side or both. In Alice's case, for instance, this d can be theoretically found once determining A_0 and A_1. However, one may question the practical feasibility of inferring it with respect to the entangled state Alice and Bob share at the same run of the ordinary Bell–CHSH experiment. We propose to measure it by employing a weak measurement [19] of the Hermitian operator $\{A_0, A_1\}/2$ on Alice's side, prior to her "strong" projective measurement. Weak measurement is known on theoretical [20] and experimental [21] grounds to asymptotically preserve entanglement, hence in the so called "weak limit" of an almost vanishing coupling constant between Alice's qubit and the measuring pointer, the back-action of the measurement would be negligible. When accumulating large enough statistics, the expectation value d can be inferred with arbitrarily high accuracy. Even though each run can be thought of as measuring weakly a pre- and post-selected system, we can take the weighted sum over all weak values [19] for generating the required expectation values. The same experimental procedure can be similarly applied to any multipartite scenario.

5. Relation to the NPA Hierarchy

The covariance C^Q is the Schur complement of

$$\mathcal{N} \overset{\text{def}}{=} \begin{bmatrix} 1 & V^T \\ V & C^Q + VV^T \end{bmatrix} \tag{16}$$

Therefore, $C^Q \succeq 0$ if and only if $\mathcal{N} \succeq 0$. This $\mathcal{N}$ may be viewed as a symmetrization of a Hermitian matrix Γ similar to those used in [3]. In particular,

$$\mathcal{N} = \frac{1}{2}\left(\Gamma + \Gamma^T\right) \tag{17}$$

where Γ is a submatrix in one of the levels of the NPA construction.

The symmetrization in Equation (17) allows entries whose values are otherwise inaccessible in the underlying experiment to be included in the derived bound. In fact, terms, e.g., $\langle\{A_0, A_1\}\rangle/2$, which are missing from Γ, have been shown in the preceding theorem to determine the extent of nonlocality. As mentioned above, bounds involving both local and nonlocal correlations are partly motivated by a possible application of weak measurements.

6. Tripartite Covariance

To examine the strength and applicability of the proposed formalism, we analyze in this section and in the next one two kinds of common generalizations of the Bell–CHSH setup. First, the covariance C^Q may be defined for any number of parties and any number of measurement devices. In the tripartite case, for example, where Alice, Bob, and Charlie each have a pair of measurement devices, the operators $X_m \overset{\text{def}}{=} A_i B_j C_k$, $m = 1 + i + 2j + 4k$, where the commuting triplets A_i, B_j, and C_k are self-adjoint. Here, C^Q is an 8×8 positive semidefinite matrix.

The tripartite covariance matrix implies bounds that may be used to characterize the set of quantum realizable three-point correlators, $\langle A_i B_j C_k \rangle$. In this respect, the results of the preceding theorems hold for any 4×4 submatrix of any matrix obtained by permuting the columns and the respective rows of C^Q. In one case, applying the reasoning of the last theorem leads to a bound tighter than Mermin's inequality [22]

$$|\langle A_0 B_0 C_0\rangle + \langle A_1 B_1 C_0\rangle + \langle A_0 B_1 C_1\rangle - \langle A_1 B_0 C_1\rangle| \le \sqrt{2(1+d)} + \sqrt{2(1-e)}, \tag{18}$$

where $d = \langle\{A_0 B_0, A_1 B_1\}\rangle/2$ and $e = \langle\{A_0 B_1, A_1 B_0\}\rangle/2$. If both pairs, (A_0, A_1) and (B_0, B_1), commute then the right hand side in Equation (18) equals the Bell limit, 2. if, on the other hand, either one of them anti-commute, in which case $d = -e$, then the right hand side in this inequality reads $2\sqrt{2(1+d)} \le 4$.

The tripartite covariance may also be composed of both two- and three-fold operators. For example, applying the first theorem to the covariance of the operators $X_1 = A_0 B_j$, $X_2 = A_1 B_j$, $X_3 = A_0 B_i C_k$, and $X_4 = A_1 B_i C_k$, yields

$$|\langle A_0 B_j\rangle\langle A_1 B_j\rangle - \langle A_0 B_i C_k\rangle\langle A_1 B_i C_k\rangle| \le$$
$$\sqrt{(1 - \langle A_0 B_j\rangle^2)(1 - \langle A_1 B_j\rangle^2)} + \sqrt{(1 - \langle A_0 B_i C_k\rangle^2)(1 - \langle A_1 B_i C_k\rangle^2)}. \tag{19}$$

which generalizes the bipartite inequality in [10,14,15]. The last theorem implies in this case

$$\left| \langle A_0 B_j \rangle + \langle A_1 B_j \rangle + \langle A_0 B_i C_k \rangle - \langle A_1 B_i C_k \rangle \right| \leq 2 + S(a, bc), \tag{20}$$

where the means of a and of bc are, respectively, $\langle \{A_0, A_1\} \rangle / 2$ and $\langle \{B_j, B_i C_k\} \rangle / 2$.

Consider now a tripartite covariance composed of only two-fold products, e.g., $X_1 = A_0 B_j$, $X_2 = A_1 B_j$, $X_3 = A_0 C_k$, and $X_4 = A_1 C_k$. By the last theorem

$$\left| \langle A_0 B_j \rangle + \langle A_1 B_j \rangle + \langle A_0 C_k \rangle - \langle A_1 C_k \rangle \right| \leq 2 + S(a, bc)$$
$$\left| \langle A_i B_0 \rangle + \langle A_i B_1 \rangle + \langle B_0 C_k \rangle - \langle B_1 C_k \rangle \right| \leq 2 + S(b, ac) \tag{21}$$
$$\left| \langle A_i C_0 \rangle + \langle A_i C_1 \rangle + \langle B_j C_0 \rangle - \langle B_j C_1 \rangle \right| \leq 2 + S(c, ab),$$

where the means of a, b, and c are, respectively, $\langle \{A_0, A_1\} \rangle / 2$, $\langle \{B_0, B_1\} \rangle / 2$, and $\langle \{C_0, C_1\} \rangle / 2$. Similarly, the means of ab, ac, and bc are, respectively, $\langle A_i B_j \rangle$, $\langle A_i C_k \rangle$, and $\langle B_j C_k \rangle$. These inequalities may be interpreted as follows. The first one, for example, suggests that the extent of nonlocality distributed between Alice–Bob and Alice–Charlie pairs is bounded by the local uncertainty at Alice site and also by the uncertainty underlying the Bob–Charlie link. The greater these uncertainties are, the stronger this nonlocality may get.

7. Further Generalization of the Covariance Matrix

The second kind of generalization refers to the natural case where Alice and Bob each have a two-level system, but now they can perform measurements in more than two incompatible bases (this is of course a very realistic scenario). For instance, when Alice and Bob may each perform three different kinds of measurements (still having ± 1 outcomes), the set of products becomes $X_k \overset{\text{def}}{=} A_i B_j$, where $k = 1 + i + 3j$, and $i, j \in \{0, 1, 2\}$. Under the assumption of local realism one finds the following Bell inequality, $|\mathcal{B}'| \leq 4$, where $\mathcal{B}' = c_1 + c_2 - c_3 + c_4 + c_5 + c_6 - c_7 + c_8$. This inequality is obtained from the well-studied I_{3322} inequality [23] by assuming ± 1 outcomes rather than $0, 1$ and by taking vanishing one-point correlators.

Let $\mathcal{C}_{123}$ be the covariance of X_1, X_2, and X_3. Similarly, let $\mathcal{C}_{456}$ and $\mathcal{C}_{78}$ be the covariances of X_4, X_5, and X_6, and of X_7 and X_8, respectively. Because $g^T C g \geq 0$, namely, $|g^T V| \leq \sqrt{g^T M g}$, for any vector g, it follows that

$$|\mathcal{B}'| \leq |c_1 + c_2 - c_3| + |c_4 + c_5 + c_6| + |c_7 - c_8| \leq$$
$$\sqrt{g_{++-}^T M_{123} g_{++-}} + \sqrt{g_{+++}^T M_{456} g_{+++}} + \sqrt{g_{+-}^T M_{78} g_{+-}} =$$
$$\sqrt{3 + 2d - 2(e + f)} + \sqrt{3 + 2d + 2(e + f)} + \sqrt{2 - 2d}, \tag{22}$$

where g_{+++}, etc., are vectors whose entries are 1 or -1, depending on the specification. Here, $d = \langle \{A_0, A_1\} \rangle / 2$, $e = \langle \{A_0, A_2\} \rangle / 2$, and $f = \langle \{A_1, A_2\} \rangle / 2$. It is straightforward to show that the maximum of the right hand side is 5, which is obtained for $e + f = 0$, and $d = 1/2$. It is worth noting that this bound coincides with numerical approximations of the bound on the original I_{3322} inequality in finite-dimensional Hilbert spaces [24].

8. Conclusions

In this paper, the analysis of a certain covariance matrix gives rise to a tight characterization of binary two-point correlators in quantum mechanics and in a general class of nonlocal theories. This formalism has further led to a natural measure of nonlocality given by the Tsallis entropy. Finally, we have discussed some generalizations of this approach and derived new bounds on tripartite two- and three-point correlators. These predictions, which often depend not only on the correlators but also on some anti-commutators might be experimentally tested with the aid of weak measurements [19], known to preserve entanglement [20,21]. That is, the nonlocal correlators can be determined as usual by performing (strong) projective measurements on the Alice and Bob sides, and at the same time

Entropy **2018**, *20*, 500

weak measurements can determine the local correlators needed for the proposed bounds. As the latter involve expectation values, rather than weak values [19], summation over all postselections should be performed. Hence, all the above seems to be experimentally testable.

Author Contributions: A.C. and E.C. both wrote the text and worked out the mathematical proofs in this paper.

Funding: A.C. acknowledges support from Israel Science Foundation Grant No. 1723/16. E.C. was supported by the Canada Research Chairs (CRC) Program.

Acknowledgments: We thank Sandu Popescu, Paul Skrzypczyk and Elie Wolfe for helpful comments and discussions.

Conflicts of Interest: The authors declare no conflict of interest.

References

1. Navascués, M.; Guryanova, Y.; Hoban, M.J.; Acin, A. Almost quantum correlations. *Nat. Commun.* **2015**, *6*, 6288. [CrossRef] [PubMed]
2. Popescu, S.; Rohrlich, D. Quantum nonlocality as an axiom. *Found. Phys.* **1994**, *24*, 379–385. [CrossRef]
3. Navascués, M.; Pironio, S.; Acín, A. A convergent hierarchy of semidefinite programs characterizing the set of quantum correlations. *New J. Phys.* **2008**, *10*, 073013. [CrossRef]
4. Pawłowski, M.; Paterek, T.; Kaszlikowski, D.; Scarani, V.; Winter, A.; Żukowski, M. Information causality as a physical principle. *Nature* **2009**, *461*, 1101–1104. [CrossRef] [PubMed]
5. Navascués, M.; Wunderlich, H. A glance beyond the quantum model. *Proc. R. Soc. A* **2010**, *466*, 881–890. [CrossRef]
6. Popescu, S. Nonlocality beyond quantum mechanics. *Nat. Phys.* **2014**, *10*, 264–270. [CrossRef]
7. Brunner, N.; Cavalcanti, D.; Pironio, S.; Scarani, V.; Wehner, S. Bell nonlocality. *Rev. Mod. Phys.* **2014**, *86*, 419. [CrossRef]
8. Goh, K.T.; Kaniewski, J.; Wolfe, E.; Vértesi, T.; Wu, X.; Cai, Y.; Liang, Y.-C.; Scarani, V. Geometry of the set of quantum correlations. *Phys. Rev. A* **2018**, *97*, 022104. [CrossRef]
9. Carmi, A.; Cohen, E. Relativistic independence bounds nonlocality. *arXiv* **2018**, arXiv:1806.03607.
10. Landau, L. Empirical two-point correlation functions. *Found. Phys.* **1988**, *18*, 449–460. [CrossRef]
11. Bell, J.S. On the Einstein Podolsky Rosen paradox. *Physics* **1964**, *1*, 195–200. [CrossRef]
12. Clauser, J.F.; Horne, M.A.; Shimony, A.; Holt, R.A. Proposed experiment to test local hidden-variable theories. *Phys. Rev. Lett.* **1969**, *23*, 880. [CrossRef]
13. Carmi, A.; Moskovich, D. Tsirelson's bound prohibits communication through a disconnected channel. *Entropy* **2018**, *20*, 151. [CrossRef]
14. Tsirel'son, B.S. Quantum analogues of the Bell inequalities. The case of two spatially separated domains. *J. Sov. Math.* **1987**, *36*, 557–570. [CrossRef]
15. Masanes, L. Necessary and sufficient condition for quantum-generated correlations. *arXiv* **2003**, arXiv:quant-ph/0309137.
16. Fine, A. Hidden variables, joint probability, and the Bell inequalities. *Phys. Rev. Lett.* **1982**, *48*, 291. [CrossRef]
17. Tsallis, C. Possible generalization of Boltzmann-Gibbs statistics. *J. Stat. Phys.* **1988**, *52*, 479–487. [CrossRef]
18. Oppenheim, J.; Wehner, S. The uncertainty principle determines the nonlocality of quantum mechanics. *Science* **2010**, *330*, 1072–1074. [CrossRef] [PubMed]
19. Aharonov, Y.; Albert, D.; Vaidman, L. How the result of a measurement of a component of the spin of a spin-1/2 particle can turn out to be 100. *Phys. Rev. Lett.* **1988**, *60*, 1351–1354. [CrossRef] [PubMed]
20. Aharonov, Y.; Cohen, E.; Elitzur, A.C. Can a future choice affect a past measurement's outcome? *Ann. Phys.* **1988**, *355*, 258–268. [CrossRef]
21. White, T.C.; Mutus, J.Y.; Dressel, J.; Kelly, J.; Barends, R.; Jeffrey, E.; Sank, D.; Megrant, A.; Campbell, B.; Chen, Y.; et al. Preserving entanglement during weak measurement demonstrated with a violation of the Bell-Leggett-Garg inequality. *NPJ Quantum Inf.* **2016**, *2*, 15022. [CrossRef]
22. Mermin, N.D. Extreme quantum entanglement in a superposition of macroscopically distinct states. *Phys. Rev. Lett.* **1990**, *65*, 1838. [CrossRef] [PubMed]

23. Collins, D.; Gisin, N. A relevant two qubit Bell inequality inequivalent to the CHSH inequality. *J. Phys. A Math. Gen.* **2004**, *37*, 1775–1787. [CrossRef]
24. Pál, F.K.; Vértesi, T. Maximal violation of a bipartite three-setting, two-outcome Bell inequality using infinite-dimensional quantum systems. *Phys. Rev. A* **2010**, *82*, 022116. [CrossRef]

entropy

MDPI

Article

Entropic Steering Criteria: Applications to Bipartite and Tripartite Systems

Ana C. S. Costa *, Roope Uola and Otfried Gühne

Naturwissenschaftlich-Technische Fakultät, Universität Siegen, 57068 Siegen, Germany;
roope.uola@gmail.com (R.U.); otfried.guehne@uni-siegen.de (O.G.)
* Correspondence: ana.sprotte@gmail.com or ana.costa@physik.uni-siegen.de; Tel.: +49-271-740-3797

Received: 23 August 2018; Accepted: 20 September 2018; Published: 5 October 2018

Abstract: The effect of quantum steering describes a possible action at a distance via local measurements. Whereas many attempts on characterizing steerability have been pursued, answering the question as to whether a given state is steerable or not remains a difficult task. Here, we investigate the applicability of a recently proposed method for building steering criteria from generalized entropic uncertainty relations. This method works for any entropy which satisfy the properties of (i) (pseudo-) additivity for independent distributions; (ii) state independent entropic uncertainty relation (EUR); and (iii) joint convexity of a corresponding relative entropy. Our study extends the former analysis to Tsallis and Rényi entropies on bipartite and tripartite systems. As examples, we investigate the steerability of the three-qubit GHZ and W states.

Keywords: steering; entropic uncertainty relation; general entropies

---

1. Introduction

The notion of steering was first introduced by Schrödinger in 1935 in order to capture the essence of the Einstein–Podolsky–Rosen argument [1]. It describes the ability of one experimenter, Alice, to remotely affect the state of another experimenter, Bob, through local actions on her system supported by classical communication. Steering is based on a quantum correlation strictly between entanglement and non-locality, meaning that not every entangled state can be used for steering and not every steerable state violates a Bell inequality [2].

Recently, it has been shown that steering plays a fundamental role in various quantum protocols and in entanglement theory. In the former, steering characterizes systems useful for one-sided device-independent quantum key distribution [3], subchannel discrimination [4] and randomness generation [5]. Concerning entanglement theory, steering has been used to find counterexamples to Peres conjecture, which was an open problem for more than fifteen years [6–8]. Steering is also known to be closely related to incompatibility of quantum measurements. Namely, any set of non-jointly measurable observables is useful for demonstrating steering [9,10], and every incompatibility problem can be mapped into a steering problem in a one-to-many manner [11–13].

The extension of steering to multipartite systems has also been proposed. In the multipartite setting the concept of steering has some ambiguity in it. Whether one is interested in the typical spooky action at a distance [14–16] or in a more detailed semi-device independent entanglement verification scheme [17,18], one ends up with two different definitions. Here we are interested in the latter scenario as it relates more closely to our approach.

To detect steerability of a given bipartite quantum state might turn into a cumbersome task. The question of steerability (with given measurements on Alice's side and tomography on Bob's side) can be formulated as a semidefinite program (SDP) [19–21] and as such one could imagine that the task is straightforward and easy to implement. Whereas SDP methods provide a powerful tool for steering detection, they are often restricted to systems with only a few measurements and small dimensions due to computational limitations. One should mention, though, that SDP methods can be used to set bounds for steering even in a scenario with a continuum of measurements [22–24]. Another way of detecting steering is through criteria based on correlations [2,25–29]. Whereas these criteria are mostly analytical and straightforward to evaluate, they are also often either not optimal or limited to qubit systems.

In Ref. [30], steering criteria are developed from entropic uncertainty relations (EURs). The criteria are based on generalized entropies, hence, forming an extension of the entropic criteria in Refs. [31,32]. The work falls into the second category of the aforementioned classification of steering criteria, and, as pointed out by the authors, the criteria of Ref. [30] manage to beat other correlation-based methods either in applicability or in detection power. In this work we extend the analysis of Ref. [30] to Rényi entropies and to tripartite steering scenarios. We discuss in detail the question of steerability with local and global measurements in the tripartite setting. Please note that recently similar efforts have been pursued in the context of Rényi entropies [33].

This work is organized as follows. First, we introduce the concept of steering for bipartite and tripartite systems in Section 2. Second, we present some useful entropies for the characterization of steering, followed by bounds of EURs in Section 3, where we also propose some bounds for Tsallis entropies, obtained from numerical investigations. We explain the criteria for the detection of steering from EURs in Section 4. In Section 5 we provide a connection to existing entanglement criteria. In Section 6 we investigate the optimal parameters from generalized entropies for the detection of steering, followed by the application of the criteria to some common examples. Finally, in Section 7 we extend the criteria to the tripartite case, and apply it to noisy GHZ and W states. We conclude the paper with some final remarks.

2. Steering

In a bipartite steering scenario, Alice and Bob share a quantum state, Alice performs local actions (measurements) on her part of the state and Bob is left with non-normalised states (or a state assemblage) depending on Alice's choice of measurement and her reported outcomes. The task for Bob is to verify if his assemblages could be prepared using a separable state or not [29]. In a more formal manner, we can assume that Alice performs a measurement A with outcome i on her part of the system, while Bob performs a measurement B with outcome j on his part. From that, they can obtain the joint probability distribution of the outcomes. If for all possible measurements A and B one can express the joint probabilities in the form

$$p(i,j|A,B) = \sum_\lambda p(\lambda)p(i|A,\lambda)p_Q(j|B,\lambda), \tag{1}$$

then the shared state is called unsteerable. Here, $p(i|A,\lambda)$ is a general probability distribution, while $p_Q(j|B,\lambda) = \mathrm{Tr}_B[B(j)\sigma_\lambda]$ is a probability distribution originating from a quantum state σ_λ. Furthermore, $B(j)$ denotes a measurement operator, i.e., $B(j) \geq 0$ and $\sum_j B(j) = \mathbb{1}$, and $\sum_\lambda p(\lambda) = 1$, where λ is a label for the hidden quantum state σ_λ. A model as in Equation (1) is called a local hidden state (LHS) model, and if it exists, Bob can explain all the results through a set of local states $\{\sigma_\lambda\}$ which is only altered by the classical information about Alice's performed measurement and the recorded outcome. Otherwise, the state is called steerable. One should notice that for a state to be unsteerable, one has to prove the existence of an LHS model for all possible measurements on Alice's side and for a tomographically complete set on Bob's side, whereas for proving steerability it suffices to find a set of measurements for Alice and Bob for which the probabilities cannot be expressed as Equation (1).

For multipartite systems, LHS models can be extended in different ways. For simplicity, let us consider the case of tripartite systems. In addition to the notation used before, we assume that Charlie performs measurements C with outcomes labelled with k. Then, one possibility is to ask if Alice can steer the state of Bob and Charlie. If for all possible measurements A, B and C the joint probability distribution can be expressed as

$$p(i,j,k|A,B,C) = \sum_\lambda p(\lambda)p(i|A,\lambda)p_Q(j|B,\lambda)p_Q(k|C,\lambda), \tag{2}$$

the system is called unsteerable from Alice to Bob and Charlie. Here, $p_Q(j|B,\lambda)p_Q(k|C,\lambda) = \mathrm{Tr}[B(j) \otimes C(k)(\sigma_\lambda^B \otimes \sigma_\lambda^C)]$, where the hidden states of Bob and Charlie are factorizable. We require the factorizability in order to distinguish the tripartite scenario from a bipartite one (i.e., Bob and Charlie being a single system) where unsteerability is defined as

$$p(i,j,k|A,B,C) = \sum_\lambda p(\lambda)p(i|A,\lambda)p_Q(j,k|B,C,\lambda), \tag{3}$$

with $p_Q(j,k|B,C,\lambda) = \mathrm{Tr}[B(j) \otimes C(k)\sigma_\lambda^{BC}]$. Please note that the factorizability requirement includes all hidden state models using separable states through a redefinition of the hidden variable space. From a physical point of view, Equation (2) corresponds to tests of full separability with untrusted Alice; whereas Equation (3) corresponds to tests of biseparability in the $A|BC$ cut with untrusted Alice.

Another possibility is to ask whether the joint probability distribution of measurements performed by Alice, Bob and Charlie, can be expressed as

$$p(i,j,k|A,B,C) = \sum_\lambda p(\lambda)p(i|A,\lambda)p(j|B,\lambda)p_Q(k|C,\lambda), \tag{4}$$

which means that the system is unsteerable from Alice and Bob to Charlie with factorizable post-processing, meaning that we assume the post-processings on one party to be independent of that of the other party. This extra assumption is one possibility to distinguish, between bipartite and tripartite scenarios. Please note that one could also require non-signalling instead of factorizability of the post-processings. In a purely bipartite scenario an unsteerable joint probability distribution would be given by

$$p(i,j,k|A,B,C) = \sum_\lambda p(\lambda)p(i,j|A,B,\lambda)p_Q(k|C,\lambda). \tag{5}$$

Similarly to the above scenario, Equation (4) corresponds to tests of full separability with untrusted Alice and Bob; whereas Equation (5) corresponds to tests of biseparability in the $AB|C$ cut with untrusted Alice and Bob.

One should notice that, in the steering scenario from Alice and Bob to Charlie, there is a difference whether Alice and Bob decide to perform global or local measurements. A simple example of this difference can be explored in the framework of super-activation of steering [34]. Here, the authors show that while one copy of a quantum state is unsteerable, many copies of the same state become steerable, in the sense that steerability is "activated". Namely, consider a state $\varrho_{ABCC'} = \varrho_{AC} \otimes \varrho_{BC'}$, where $\varrho_{BC'}$ is a copy of ϱ_{AC}, and ϱ_{AC} is unsteerable, but its steerability can be super-activated (where only two copies is already enough [34]). For this state, local measurements give an unsteerable state assemblage, whereas, because of super-activation, it is steerable with global measurements.

3. Entropies and Entropic Uncertainty Relations

3.1. Entropies

Let us state some basic facts about entropies. For a general probability distribution $\mathcal{P} = (p_1,\ldots,p_N)$, the Shannon entropy is defined as [35]

$$S(\mathcal{P}) = -\sum_i p_i \ln(p_i). \tag{6}$$

As a possible generalized entropy, we consider the so-called Tsallis entropy [36,37] which depends on a parameter $0 < q \neq 1$. It is given by

$$S_q(\mathcal{P}) = -\sum_i p_i^q \ln_q(p_i),$$

(7)

where the q-logarithm is defined as $\ln_q(x) = (x^{1-q} - 1)/(1-q)$. Another generalization of Shannon entropy is known as Rényi entropy [38], which is defined depending on a parameter $0 < r \neq 1$ as

$$\tilde{S}_r(\mathcal{P}) = \frac{1}{1-r} \ln \left[\sum_i p_i^r \right].$$

(8)

The above entropies have the following properties [35–38]:

1. The entropies S, S_q and $\tilde{S}_r$ are positive and they are zero if and only if the probability distribution is concentrated at one value (k), i.e., $p_i = \delta_{ik}$.
2. In the limit of $q \to 1$ and $r \to 1$, the Tsallis and Rényi entropies converge to the Shannon entropy, and both decrease monotonically in q and r.
3. The Rényi entropy is a monotonous function of the Tsallis entropy:

$$\tilde{S}_r(\mathcal{P}) = \frac{\ln[1 + (1-r)S_{q=r}(\mathcal{P})]}{1-r}.$$

(9)

4. Shannon and Tsallis entropy are concave functions in $\mathcal{P}$, i.e., they obey the relation

$$f(\lambda \mathcal{P}_1 + (1-\lambda)\mathcal{P}_2) \geq \lambda f(\mathcal{P}_1) + (1-\lambda)f(\mathcal{P}_2),$$

(10)

where $f = S$ for Shannon entropy, and $f = S_q$ for Tsallis entropy. The Rényi entropy is concave if $r \in (0;1)$, and for other values of r it is neither convex nor concave.

5. In the limit of $r \to \infty$, the Rényi entropy is known as min-entropy

$$\lim_{r \to \infty} \tilde{S}_r(\mathcal{P}) = -\ln \max_i(p_i).$$

(11)

6. For two independent distributions, $\mathcal{P}$ and $\mathcal{Q}$, Shannon and Rényi entropies are additive, i.e.,

$$S(\mathcal{P}, \mathcal{Q}) = S(\mathcal{P}) + S(\mathcal{Q}),$$

(12)

$$\tilde{S}_r(\mathcal{P}, \mathcal{Q}) = \tilde{S}_r(\mathcal{P}) + \tilde{S}_r(\mathcal{Q}),$$

(13)

whereas Tsallis entropy is pseudo-additive, i.e.,

$$S_q(\mathcal{P}, \mathcal{Q}) = S_q(\mathcal{P}) + S_q(\mathcal{Q}) + (1-q)S_q(\mathcal{P})S_q(\mathcal{Q}).$$

(14)

3.2. Relative Entropies

The relative entropy, also known as Kullback–Leibler divergence [35], for two probability distributions $\mathcal{P}$ and $\mathcal{Q}$ is given by

$$D(\mathcal{P}||\mathcal{Q}) = \sum_i p_i \ln \left(\frac{p_i}{q_i} \right).$$

(15)

For Tsallis and Rényi entropies the relative entropy is defined as [38–40]

$$D_q(\mathcal{P}||\mathcal{Q}) = -\sum_i p_i \ln_q \left(\frac{q_i}{p_i} \right), \qquad \tilde{D}_r(\mathcal{P}||\mathcal{Q}) = \frac{1}{r-1} \ln \left(\sum_i p_i^r q_i^{1-r} \right),$$

(16)

respectively. The Rényi relative entropy is also known as Rényi divergence.

Here, we discuss two properties which are essential in this work: first, the relative entropy is additive for independent distributions, that is if $\mathcal{P}_1, \mathcal{P}_2$ are two probability distributions with the joint distribution $\mathcal{P}(x,y) = \mathcal{P}_1(x)\mathcal{P}_2(y)$, and the same for $\mathcal{Q}_1, \mathcal{Q}_2$, then one has

$$D(\mathcal{P}||\mathcal{Q}) = D(\mathcal{P}_1||\mathcal{Q}_1) + D(\mathcal{P}_2||\mathcal{Q}_2), \tag{17}$$

and the same holds for the generalized Rényi relative entropy,

$$\tilde{D}_r(\mathcal{P}||\mathcal{Q}) = \tilde{D}_r(\mathcal{P}_1||\mathcal{Q}_1) + \tilde{D}_r(\mathcal{P}_2||\mathcal{Q}_2). \tag{18}$$

However, for the generalized Tsallis relative entropy, we have [40]

$$D_q(\mathcal{P}||\mathcal{Q}) = D_q(\mathcal{P}_1||\mathcal{Q}_1) + D_q(\mathcal{P}_2||\mathcal{Q}_2) + (q-1)D_q(\mathcal{P}_1||\mathcal{Q}_1)D_q(\mathcal{P}_2||\mathcal{Q}_2),$$

where the additional term is due to the pseudo-additivity of the generalized entropy.

Second, the relative entropy is jointly convex. This means that for two pairs of distributions $\mathcal{P}_1, \mathcal{Q}_1$ and $\mathcal{P}_2, \mathcal{Q}_2$ one has

$$D[\lambda\mathcal{P}_1 + (1-\lambda)\mathcal{P}_2||\lambda\mathcal{Q}_1 + (1-\lambda)\mathcal{Q}_2] \le \lambda D(\mathcal{P}_1||\mathcal{Q}_1) + (1-\lambda)D(\mathcal{P}_2||\mathcal{Q}_2). \tag{19}$$

The generalized Tsallis relative entropy is also jointly convex for all values of q, while the generalized Rényi relative entropy is jointly convex only for $r \in (0;1)$ (see Theorem 11 in Ref. [41]).

3.3. Entropic Uncertainty Relations

Entropies are useful for the investigation of uncertainty relations [42]. Entropic uncertainty relations (or EURs for short) can be easily explained with an example. Consider the Pauli measurements σ_x and σ_z on a single qubit. For any quantum state these measurements give rise to a two-valued probability distribution and to the corresponding entropy $S(\sigma_m)$ for $m = x, z$. The fact that σ_x and σ_z do not share a common eigenstate can be expressed as [43]

$$S(\sigma_x) + S(\sigma_z) \ge \ln(2), \tag{20}$$

where the lower bound does not depend on the state. These type of relations can be extended to more measurements and other entropies, and the search for the optimal bounds is an active field of research.

In a general way, if one performs m measurements, the bounds of an EUR can be estimated in the following way

$$\sum_m S(X_m) \ge \min_\varrho \sum_m S(X_m)_\varrho = \mathcal{B}, \tag{21}$$

where the minimization, due to the concavity of the entropy, involves all pure (single system) states. Various analytical entropic uncertainty bounds are known for Shannon, Tsallis and Rényi entropies, and we introduce some of them in this section, together with new bounds for Tsallis entropy obtained from numerical investigations. These bounds will be useful in later sections, where we develop steering criteria based on the relative entropy between two probability distributions.

For the estimation of the bounds for EURs we consider mutually unbiased bases (MUBs) [44]. Two orthonormal bases are mutually unbiased if the absolute value of the overlap between any vector from one basis with any vector from the other basis is equal to $1/d$. For a given dimension d, it is simple to construct a pair of MUBs through, for example, the discrete Fourier transform. If d is a prime or power of a prime, the existence of $d + 1$ MUBs (i.e., a complete set of MUBs) is known. However, the number of MUBs existing in other than prime and power of prime dimensions is a long standing open problem [45].

For the Shannon entropy and a complete set of MUBs (provided that they exist), the bounds for EURs for dimension d were analytically derived in Ref. [46] and are given by

$$
\mathcal{B} =
\begin{cases}
(d+1)\ln\left(\frac{d+1}{2}\right), & d \text{ odd} \\[2mm]
\frac{d}{2}\ln\left(\frac{d}{2}\right) + \left(\frac{d}{2}+1\right)\ln\left(\frac{d}{2}+1\right), & d \text{ even.}
\end{cases}
\tag{22}
$$

Later, a bound was proved in Ref. [47] for m MUBs (which coincides with the above bound for a complete set),

$$
\mathcal{B} = m\ln(K) + (K+1)\left(m - K\frac{d+m-1}{d}\right)\ln\left(1 + \frac{1}{K}\right),
\tag{23}
$$

where $K = \left\lfloor \frac{md}{d+m-1} \right\rfloor$ and $\lfloor \cdot \rfloor$ is the floor function. Please note that steering with MUBs can be also attacked using techniques from the field of joint measurability [48–50].

We are also interested in EURs not only for single systems, but for composite ones as well. For bipartite systems we have the following bound

$$
\sum_m S(X_m^A, X_m^B) \geq \min_{\varrho_{AB}} \sum_m S(X_m^A, X_m^B)_{\varrho_{AB}} = \mathcal{C},
\tag{24}
$$

where the minimization involves all pure single system states. Here, $S(X_m^A, X_m^B)$ is the Shannon entropy of the probability distribution $p_{ij}^{(m)} = \langle i_m|\langle j_m|\varrho_{AB}|i_m\rangle|j_m\rangle$, with d^2 outcomes, where the MUBs $\{|i_m\rangle\}_m, \{|j_m\rangle\}_m$ work as the eigenvectors of the measurement $X_m^{A(B)}$. Please note that we use the symbol $\mathcal{B}$ for the bounds on single systems, while $\mathcal{C}$ is used for composite ones.

One should note that there might be a difference between the bounds obtained from separable and entangled states. A simple example is given by a two-qubit system and Pauli measurements. Optimizing over all two-qubit states and considering two Pauli measurements, we have

$$
S(\sigma_x, \sigma_x) + S(\sigma_y, \sigma_y) \geq 2\ln(2).
\tag{25}
$$

This bound is already reached with separable states [51]. Meanwhile, if one considers separable states and three Pauli measurements (which represent a complete set of MUBs for two-dimensional systems), the bound is

$$
\sum_m S(\sigma_m, \sigma_m)_{\varrho_{sep}} \geq 4\ln(2),
\tag{26}
$$

where $m = \{x, y, z\}$. However, if one considers the maximally entangled state $\varrho_{ent} = |\psi^-\rangle\langle\psi^-|$, the following value is reached

$$
\sum_m S(\sigma_m, \sigma_m)_{\varrho_{ent}} = 3\ln(2),
\tag{27}
$$

for the same measurements. Please note that, for separable states, the bound in Equation (25) follows from additivity of Shannon entropy and Equation (20), whereas the bound in Equation (26) follows from additivity and the bound in Ref. [46]. Moreover, the additivity of EUR for Shannon entropy is discussed in Ref. [51].

An analytical bound for separable states ($\varrho = \sum_j p_j \varrho_j^A \otimes \varrho_j^B$, with Hilbert space dimensions d_A and d_B) and m MUBs performed in each system is given by [47]

$$\begin{aligned}
\mathcal{C} \;=\;& m\ln(K_A) + m\ln(K_B) + (K_A+1)\left(m - K_A\frac{d_A+m-1}{d_A}\right)\ln\left(1+\frac{1}{K_A}\right) \\
&+(K_B+1)\left(m - K_B\frac{d_B+m-1}{d_B}\right)\ln\left(1+\frac{1}{K_B}\right),
\end{aligned}$$

$$(28)$$

with $K_{A(B)}$ defined as above. Please note that this bound is the sum of the bounds (23) for both subsystems. Here, this bound also holds because of concavity and additivity of Shannon entropy.

Now, let us present the bounds for generalized entropies. For the Tsallis entropy and m MUBs it has been shown in Ref. [52] that, for $q \in (0;2]$, the bound is given by

$$\mathcal{B}^{(q)} = m\ln_q\left(\frac{md}{d+m-1}\right).$$

$$(29)$$

For $q \to 1$, this bound is not optimal for even dimensions, so in this case it is more appropriate to consider the bounds given in Equation (22).

In Ref. [53], a bound for the Tsallis entropy and two-qubit systems was analytically derived, and for every $q \in [2n-1, 2n], n \in \mathbb{N}$, the bound is

$$\mathcal{B}^{(q)} = \ln_q(2),$$

$$(30)$$

for two-measurement settings composed by Pauli operators, which are MUBs in dimension 2. Numerically, these bounds seem also to hold for other values of q, except $q \in (2;3)$. For three measurement settings, one can obtain numerically the following bound

$$\mathcal{B}^{(q)} = 2\ln_q(2),$$

$$(31)$$

which is also not optimal for $q \in (2;3)$ (see Ref. [53]). Extending these bounds for arbitrary (finite) dimensions and m mutually unbiased measurements, numerical investigations suggest that, for $q \geq 2$,

$$\mathcal{B}^{(q)} = (m-1)\ln_q(d).$$

$$(32)$$

To be more precise, the above function seems to match the numerically calculated optimal values for small values of q, d and m.

Now, if one considers two-qubit systems, we introduce here the bounds for Tsallis entropy (with $q > 1$), obtained from numerical investigation by minimizing over all pure states. They are given by

$$\mathcal{C}^{(q)} = \ln_q(4)$$

$$(33)$$

for two Pauli measurements, where this bound is already reached by separable states, and

$$\mathcal{C}^{(q)} = \begin{cases} 3\ln_q(2), & 1 \leq q \leq 2 \\ 2\ln_q(4) & q \geq 2 \end{cases}$$

$$(34)$$

for three Pauli measurements. Here, in the range of $1 \leq q \leq 2$ the bound gets lower due to entanglement. In the range $q \geq 2$ separable states give the best bounds for this setting of measurements. In Figure 1 we show these results. All these bounds were obtained numerically and as their analytical proof remains an open question, we use these conjectured bounds in our calculations.

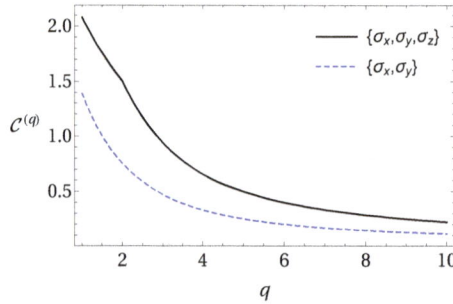

Figure 1. Numerical lower bounds for composite systems in Equations (33) and (34) in terms of the parameter q.

Regarding the bounds for Rényi entropy, we have the following scenarios [52]: in the range $r \in (0; 2)$ the bounds are independent of r and they equal the bounds for Shannon entropy; and for $r \in [2; \infty)$ the state-independent bounds are

$$\mathcal{B}_r^{(r)} = \frac{mr}{2(r-1)} \ln \left(\frac{md}{d+m-1} \right). \tag{35}$$

4. Entropic Steering Criteria

In this section we present the detailed derivation of the generalized entropic steering criteria proposed in Ref. [30]. Here we show the results for Shannon and Tsallis entropies, as has been made in Ref. [30], and also extend the criteria for Rényi entropy. Please note that the proof is not at all restricted to these functions as it can be applied to all functions which satisfy the following properties: (i) (pseudo-)additivity for independent distributions; (ii) state independent EUR; and (iii) joint convexity of the relative entropy. In the following we present our proof for specific entropies, and the method becomes clear from its application to each of them.

4.1. Entropic Steering Criteria for Shannon Entropy

The starting point of our method is to consider the relative entropy (15) between two distributions, i.e.,

$$F(A, B) = -D(A \otimes B || A \otimes \mathbb{I}). \tag{36}$$

Here $A \otimes B$ denotes the joint probability distribution $p(i, j | A, B)$, which we further denote by p_{ij}, A is the marginal distribution $p(i|A)$, which we denote by p_i, and $\mathbb{I}$ is a uniform distribution with $q_j = 1/N$ for all outcomes $j \in \{1, \cdots, N\}$. As the relative entropy is jointly convex, $F(A, B)$ is concave in the probability distribution $A \otimes B$. Then, we get

$$F(A, B) = -\sum_{ij} p_{ij} \ln \left(\frac{p_{ij}}{p_i / N} \right) = S(A, B) - S(A) - \ln(N) = S(B|A) - \ln(N), \tag{37}$$

where $S(B|A)$ is the Shannon conditional entropy. On the other hand, considering a product distribution $p(i|A, \lambda) p_Q(j|B, \lambda)$ with a fixed λ and using the property from Equation (17), we have

$$F^{(\lambda)}(A, B) = -D[p(i|A, \lambda)||p(i|A, \lambda)] - D[p_Q(j|B, \lambda)||\mathbb{I}] = -D[p_Q(j|B, \lambda)||\mathbb{I}] = S^{(\lambda)}(B) - \ln(N). \tag{38}$$

The term $S^{(\lambda)}(B)$ on the right-hand side of this equation depends on probability distributions taken from the quantum state σ_λ. For a given set of measurements $\{B_m\}$, such distributions typically obey an EUR

$$\sum_m S^{(\lambda)}(B_m) \geq \mathcal{B}_B, \tag{39}$$

where $\mathcal{B}_B$ is some entropic uncertainty bound for the observables B_m. Finally, since S is concave, the same bound holds for convex combinations of product distributions $p(i|A,\lambda)p_Q(j|B,\lambda)$ from Equation (1). Connecting this to Equations (37) and (38) we have, for a set of measurements $\{A_m \otimes B_m\}_m$,

$$\sum_m S(B_m|A_m) \geq \mathcal{B}_B, \tag{40}$$

which means that any nonsteerable quantum system obeys this relation. In this way EURs can be used to derive steering criteria. The intuition behind these criteria is based on the interpretation of Shannon conditional entropy. In Equation (40), one can see that the knowledge Alice has about Bob's outcomes is bounded. If this inequality is violated, then the system is steerable, meaning that Alice can do better predictions than those allowed by an EUR.

This criterion is more general than the one in Ref. [32], since our proof can easily also be extended to other generalized entropies, as we show in the following.

4.2. Entropic Steering Criteria for Generalized Entropies

4.2.1. Tsallis Entropy

Now we can apply the machinery derived above and consider the quantity $F_q(A,B) = -D_q(A \otimes B||A \otimes \mathbb{I})$. Using the definition of the generalized relative entropy, we have

$$F_q(A,B) = \sum_{i,j} p_{ij} \ln_q \left(\frac{p_i/N}{p_{ij}} \right) = \frac{x}{1-q} + (1+x)\left\{ S_q(B|A) + (1-q)C(A,B) \right\}, \tag{41}$$

where $S_q(B|A) = S_q(A,B) - S_q(A)$ is the conditional Tsallis entropy [54], $x = N^{q-1} - 1$, and

$$C(A,B) = \sum_i p_i^q [\ln_q(p_i)]^2 - \sum_{i,j} p_{ij}^q \ln_q(p_i) \ln_q(p_{ij}), \tag{42}$$

is the correction term.

Now, considering the property from Equation (19) and a product distribution $p(i|A,\lambda)p_Q(j|B,\lambda)$ with a fixed λ one gets

$$F_q^{(\lambda)}(A,B) = \frac{x}{1-q} + (1+x)S_q^{(\lambda)}(B). \tag{43}$$

It follows by direct calculation that if the measurements $\{B_m\}_m$ obey an EUR

$$\sum_m S_q(B_m) \geq \mathcal{B}_B^{(q)} \tag{44}$$

then one has the steering criterion

$$\sum_m \left[S_q(B_m|A_m) + (1-q)C(A_m,B_m) \right] \geq \mathcal{B}_B^{(q)}. \tag{45}$$

From Equation (45) it is easy to see that if we consider $q \to 1$, we arrive at Equation (40). Note that one can rewrite Equation (45) in terms of probabilities as

$$\frac{1}{q-1}\left[\sum_k \left(1 - \sum_{ij}\frac{(p_{ij}^{(m)})^q}{(p_i^{(m)})^{q-1}}\right)\right] \geq \mathcal{B}_B^{(q)}. \tag{46}$$

Here, $p_{ij}^{(m)}$ is the probability of Alice and Bob for outcome (i, j) when measuring $A_m \otimes B_m$, and $p_i^{(m)}$ are the marginal outcome probabilities of Alice's measurement A_m. This form of the criterion is straightforward to evaluate.

4.2.2. Rényi Entropy

If one considers the quantity $\tilde{F}_r(A, B) = -\tilde{D}_r(A \otimes B || A \otimes \mathbb{I})$ with the measurements B_m obeying an EUR

$$\sum_m \tilde{S}_r(B_m) \geq \check{\mathcal{B}}_B^{(r)}, \tag{47}$$

we have the following steering criterion for Rényi entropy

$$\frac{1}{1-r}\sum_m \ln\left[\sum_{i,j}(p_{ij}^{(m)})^r\,(p_i^{(m)})^{1-r}\right] \geq \check{\mathcal{B}}_B^{(r)}. \tag{48}$$

Please note that for the range $r \in (0; 1)$ the bound is independent of r, and it is the same as the bound for Shannon entropy [52]. Unlike the other entropies, we cannot write the result in terms of Rényi conditional entropies, given its definition is not clear in the literature (see discussion in Ref. [55]).

5. Connection to Existing Entanglement Criteria

At this point, it is interesting to connect our approach with the entanglement criteria derived from EURs [53]. In Ref. [53], it has been shown that for separable states the following inequality

$$S_q(A_1 \otimes B_1) + S_q(A_2 \otimes B_2) \geq \mathcal{B}_B^{(q)} \tag{49}$$

holds. Here, A_1 and A_2 (B_1 and B_2) are observables on Alice's (Bob's) laboratory, and Bob's observables obey an EUR $S_q(B_1) + S_q(B_2) \geq \mathcal{B}_B^{(q)}$. Differently from our approach, $S_q(A_m \otimes B_m)$ is the entropy of the probability distribution of the outcomes of the *global* observable $A_m \otimes B_m$. Please note that for a degenerate $A_m \otimes B_m$ the probability distribution differs from the local ones. For instance, measuring $\sigma_z \otimes \sigma_z$ gives four possible local probabilities $p_{++}, p_{+-}, p_{-+}, p_{--}$, but for the evaluation of $S(A_m \otimes B_m)$ one combines them as $q_+ = p_{++} + p_{--}$ and $q_- = p_{+-} + p_{-+}$, as these correspond to the global outcomes.

There are some interesting connections between our derivation of steering inequalities and this entanglement criterion. First, the proof in Ref. [53] is based on EURs for Bob's observables (the same as our criteria), and this is the only quantum restriction in the criterion, so Equation (49) is a steering inequality, meaning that all probability distributions of the form in Equation (1) fulfil it. Second, in Ref. [53] it was observed that the criterion is strongest for values $2 \leq q \leq 3$, which seems to be the case also for our criteria (shown later). Third, for special scenarios (e.g., Bell-diagonal two-qubit states and Pauli measurements), Equation (49) and Equations (40) and (45) give the same results. However, it does not hold for more general scenarios.

The approach of Ref. [53] has been slightly improved in Ref. [56], where the main idea is to recombine the probability distribution in a different way (see below). Also, the criteria in Ref. [56] are more general in the sense that can be applied to any symmetric and concave function.

Similar to the case of Ref. [53], the criteria from Ref. [56] can be also applied to steering. To see this, let us first explain the main ideas in [56]. In this work, they consider concave and symmetrical (i.e., invariant under the permutation of variables) functions $f : \mathbb{R}^n \longrightarrow \mathbb{R}$. For simplicity, define

$$f(\rho_0, e) = f(\langle e_1|\rho_0|e_1\rangle, \ldots, \langle e_n|\rho_0|e_n\rangle),\tag{50}$$

where $e = \{|e_i\rangle | i = 1, \ldots, n\}$ is an orthonormal basis of the n-dimensional Hilbert space in which the state ρ_0 acts. Then, one can construct a probability matrix $P = (p_{ij})$, where the elements are defined as $p_{ij} = \langle e_i^A|\langle e_j^B|\rho|e_j^B\rangle|e_i^A\rangle$, where $e_A = \{|e_i^A\rangle | i = 1, \cdots, n_A\}$ and $e_B = \{|e_j^B\rangle | j = 1, \cdots, n_B\}$ are orthonormal bases of the $n_{A(B)}$-dimensional Hilbert space $\mathcal{H}_{A(B)}$.

Then, define a permutation matrix $Q = (q_{ij})$, where $\{q_{i1}, \ldots, q_{in_B}\}$ is a permutation of an n_B-element set $S = \{s_1, \ldots, s_{n_B}\}$ for $i = 1, \ldots, n_A$. For example, if we consider the case of $n_A = n_B = 3$, three examples of possible constructions of Q are

$$\begin{pmatrix} s_1 & s_1 & s_1 \\ s_2 & s_2 & s_2 \\ s_3 & s_3 & s_3 \end{pmatrix}, \quad \begin{pmatrix} s_1 & s_1 & s_3 \\ s_2 & s_3 & s_1 \\ s_3 & s_2 & s_2 \end{pmatrix}, \quad \begin{pmatrix} s_3 & s_1 & s_2 \\ s_1 & s_3 & s_3 \\ s_2 & s_2 & s_1 \end{pmatrix}.\tag{51}$$

Now, define

$$f(\varrho_{AB}, e_A, e_B, Q) = f\left(\sum_{ij}\delta(q_{ij}, s_1)p_{ij}, \sum_{ij}\delta(q_{ij}, s_2)p_{ij}, \cdots, \sum_{ij}\delta(q_{ij}, s_{n_A})p_{ij}\right),\tag{52}$$

where $\delta(a, b)$ is the Kronecker function. Here, the argument of this function is the combination of the probabilities given a permutation matrix Q. If we take the third example in Equation (51), we have

$$f(\varrho_{AB}, e_A, e_B, Q) = f(p_{21} + p_{12} + p_{33}, p_{31} + p_{32} + p_{13}, p_{11} + p_{22} + p_{23}).\tag{53}$$

Here one can see that the combination of the probabilities will depend on the permutation matrix Q.

Given the above definitions, the authors prove the following bound for product states ($\varrho_{AB} = \varrho_A \otimes \varrho_B$),

$$f(\varrho_{AB}, e_A, e_B, Q) \geq f(\varrho_B, e_B),\tag{54}$$

holding for any permutation matrix Q and any concave symmetrical function f. The bound is an entanglement criterion for pure states. Here, one can notice that the right-hand side of Equation (54) is independent of the space $\mathcal{H}_A$, giving some hint that the criterion actually detects steerability of the state.

Using the notation $e_k^A = \{|e_{i_k}^A\rangle | i = 1, \ldots, n_A\}$ and $e_k^B = \{|e_{j_k}^B\rangle | j = 1, \ldots, n_B\}$ for different bases of $\mathcal{H}_{A(B)}$, the authors prove that for any separable state ϱ_{AB}

$$\sum_k f_k(\varrho_{AB}, e_k^A, e_k^B, Q_k) \geq \min_{|\psi\rangle \in \mathcal{H}_B} \sum_k f_k(|\psi\rangle\langle\psi|, e_k^B)\tag{55}$$

holds for arbitrary symmetrical concave functions f_k, permutation matrices Q_k and bases $e_k^{A(B)}$. Equation (55) is a general entanglement criterion based on symmetrical concave functions f_k. In order to find the optimal criteria, an optimization over all possible permutation matrices should be performed. A specific criterion is given for the case where f_k is replaced by the Shannon entropy, and the bound in Equation (55) is related to EURs.

Now we show that the above entanglement criterion is actually a steering criterion, given that Equation (55) can be obtained if one considers an LHS model. Note first that one can include general measurements into the above considerations by defining

$$f(\varrho_0, M) := f(\text{Tr}[\varrho_0 M_i]_i), \quad \text{for} \quad i = 1, \cdots, n,\tag{56}$$

where the operators $\{M_i\}_i$ form a positive operator valued measure (POVM) and ϱ_0 is a quantum state. Then, taking an unsteerable state ρ_{AB} and labelling by N_i the POVM elements of Alice's measurements, one has for a fixed hidden variable λ

$$p_{ij}(\lambda) = p(i|N,\lambda)\mathrm{Tr}[M_j\rho_\lambda^B]. \tag{57}$$

Hence,

$$
\begin{aligned}
f\left(\sum_{ij}\delta(q_{ij},s_1)p_{ij}(\lambda),\cdots\right) &= f\left(\sum_i p(i|N,\lambda)\sum_j \delta(q_{ij},s_1)\mathrm{Tr}[M_j\rho_\lambda^B],\cdots\right) \\
&\geq \sum_i p(i|N,\lambda)f\left(\sum_j \delta(q_{ij},s_1)\mathrm{Tr}[M_j\rho_\lambda^B],\cdots\right) \\
&= \sum_i p(i|N,\lambda)f\left(\mathrm{Tr}[M_1\rho_\lambda^B],\cdots\right) \\
&= f(\varrho_\lambda^B,M). \tag{58}
\end{aligned}
$$

On the second line we use concavity of the function f, and in the third line we use symmetry. Taking the sum over all hidden variables λ gives

$$f(\varrho,M,N,Q) := f\left(\sum_{ij}\sum_\lambda \delta(q_{ij},s_1)p(\lambda)p_{ij}(\lambda),\cdots\right) \geq \sum_\lambda p(\lambda)f(\varrho_\lambda^B,M) \geq \min_{\varrho\in\mathcal{H}_B}f(\varrho^B,M). \tag{59}$$

Considering more measurements one has

$$\sum_k f_k(\varrho,M_k,N_k,Q_k) = \sum_\lambda p(\lambda)\sum_k f(\varrho_\lambda^B,M_k) \geq \min_{|\psi\rangle\in\mathcal{H}_B}\sum_k f_k(|\psi\rangle\langle\psi|,M_k), \tag{60}$$

which is exactly the same criteria of Equation (55). This means that the entanglement criteria proposed in Ref. [56] are actually steering criteria.

6. Applications

6.1. Optimal Values of q and r for Steering Detection

In this section we investigate the dependence of our steering criteria on the parameters q and r appearing in Tsallis and Rényi entropies. Also a comparison between the criteria obtained from Tsallis and Rényi entropy (with Shannon entropy as a special case) is presented for specific examples. We base our calculations on numerics for the cases where the optimal (analytical) uncertainty bounds are not known.

Let us first consider the case of qubit systems. For this analysis, consider three noisy two-qubit entangled states, $\varrho_{ex}^{(2)}(w) = w\varrho_x^{(2)} + (1-w)\mathbb{1}/4$ with $x = 1,2,3$ where $\rho_1^{(2)} = |\psi^-\rangle\langle\psi^-|$, $\rho_2^{(2)}$ and $\rho_3^{(2)}$ are two example states, given by

$$\rho_2^{(2)} = \frac{1}{4}\begin{pmatrix} 0.14 & 0.09-0.18i & -0.12+0.17i & -0.06 \\ 0.09+0.18i & 1.58 & -1.72 & -0.12+0.17i \\ -0.12-0.17i & -1.72 & 1.98 & 0.09-0.18i \\ -0.06 & -0.12-0.17i & 0.09+0.18i & 0.3 \end{pmatrix},$$

$$\rho_3^{(2)} = \frac{1}{4}\begin{pmatrix} 0.06 & -0.13 & 0.16+0.02i & -0.02 \\ -0.13 & 1.74 & -1.82 & 0.16+0.02i \\ 0.16-0.02i & -1.82 & 1.96 & -0.13 \\ -0.02 & 0.16-0.02i & -0.13 & 0.24 \end{pmatrix},$$

which give a fair violation of the criteria. Please note that this behaviour is typical not only for these states. For all the states that we tried a similar plot was obtained.

Here we will focus on the Pauli measurements $\{\sigma_x, \sigma_y, \sigma_z\}$. In Figure 2 we show the critical value of white noise w for the violation of the generalized entropic criteria from Equations (45) and (48).

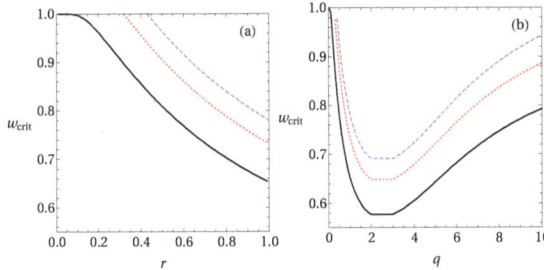

Figure 2. The critical value w for noisy two-qubit entangled states $\varrho_{e_x}^{(2)}(w)$ for the detection of steering. Solid black line corresponds to Werner states ($\varrho_{e_1}^{(2)}(w)$), and the dashed blue and dotted red lines correspond to $\varrho_{e_2}^{(2)}(w)$ and $\varrho_{e_3}^{(2)}(w)$, respectively, with **(a)** the criteria based on Rényi entropy [Equation (45)] and **(b)** on Tsallis entropy [Equation (48)].

From these simple examples, one is able to extract some hint about the optimal values of q and r that best identify steerability of the state. If one considers the criteria based on Rényi entropy, one notices in Figure 2a that the smallest critical value of white noise occurs for $r \rightarrow 1$, which corresponds to the criteria based on Shannon entropy. Meanwhile, in Figure 2b, the best criteria from Tsallis entropy are the ones for $q = 2$ and $q = 3$, which give an improvement to the Shannon-based criteria. Please note that within the interval $q \in [2; 3]$ the line seems to be flat, meaning that any q in this interval could be considered as an optimal value for the detection of steering from generalized entropies. However, this statement does not hold in general, as one can see in Figure 3. It is true for the case of Werner states, whereas for the other considered states the optimal parameter values are $q = 2$ and $q = 3$ only.

It is worth mentioning that the criteria for $q = 2$ and $q = 3$, in the case of $d = 2$, are analytically the same. Also, for these values of q, they can be connected to the variance criteria from Refs. [57,58]. To see this, consider an observable A with eigenvalues ± 1 and corresponding outcome probabilities $p_\pm$. The variance of the observable A is given by

$$\delta^2(A) = 1 - \langle A \rangle^2 = 1 - (p_+ - p_-)^2 = 2(1 - p_+^2 - p_-^2) \sim S_2(A). \tag{61}$$

The same relation can be found for $q = 3$. This equivalence between variances and Tsallis entropies with $q = 2$ and $q = 3$ can be extended to the related steering criteria.

For the two-qubit Werner state, it is known that the optimal white noise threshold (w_{crit}) is $1/\sqrt{3}$ for three (orthogonal) projective measurements [59]. Interestingly, the criteria based on Tsallis entropy achieve these values with $q = 2$ and $q = 3$.

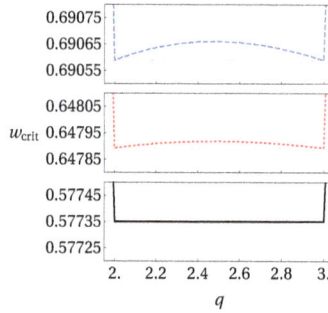

Figure 3. Zoom-in of Figure 2, for the interval $q \in [2;3]$. Solid black line corresponds to Werner states, and the dashed blue and dotted red lines correspond to two random entangled states.

It is interesting to check whether the same optimal values of r and q also hold for some higher dimensional states. For this, consider noisy two-qutrit entangled states $\varrho_{ex}^{(3)}(w) = w|\psi_x\rangle\langle\psi_x| + (1-w)\mathbb{1}/9$, where $|\psi_x\rangle = \frac{1}{\sqrt{2+x^2}}(|00\rangle + x|11\rangle + |22\rangle)$. In Figure 4, we analyse the states with $x = \{0.2, 0.5, 1\}$ when Alice performs a complete set of MUBs. One can see in Figure 4a that for these states our criteria based on Rényi entropy are weaker than the ones based on Shannon entropy, similar to the case of two-qubit states. Interestingly, in Figure 4b the optimal q for the detection of steering using Tsallis entropy is only $q = 2$ (and not $q = 2$ and $q = 3$ as in the two-qubit case).

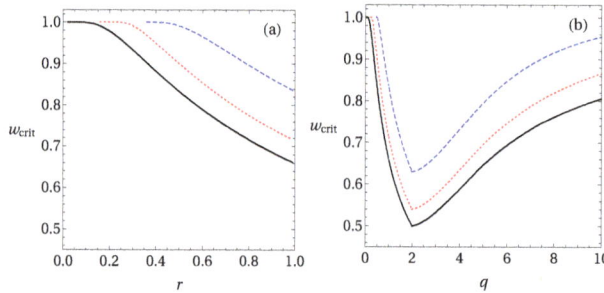

Figure 4. The critical value w for noisy two-qutrit entangled states $\varrho_{ex}^{(3)}(w)$ for the detection of steering. The solid black line corresponds to the state with $x = 1$, the dotted red line with $x = 0.5$, and the dashed blue line with $x = 0.2$, with (**a**) the criteria based on Rényi entropy (45) and (**b**) on Tsallis entropy (48).

We close this section with the conjecture that the criterion obtained from Tsallis entropy with $q = 2$ is the best one to detect steerable states using the method proposed in this work for arbitrary (finite) dimensions. Our criteria based on Rényi entropy seems to be weaker than the one based on Shannon entropy (see Figures 2 and 4) and, hence, we will not consider it further. Moreover, we will focus mainly on the criterion based on Tsallis entropy with $q = 2$, but we also discuss results for different values of q.

6.2. Isotropic States

The generalized entropic steering criteria are interesting for many scenarios, especially in the case of higher dimensional systems. Here, we address this scenario by applying our criteria to d-dimensional isotropic states [60]

$$\varrho_{iso} = \alpha |\phi_d^+\rangle\langle\phi_d^+| + \frac{1-\alpha}{d^2}\mathbb{1}, \tag{62}$$

where $|\phi^+\rangle = (1/\sqrt{d})\sum_{i=0}^{d-1}|i\rangle|i\rangle$ is a maximally entangled state. These states are known to be entangled for $\alpha > 1/(d+1)$ and separable otherwise. To detect steering via our entropic criteria, we consider as measurements m MUBs in dimension d (provided that they exist).

The marginal probabilities for this class of states are $p_i = 1/d$ for all i and the joint probabilities are $p_{ii} = [1+(d-1)\alpha]/d^2$ (occurring d times), and $p_{ij} = (1-\alpha)/d^2$ (for $i \neq j$ and occurring $d(d-1)$ times). Please note that since isotropic states are invariant under local unitary operators of the form $U \otimes U^*$ Ref. [60], we choose Bob's measurements to be the conjugates of Alice's measurements. Inserting these probabilities in Equation (46), the condition for non-steerability reads

$$\frac{m}{q-1}\left\{1-\frac{1}{d^q}[(1+(d-1)\alpha)^q+(d-1)(1-\alpha)^q]\right\} \geq B_B^{(q)}, \tag{63}$$

where $B_B^{(q)}$ is given in Equation (29) and (32) [in the limit of $q \to 1$, we use the bounds from Equation (22)]. One can see that Equation (63) is valid for any dimension d, and depends only on the parameter q and the number of MUBs m.

Numerical investigations suggest that the criterion is strongest for $q = 2$, as one can see in Figure 5. For this value of q the violation of Equation (63) occurs for $\alpha > 1/\sqrt{m}$. For a complete set of MUBs ($m = d+1$) (with d being a power of a prime) the violation happens for $\alpha > 1/\sqrt{d+1}$. For example, if we consider $d = 2$ (qubits), isotropic states are equivalent to Werner states [61]. For a complete set of MUBs the violation of our criteria occurs for $\alpha > 1/\sqrt{3} \approx 0.577$, which is known to be the optimal threshold [59] for three MUBs.

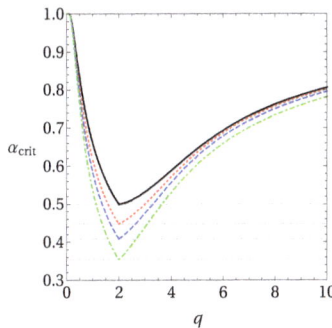

Figure 5. The critical value of white noise α of states in Equation (62) as function of the Tsallis parameter q, considering a complete set of MUBs. Here, the solid black line corresponds to $d = 3$, the dotted red line to $d = 4$, the dashed blue line to $d = 5$, and the dot-dashed green line to $d = 7$. The optimal value for the detection of steerability is given by $q = 2$.

Now, we are able to compare our results with two others which investigated steering for the class of isotropic states and MUBs. In Ref. [62], a steering inequality has been presented which is violated for $\alpha > (d^{3/2} - 1)/(d^2 - 1)$, whereas in Ref. [63] the authors used semi-definite programming for this task. In Figure 6, we show this comparison. Please note that we present only the results for $q = 2$, which is the conjectured optimal value (Figure 5). From Figure 6, one sees that our criterion is stronger than the one from Ref. [62]. For $3 \leq d \leq 5$ a better threshold than ours was obtained in Ref. [63], but it is worth mentioning that our criteria directly use probability distributions from a few measurements, without the need of performing full tomography on Bob's conditional state. In addition the numerical approach becomes computationally more demanding when increasing the number of variables.

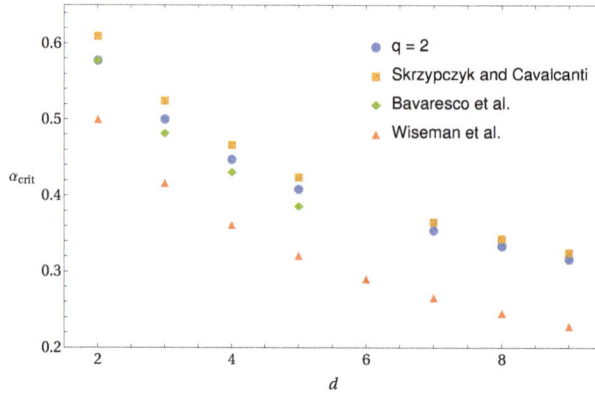

Figure 6. The critical value of white noise α for different dimensions d, considering a complete set of MUBs. In this plot, blue circles correspond to our criterion in Equation (63) for $q = 2$. The yellow squares correspond to the results for the inequality presented in Ref. [62] and the green diamonds in Ref. [63], where α_{crit} was calculated via SDP (numerical method). Below the red triangles the existence of an LHS model for all projective measurements (i.e. infinite amount of measurements instead of $d + 1$ MUBs) is known [2]. Please note that Ref. [2] is given for comparison, this is not a steering criterion, but a bound on any criterion.

6.3. General Two-Qubit States

Let us now consider the application of our method to general two-qubit states. Any two-qubit state can, after application of local unitaries, be written as

$$\varrho_{AB} = \frac{1}{4}\left[\mathbb{1} \otimes \mathbb{1} + (\vec{a}\vec{\sigma}) \otimes \mathbb{1} + \mathbb{1} \otimes (\vec{b}\vec{\sigma}) + \sum_{i=1}^{3} c_i \sigma_i \otimes \sigma_i\right], \tag{64}$$

where $\vec{a}, \vec{b}, \vec{c} \in \mathbb{R}^3$ are vectors with norm less than one, $\vec{\sigma}$ is a vector composed of the Pauli matrices and $(\vec{a}\vec{\sigma}) = \sum_i a_i \sigma_i$. We assume that Alice performs projective measurements with effects $P_m^A = [\mathbb{1} + \mu_m(\vec{u}_m\vec{\sigma})]/2$ and Bob with effects $P_m^B = [\mathbb{1} + \nu_m(\vec{v}_m\vec{\sigma})]/2$, where $\mu_m, \nu_m = \pm 1$ and $\{\vec{u}, \vec{v}\}$ are unit vectors in $\mathbb{R}^3$. We have the following probabilities:

$$p(\mu_m) = \text{Tr}[(P_m^A \otimes \mathbb{1})\varrho_{AB}] = \frac{1}{2}(1 + \mu_m(\vec{a}\vec{u}_m)),$$

$$p(\mu_m, \nu_m) = \text{Tr}[(P_m^A \otimes P_m^B)\varrho_{AB}] = \frac{1}{4}(1 + \mu_m(\vec{a}\vec{u}_m) + \nu_m(\vec{b}\vec{v}_m) + \mu_m\nu_m T_m),$$

where $T_m = \sum_{i=1}^{3} c_i u_{im} v_{im}$. Now Equation (46) can be written as

$$\sum_m\left[1 - \sum_{\mu_m,\nu_m} \frac{[1 + \mu_m(\vec{a}\vec{u}_m) + \nu_m(\vec{b}\vec{v}_m) + \mu_m\nu_m T_m]^q}{2^{q+1}[1 + \mu_m(\vec{a}\vec{u}_m)]^{q-1}}\right] \geq (q-1)\mathcal{B}_B^{(q)}. \tag{65}$$

The optimization over measurements in this criterion for a general two-qubit state is involving. We will focus on the simple case of Pauli measurements, meaning that $\vec{u}_m = \vec{v}_m = \{(1,0,0)^T,(0,1,0)^T,(0,0,1)^T\}$ and $q = 2$. Then we have the following inequality

$$\sum_{i=1}^{3}\left[\frac{1-a_i^2-b_i^2-c_i^2+2a_ib_ic_i}{2(1-a_i^2)}\right] \geq 1, \tag{66}$$

the violation of which implies steerability.

Now, we can compare our criteria with other proposals for the detection of steerable two-qubit states using three measurements. The criterion from [53] (see Equation (49)) proves steerability if $\sum_{i=1}^{3} c_i^2 > 1$, and from the linear criteria [2,64] steerability follows if $(\sum_{i=1}^{3} c_i^2)^{1/2} > 1$. Not surprisingly, Equation (66) is stronger, since it uses more information about the state. The claim can be made hard by analyzing 10^6 (Hilbert-Schmidt) random two-qubit states [65]. 94.34% of the states do not violate any of the criteria, 3.81% are steerable according to all criteria, 1.85% violate only criterion (66), and none of the states violates the linear criteria without violating (66).

A special case of two-qubit states are the Bell diagonal ones, which can be obtained if we set $\vec{a} = \vec{b} = 0$ in Equation (64). For this class of states it is easy to see that the three criteria are equivalent. Note, moreover, that a necessary and sufficient condition for steerability of this class of states with all projective measurements has recently been found [27].

6.4. One-Way Steerable States

As an example of weakly steerable states that can be detected with our methods we take one-way steerable states, i.e., states that are steerable from Alice to Bob but not the other way around. More specifically, we consider the family of states given as

$$\varrho_{AB} = \beta|\psi(\theta)\rangle\langle\psi(\theta)| + (1-\beta)\frac{1}{2}\otimes\varrho_B^\theta, \tag{67}$$

where $|\psi(\theta)\rangle = \cos(\theta)|00\rangle + \sin(\theta)|11\rangle$ and $\varrho_B^\theta = \mathrm{Tr}_A[|\psi(\theta)\rangle\langle\psi(\theta)|]$. It is known that states with $\theta \in [0, \pi/4]$ and $\cos^2(2\theta) \geq (2\beta - 1)((2-\beta)\beta^3)$ are not steerable from Bob to Alice considering all possible projective measurements [28], while Alice can steer Bob whenever $\beta > 1/2$.

Considering two measurement settings, we have that this state is one-way steerable for $1/\sqrt{2} < \beta \leq \beta_{max}^{(2)}$ with $\beta_{max}^{(2)} = [1+\sin^2(2\theta)]^{-1/2}$, and for three measurement settings, this state is one way-steerable for $1/\sqrt{3} < \beta \leq \beta_{max}^{(3)}$ with $\beta_{max}^{(3)} = [1+2\sin^2(2\theta)]^{-1/2}$ [66]. For our entropic steering criterion (45) with $q = 2$ we find that this state is one-way steerable in the range

$$\frac{\sqrt{1+\tan^2(\theta)}}{1+\tan(\theta)} < \beta \leq \beta_{max}^{(2)}, \tag{68}$$

for two Pauli measurements (σ_x, σ_z), and

$$\frac{1}{2\cos(2\theta)}\sqrt{3-\sqrt{1+8\sin^2(2\theta)}} < \beta \leq \beta_{max}^{(3)}, \tag{69}$$

for three Pauli measurements $(\sigma_x, \sigma_y, \sigma_z)$. For any θ this gives a non-empty interval of β for which our criterion detects these weakly steerable states. In Figure 7, we show the range of one-way steerability considering two and three measurement settings.

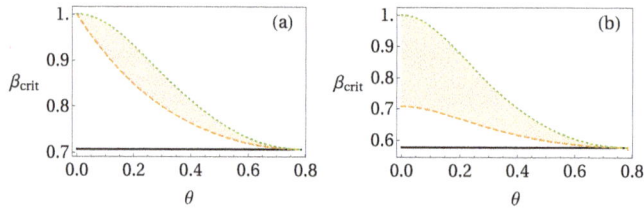

Figure 7. One-way steerability of states (67) for (**a**) two and (**b**) three measurement settings. The shaded area is the region where our criterion detects these weakly steerable states.

6.5. Bound Entangled States

It is also interesting to investigate whether the entropic steering criteria from generalized entropies are able to detect steerability of bound entangled states, which is related to the stronger version of Peres conjecture [19,67,68]. The conjecture states the possibility of constructing local models for bound entangled states and it was proven wrong in Refs. [6,7].

For this task, we investigated the following class of states presented in [6]

$$\varrho_{BES} = \lambda_1 |\psi_1\rangle\langle\psi_1| + \lambda_2 |\psi_2\rangle\langle\psi_2| + \lambda_3 (|\psi_3\rangle\langle\psi_3| + |\tilde\psi_3\rangle\langle\tilde\psi_3|), \tag{70}$$

with the following normalized states

$$
\begin{aligned}
|\psi_1\rangle &= (|12\rangle + |21\rangle)/\sqrt{2}, \\
|\psi_2\rangle &= (|00\rangle + |11\rangle - |22\rangle)/\sqrt{3}, \\
|\psi_3\rangle &= m_1|01\rangle + m_2|10\rangle + m_3(|11\rangle + |22\rangle), \\
|\tilde\psi_3\rangle &= m_1|02\rangle - m_2|20\rangle + m_3(|21\rangle - |12\rangle),
\end{aligned}
\tag{71}
$$

where $m_{1(2)} \geq 0$ and $m_3 = \sqrt{(1 - m_1^2 - m_2^2)/2}$. This class of states has a positive partial transpose if the eigenvalues are fixed as

$$
\begin{aligned}
\lambda_1 &= 1 - (2 + 3m_1 m_2)/N, \\
\lambda_2 &= 3m_1 m_2/N, \\
\lambda_3 &= 1/N,
\end{aligned}
\tag{72}
$$

with $N = 4 - 2m_1^2 + m_1 m_2 - 2m_2^2$ and $m_1^2 + m_2^2 + m_1 m_2 \leq 1$. In Ref. [6], the authors show that this class of states is steerable for certain measurements. Now, to check whether the generalized entropic criteria are also able to detect the steerability of such states, consider that we perform the following two MUBs on Alice's and Bob's system [6]:

$$
\begin{aligned}
M_1^1 &= [1/\sqrt{3}, -1/\sqrt{6}, -1/\sqrt{2}], \\
M_2^1 &= [1/\sqrt{3}, -1/\sqrt{6}, 1/\sqrt{2}], \\
M_3^1 &= [1/\sqrt{3}, \sqrt{2/3}],
\end{aligned}
\tag{73}
$$

for measurement $m = 1$, and

$$
\begin{aligned}
M_1^2 &= [1, 0, 0], \\
M_2^2 &= [0, q/\sqrt{2}, iq/\sqrt{2}], \\
M_3^2 &= [0, q^*/\sqrt{2}, -iq^*/\sqrt{2}],
\end{aligned}
\tag{74}
$$

for measurement $m = 2$. These rotated MUBs are given by the symmetry of the above class of states. Since they are MUBs, the bound $\mathcal{C}^{(2)} = 1$ holds.

In Figure 8, one can see that no violation for this specific class of bound entangled states occurs (given the above measurements). Surprisingly, performing more measurements makes no difference for the detection of steerability using our entropic steering criterion. This situation can be explained by the symmetry of such states, i.e., with the addition of more mutually unbiased measurements the entropic uncertainty bound increases with the same rate as the l.h.s. of criterion (45). The same result can also be obtained by a numerical optimization over random unitaries applied in the standard MUBs, where we use the parametrization given in Ref. [69]. In this sense, it remains as an open question if the criterion is able to detect steerable bound entangled states.

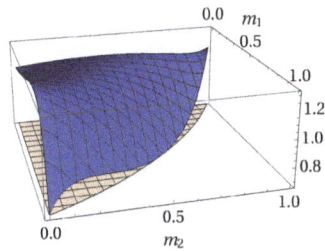

Figure 8. Plot of Equation (45) in terms of m_1 and m_2 with $q = 2$ (blue curve) for ϱ_{BES} in the region $m_1^2 + m_2^2 + m_1 m_2 \leq 1$. The opaque flat plot is the entropic uncertainty bound for $q = 2$ and the measurements given by Equations (73) and (74). From the plot one can see that there is no violation of Equation (45) for any state in this family.

7. Multipartite Scenario

In this section we extend the generalized entropic criteria to the case of tripartite systems. For such systems, one can consider two different steering scenarios: either Alice tries to steer Bob and Charlie or Alice and Bob try to steer Charlie. In the latter scenario, one should notice that there is a difference regarding the kind of measurements Alice and Bob perform: local or global ones. In this section we consider all these cases and derive generalized multipartite steering criteria from the Tsallis entropy. This specific choice of entropy is given by the examples presented in the previous sections, where the criteria based on Rényi entropy were found weak in comparison to the one based on Shannon entropy, which, by extension, is included in the Tsallis entropy.

A proposal for multipartite steering using EURs based on Shannon entropy has been recently introduced in Ref. [18]. Here, we derive our criteria from a different perspective considering a general approach via Tsallis entropy.

7.1. Steering from Alice to Bob and Charlie

Let us first focus on the scenario where Alice tries to steer Bob and Charlie. Consider the quantity

$$F_q(A, B, C) = -D_q(A \otimes B \otimes C || A \otimes \mathbb{I}_B \otimes \mathbb{I}_C), \qquad (75)$$

where $\mathbb{I}_{B(C)}$ are equal distributions with $p_j = 1/N_B$ and $p_k = 1/N_C$, respectively. Writing this in terms of probabilities gives [see also Equations (41) and (42)]

$$
\begin{aligned}
F_q(A, B, C) &= \sum_{i,j,k} p_{ijk} \ln_q \left(\frac{p_i/N_{BC}}{p_{ijk}} \right) \\
&= \frac{x_{BC}}{1-q} + (1 + x_{BC})[S_q(A, B, C) - S_q(A) + (1-q)T_q^{(1)}(A, B, C)], \qquad (76)
\end{aligned}
$$

where $N_{BC} = N_B N_C$, $x_{BC} = (N_B N_C)^{q-1} - 1$ and

$$T_q^{(1)}(A, B, C) = \sum_i p_i^q (\ln_q(p_i))^2 - \sum_{i,j,k} p_{ijk}^q \ln_q(p_i) \ln_q(p_{ijk}), \tag{77}$$

is the correction term.

Now, from the LHS model (2), the probability distribution $p(i|A, \lambda) p_Q(j|B, \lambda) p_Q(k|C, \lambda)$ with a fixed λ yields

$$F_q^{(\lambda)}(A, B, C) = \frac{x_{BC}}{1-q} + (1 + x_{BC}) S_q^{(\lambda)}(B, C), \tag{78}$$

with $S_q^{(\lambda)}(B, C) = S_q^{(\lambda)} + S_q^{(\lambda)}(C) + (1-q) S_q^{(\lambda)}(B) S_q^{(\lambda)}(C)$. For a given set of measurements $B_m \otimes C_m$ one has an EUR

$$\sum_m S_q^{(\lambda)}(B_m, C_m) \geq C_{BC}^{(q)}, \tag{79}$$

where $C_{BC}^{(q)}$ is some entropic uncertainty bound for the observables $B_m \otimes C_m$. Since S_q is a concave function, the same bound holds for convex combinations of product distributions $p(i|A, \lambda) p_Q(j|B, \lambda) p_Q(k|C, \lambda)$. Connecting the above results, the generalized multipartite steering criteria from Alice to Bob and Charlie are given by

$$\sum_m [S_q(B_m, C_m|A_m) + (1-q) T_q^{(1)}(A_m, B_m, C_m)] \geq C_{BC}^{(q)}, \tag{80}$$

where $S_q(B_m, C_m|A_m) = S_q(B_m, C_m) - S_q(A_m)$ is the conditional Tsallis entropy. In terms of probabilities, these criteria can be written as

$$\frac{1}{q-1} \left[\sum_m \left(1 - \sum_{i,j,k} \frac{(p_{ijk}^{(m)})^q}{(p_i^{(m)})^{q-1}} \right) \right] \geq C_{BC}^{(q)}. \tag{81}$$

Note here that we define tripartite steering from Alice to Bob and Charlie from the LHS model given in Equation (2), and in this case we should consider the EUR bounds for separable states, see for example Equations (25), (26), (28), (33) and (34), for the case of qubits and Pauli measurements. Moreover, if we consider the bound where we allow the state of Bob and Charlie to be entangled, which leads effectively to the scenario of bipartite steering, the bound for three measurement settings changes for Shannon entropy (see Equation (27)). For Tsallis entropy, the EUR bound differs by non-separable states in the range of $1 \leq q < 2$, for three measurement settings (see Figure 1). These different scenarios will be discussed further for some class of states in the next section.

7.2. Steering from Alice and Bob to Charlie

Let us now consider the scenario where Alice and Bob try to steer Charlie. Here, we follow the definition of tripartite steering given through Equation (4). To start with, consider the quantity

$$F_q(A, B, C) = -D_q(A \otimes B \otimes C || A \otimes B \otimes \mathbb{I}_C), \tag{82}$$

where $\mathbb{I}_C$ represents a uniform distribution with $p_k = 1/N_C$. In terms of probabilities one gets

$$\begin{aligned}
F_q(A, B, C) &= \sum_{i,j,k} p_{ijk} \ln_q \left(\frac{p_{ij}/N_C}{p_{ijk}} \right) \\
&= \frac{x_C}{1-q} + (1 + x_C)[S_q(A, B, C) - S_q(A, B) + (1-q) T_q^{(2)}(A, B, C)], \tag{83}
\end{aligned}$$

where $x_C = N_C^{q-1} - 1$ and

$$T_q^{(2)}(A, B, C) = \sum_{i,j} p_{ij}^q (\ln_q(p_{ij}))^2 - \sum_{i,j,k} p_{ijk}^q \ln_q(p_{ij}) \ln_q(p_{ijk}), \tag{84}$$

is the correction term.

Assuming that one has the LHS model from Equation (4) and considering the probability distribution $p(i|A, \lambda)p(j|B, \lambda)p_Q(k|C, \lambda)$ with a fixed λ one gets

$$F_q^{(\lambda)}(A, B, C) = \frac{x_C}{1-q} + (1 + x_C)S_q^{(\lambda)}(C). \tag{85}$$

For a given set of measurements $\{C_m\}_m$ one has an EUR

$$\sum_m S_q^{(\lambda)}(C_m) \geq \mathcal{B}_C^{(q)}, \tag{86}$$

where $\mathcal{B}_C^{(q)}$ is some entropic bound for the observables $\{C_m\}_m$. Since S_q is concave function, the same bound holds for convex combinations of product distributions $p(i|A, \lambda)p(j|B, \lambda)p_Q(k|C, \lambda)$. Connecting the above results, the generalized multipartite steering criteria from Alice to Bob and Charlie are given by

$$\sum_m [S_q(A_m, B_m, C_m) - S_q(A_m, B_m) + (1-q)T_q^{(2)}(A_m, B_m, C_m)] \geq \mathcal{B}_C^{(q)}. \tag{87}$$

In terms of probabilities, these criteria can be written as

$$\frac{1}{q-1} \left[\sum_m \left(1 - \sum_{i,j,k} \frac{(p_{ijk}^{(m)})^q}{(p_{ij}^{(m)})^{q-1}} \right) \right] \geq \mathcal{B}_C^{(q)}. \tag{88}$$

In this scenario, this framework is not able to distinguish between bipartite and tripartite steering. This comes from the fact that if we consider the LHS model given in (5), with product distributions $p(i, j|A, B, \lambda)p_Q(k|C, \lambda)$, we obtain the same criteria.

7.3. Applications

For the application of the multipartite entropic steering criteria, we consider systems of three qubits with Pauli measurements. We focus our discussion on GHZ and W states. A noisy GHZ state is defined as

$$\rho_{GHZ} = \gamma|GHZ\rangle\langle GHZ| + \frac{1-\gamma}{8}\mathbb{1}, \tag{89}$$

where $|GHZ\rangle = \frac{1}{\sqrt{2}}(|000\rangle + |111\rangle)$. This state is known to be not fully separable iff $\gamma > 1/5$ [70,71] and to be Bell nonlocal for $\gamma > 1/2$ for two and three measurements per site [72]. A noisy W state reads

$$\rho_W = \delta|W\rangle\langle W| + \frac{1-\delta}{8}\mathbb{1}, \tag{90}$$

where $|W\rangle = \frac{1}{\sqrt{3}}(|100\rangle + |010\rangle + |001\rangle)$, being entangled for $\delta > \sqrt{3}/(8 + \sqrt{3}) \approx 0.178$ and fully separable for $\delta \leq 0.177$ [73]. This state is Bell nonlocal for $\delta > 0.6442$ for two measurements per site and $\delta > 0.6048$ for three measurements [72]. Here, we are interested in the critical amount of white noise for the violation of criteria (80) and (87) with the aforementioned of measurements (together with an optimization over local unitaries).

Let us start discussing the results for the scenario of steering from Alice to Bob and Charlie. As mentioned above, we can distinguish the results into two different steering scenarios-bipartite and tripartite-depending on the considered LHS model and, consequently, the associated entropic bounds.

For the case of noisy GHZ states we have the following results for two measurement settings. Considering that Bob and Charlie always perform the same measurements (restriction given by the EUR bounds), violation of the criterion (80) is found for

$$
\begin{aligned}
A_1 &= B_1 = C_1 = \sigma_x, \\
A_2 &= B_2 = C_2 = \sigma_z,
\end{aligned}
\tag{91}
$$

with $\gamma > \gamma_{crit}^{(1)} \approx 0.8631$ and $\gamma > \gamma_{crit}^{(2)} \approx 0.866$, where the notation $\gamma^{(q)}$ is used to distinguish between Shannon and Tsallis entropies. For three measurement settings, we choose the measurements as

$$
\begin{aligned}
A_1 &= A_2 = B_1 = C_1 = \sigma_x, \\
B_2 &= C_2 = \sigma_y, \\
A_3 &= B_3 = C_3 = \sigma_z,
\end{aligned}
\tag{92}
$$

and the state is steerable from Alice to Bob and Charlie for $\gamma > \gamma_{crit}^{(1)} \approx 0.7642$ (for the bound (26)) and $\gamma > \gamma_{crit}^{(1)} \approx 0.909$ (for the bound (34) with $q \to 1$). Using the criteria from Tsallis entropy (for the bound (34)) $\gamma > \gamma_{crit}^{(2)} \approx 0.775$. Please note that the best noise threshold is obtained using Shannon entropy and the bound for separable states, which leads to a "truly" tripartite steering scenario. In other words, the criteria obtained from Shannon entropy is sensitive to this distinction and demonstrates that is "easier" for Alice to steer Bob and Charlie if they share a separable state. In contrast, the criteria from Tsallis entropy with $q = 2$ is not sensitive (indifferent) within these different scenarios.

For the noisy W states, we have the following results for two-measurement settings. The optimal measurements are the ones given by Equation (91). Violation of the criteria occurs for $\delta > \delta_{crit}^{(1)} \approx 0.9814$, and no violation was found for $q = 2$. Considering three-measurement setting, the optimal set of measurements is

$$
\begin{aligned}
A_1 &= B_1 = C_1 = \sigma_x, \\
A_2 &= B_2 = C_2 = \sigma_y, \\
A_3 &= B_3 = C_3 = \sigma_z,
\end{aligned}
\tag{93}
$$

and there is no violation for the criteria with the bound (27), but $\delta > \delta_{crit}^{(1)} \approx 0.8523$ for the criteria with the bound (26), and $\delta > \delta_{crit}^{(2)} \approx 0.8366$ for the bound (34). The best threshold for steerability occurs for the criterion based on Tsallis entropy (contrary to the results found for noisy GHZ states), although the criterion does not distinguish between bipartite and tripartite LHS models.

Now, consider steering from Alice and Bob to Charlie. As mentioned above, in this scenario we have no distinction between bipartite and tripartite steering, since both models lead to the same criteria. However, it is possible to explore the difference between performing local and global measurements.

Let us first discuss the results for local measurements. For noisy GHZ states and two measurement settings we use the measurements from Equation (91). Steerability from Alice and Bob to Charlie occurs for $\gamma > \gamma_{crit}^{(1)} \approx 0.7476$ and $\gamma > \gamma_{crit}^{(2)} \approx 0.6751$. For three measurement settings, considering the measurements from Equation (92), one has $\gamma > \gamma_{crit}^{(1)} \approx 0.6247$ and $\gamma > \gamma_{crit}^{(2)} \approx 0.5514$.

For noisy W states we use the measurements

$$
\begin{aligned}
A_1 &= C_1 = \sigma_x, \\
A_2 &= B_1 = B_2 = C_2 = \sigma_z.
\end{aligned}
\tag{94}
$$

Here steering occurs for $\delta > \delta_{crit}^{(1)} \approx 0.818$ and $\delta > \delta_{crit}^{(2)} \approx 0.75$. For three measurement settings we take

$$A_1 = C_1 = \sigma_x,$$
$$A_2 = C_2 = \sigma_y,$$
$$A_3 = B_1 = B_2 = B_3 = C_3 = \sigma_z. \tag{95}$$

The corresponding thresholds are $\delta_{crit}^{(1)} \approx 0.698$ and $\delta_{crit}^{(2)} \approx 0.623$. In this scenario, one can notice that increasing the number of measurements and choosing $q = 2$, one is able to detect more steering for both families of states-in the same way as in bipartite steering.

Now, let us explore the scenario where Alice and Bob perform global measurements. For this, we consider MUBs in dimension 4 for the global measurements and Pauli measurements to be performed in Charlie system. A possible set of MUBs in dimension 4 is given by

$$M_1 = \begin{pmatrix} 1 & 0 & 0 & 0 \\ 0 & 1 & 0 & 0 \\ 0 & 0 & 1 & 0 \\ 0 & 0 & 0 & 1 \end{pmatrix}, \quad M_2 = \frac{1}{2}\begin{pmatrix} 1 & 1 & 1 & 1 \\ 1 & 1 & -1 & -1 \\ 1 & -1 & -1 & 1 \\ 1 & -1 & 1 & -1 \end{pmatrix}, \quad M_3 = \frac{1}{2}\begin{pmatrix} 1 & 1 & 1 & 1 \\ -1 & -1 & 1 & 1 \\ -i & i & i & -i \\ -i & i & -i & i \end{pmatrix},$$

$$M_4 = \frac{1}{2}\begin{pmatrix} 1 & 1 & 1 & 1 \\ -i & -i & i & i \\ -i & i & i & -i \\ -1 & 1 & -1 & 1 \end{pmatrix}, \quad M_5 = \frac{1}{2}\begin{pmatrix} 1 & 1 & 1 & 1 \\ -i & -i & i & i \\ -1 & 1 & -1 & 1 \\ -i & i & i & -i \end{pmatrix}. \tag{96}$$

From this set, we can choose within five measurements, while for the measurements of Charlie's system we can choose within three Pauli measurements. The task is to find the optimal combination which shows the best threshold for steerability in this scenario.

Considering the noisy GHZ states, the optimal two-measurement choice (from the given set) is $(AB)_1 = M_1$, $(AB)_2 = M_2$, and $C_1 = \sigma_z$, $C_2 = \sigma_x$, and the optimal three-measurement setting is the same as the two-measurement setting, with the addition of the third measurement $(AB)_3 = M_3$ and $C_3 = \sigma_y$, which gives the same noise threshold found in the scenario of local measurements. Hence, the criterion is not able to detect a difference between local and global measurements for this specific family of states and set of measurements.

However, this is not the case for the noisy W states. The optimal two-measurement setting (from the given set) is $(AB)_1 = M_1$, $(AB)_2 = M_2$, and $C_1 = \sigma_z$, $C_2 = \sigma_x$, with the noise threshold $\delta_{crit}^{(1)} \approx 0.8571$ and $\delta_{crit}^{(2)} \approx 0.7802$. The optimal three-measurement setting is the same as the two-measurement setting, with the addition of the third measurement $(AB)_3 = M_4$ and $C_3 = \sigma_y$, with the noise threshold $\delta_{crit}^{(1)} \approx 0.7414$ and $\delta_{crit}^{(2)} \approx 0.6548$. These results show that for noisy W states, local measurements are able to detect steerability with smaller noise threshold while compared to global ones. This result shows that the standard MUBs are not a good choice of global measurements, since they should reveal steerability with lower thresholds while compared to local ones.

Now, we are able to compare our results to the literature. For example, in the case of Shannon entropy and two measurement settings, we obtain the same results as the ones presented in Ref. [18] for noisy GHZ and W states and scenarios of steering from Alice to Bob and Charlie and Alice and Bob to Charlie. In the latter case, we were able to find a smaller threshold considering Tsallis entropy and $q = 2$. However, if we compare our results with the ones in Ref. [17], our noise thresholds are bigger for all scenarios, and the same happens if we compare them with the nonlocality thresholds presented in Ref. [72].

8. Conclusions

In this work we have extended to several directions the straightforward technique for the construction of strong steering criteria from EURs [30]. These criteria are easy to implement using a finite set of measurement settings only, and do not need the use of semi-definite programming and full tomography on Bob's conditional states. We also show that they can be extended to multipartite systems, where different steering scenarios can be identified and evaluated.

For future work, several directions seem promising. First, considering EURs in the presence of quantum memory [74] might improve the criteria. Second, connecting our results to measurement uncertainty relations for discrete observables [75]. Third, making quantitative statements about steerability from steering criteria. Recently, some attempts in this direction have been pursued [76].

Author Contributions: All authors contributed equally to this work.

Funding: This work was supported by the DFG, the ERC (Consolidator Grant No. 683107/TempoQ) and the Finnish Cultural Foundation.

Acknowledgments: We thank Chau Nguyen for the discussions about global versus local measurements, and for pointing us out the example of super-activation of steering. We also thank to Yichen Huang for bringing our attention to Ref. [56], and Alberto Riccardi and René Schwonnek for discussions. We are also thankful for the comments on the earlier version of the manuscript by C. Jebaratnam.

Conflicts of Interest: The authors declare no conflict of interest.

References

1. Schrödinger, E. *Eine Entdeckung Von Ganz Außerordentlicher Tragweite*; Springer: Berlin, Germany, 2011; p. 551.
2. Wiseman, H.M.; Jones, S.J.; Doherty, A.C. Steering, Entanglement, Nonlocality, and the Einstein-Podolsky-Rosen Paradox. *Phys. Rev. Lett.* **2007**, *98*, 140402. [CrossRef] [PubMed]
3. Branciard, C.; Cavalcanti, E.G.; Walborn, S.P.; Scarani, V.; Wiseman, H.M. One-sided device-independent quantum key distribution: Security, feasibility, and the connection with steering. *Phys. Rev. A* **2012**, *85*, 010301. [CrossRef]
4. Piani, M.; Watrous, J. Necessary and Sufficient Quantum Information Characterization of Einstein-Podolsky-Rosen Steering. *Phys. Rev. Lett.* **2015**, *114*, 060404. [CrossRef] [PubMed]
5. Law, Y.Z.; Thinh, L.P.; Bancal, J.-D.; Scarani, V. Quantum randomness extraction for various levels of characterization of the devices. *J. Phys. A* **2014**, *47*, 424028. [CrossRef]
6. Moroder, T.; Gittsovich, O.; Huber, M.; Gühne, O. Steering Bound Entangled States: A Counterexample to the Stronger Peres Conjecture. *Phys. Rev. Lett.* **2014**, *113*, 050404. [CrossRef] [PubMed]
7. Vértesi, T.; Brunner, N. Disproving the Peres conjecture by showing Bell nonlocality from bound entanglement. *Nat. Commun.* **2014**, *5*, 5297. [CrossRef] [PubMed]
8. Yu, S.; Oh, C.H. Family of nonlocal bound entangled states. *Phys. Rev. A* **2017**, *95*, 032111. [CrossRef]
9. Quintino, M.T.; Vértesi, T.; Brunner, N. Joint Measurability, Einstein-Podolsky-Rosen Steering, and Bell Nonlocality. *Phys. Rev. Lett.* **2014**, *113*, 160402. [CrossRef] [PubMed]
10. Uola, R.; Moroder, T.; Gühne, O. Joint Measurability of Generalized Measurements Implies Classicality. *Phys. Rev. Lett.* **2014**, *113*, 160403. [CrossRef] [PubMed]
11. Uola, R.; Budroni, C.; Gühne, O.; Pellonpää, J.-P. One-to-One Mapping between Steering and Joint Measurability Problems. *Phys. Rev. Lett.* **2015**, *115*, 230402. [CrossRef] [PubMed]
12. Uola, R.; Lever, F.; Gühne, O.; Pellonpää, J.-P. Unified picture for spatial, temporal, and channel steering. *Phys. Rev. A* **2018**, *97*, 032301. [CrossRef]
13. Kiukas, J.; Budroni, C.; Uola, R.; Pellonpää, J.-P. Continuous-variable steering and incompatibility via state-channel duality. *Phys. Rev. A* **2017**, *96*, 042331. [CrossRef]
14. He, Q.Y.; Drummond, P.D.; Reid, M.D. Entanglement, EPR steering, and Bell-nonlocality criteria for multipartite higher-spin systems. *Phys. Rev. A* **2011**, *83*, 032120. [CrossRef]
15. Cavalcanti, E.G.; He, Q.Y.; Reid, M.D.; Wiseman, H.M. Unified criteria for multipartite quantum nonlocality. *Phys. Rev. A* **2011**, *84*, 032115. [CrossRef]
16. He, Q.Y.; Reid, M.D. Genuine Multipartite Einstein-Podolsky-Rosen Steering. *Phys. Rev. Lett.* **2013**, *111*, 250403. [CrossRef] [PubMed]

17. Cavalcanti, D.; Skrzypczyk, P.; Aguilar, G.H.; Nery, R.V.; Souto Ribeiro, P.H.; Walborn, S.P. Detection of entanglement in asymmetric quantum networks and multipartite quantum steering. *Nat. Commun.* **2015**, *6*, 7941. [CrossRef] [PubMed]
18. Riccardi, A.; Macchiavello, C.; Maccone, L. Multipartite steering inequalities based on entropic uncertainty relations. *Phys. Rev. A* **2018**, *97*, 052307. [CrossRef]
19. Pusey, M.F. Negativity and steering: A stronger Peres conjecture. *Phys. Rev. A* **2013**, *88*, 032313. [CrossRef]
20. Cavalcanti, D.; Skrzypczyk, P. Quantum steering: A review with focus on semidefinite programming. *Rep. Prog. Phys.* **2017**, *80*, 024001. [CrossRef] [PubMed]
21. Kogias, I.; Skrzypczyk, P.; Cavalcanti, D.; Acín, A.; Adesso, G. Hierarchy of Steering Criteria Based on Moments for All Bipartite Quantum Systems. *Phys. Rev. Lett.* **2015**, *115*, 210401. [CrossRef] [PubMed]
22. Fillettaz, M.; Hirsch, F.; Designolle, S.; Brunner, N. Algorithmic construction of local models for entangled quantum states: Optimization for two-qubit states. *arXiv* **2018**, arXiv:1804.07576.
23. Hirsch, F.; Quintino, M.T.; Vértesi, T.; Pusey, M.F.; Brunner, N. Algorithmic Construction of Local Hidden Variable Models for Entangled Quantum States. *Phys. Rev. Lett.* **2016**, *117*, 190402. [CrossRef] [PubMed]
24. Cavalcanti, D.; Guerini, L.; Rabelo, R.; Skrzypczyk P. General Method for Constructing Local Hidden Variable Models for Entangled Quantum States. *Phys. Rev. Lett.* **2016**, *117*, 190401. [CrossRef] [PubMed]
25. Cavalcanti, E.G.; Foster, C.J.; Fuwa, M.; Wiseman, H.M. Analog of the Clauser–Horne–Shimony–Holt inequality for steering. *J. Opt. Soc. Am. B* **2015**, *32*, A74–A81. [CrossRef]
26. Jevtic, S.; Pusey, M.; Jennings, D.; Rudolph, T. Quantum Steering Ellipsoids. *Phys. Rev. Lett.* **2014**, *113*, 020402. [CrossRef] [PubMed]
27. Nguyen, H.C.; Vu, T. Necessary and sufficient condition for steerability of two-qubit states by the geometry of steering outcomes. *Europhys. Lett.* **2016**, *115*, 10003. [CrossRef]
28. Bowles, J.; Hirsch, F.; Quintino, M.T.; Brunner, N. Sufficient criterion for guaranteeing that a two-qubit state is unsteerable. *Phys. Rev. A* **2016**, *93*, 022121. [CrossRef]
29. Moroder, T.; Gittsovich, O.; Huber, M.; Uola, R.; Gühne, O. Steering Maps and Their Application to Dimension-Bounded Steering. *Phys. Rev. Lett.* **2016**, *116*, 090403. [CrossRef] [PubMed]
30. Costa, A.C.S; Uola, R.; Gühne, O. Steering criteria from general entropic uncertainty relations. *arXiv* **2017**, arXiv:1710.04541.
31. Walborn, S.P.; Salles, A.; Gomes, R.M.; Toscano, F.; Souto Ribeiro, P.H. Revealing Hidden Einstein-Podolsky-Rosen Nonlocality. *Phys. Rev. Lett.* **2011**, *106*, 130402. [CrossRef] [PubMed]
32. Schneeloch, J.; Broadbent, C.J.; Walborn, S.P.; Cavalcanti, E.G.; Howell, J.C. Einstein-Podolsky-Rosen steering inequalities from entropic uncertainty relations. *Phys. Rev. A* **2013**, *87*, 062103. [CrossRef]
33. Kriváchy, T.; Fröwis, F.; Brunner, N. Tight steering inequalities from generalized entropic uncertainty relations. *arXiv* **2018**, arXiv:1807.09603.
34. Quintino, M.T.; Brunner, N.; Huber, M. Superactivation of quantum steering. *Phys. Rev. A* **2016**, *94*, 062123. [CrossRef]
35. Cover, T.M.; Thomas, J.A. *Elements of Information Theory*, 2nd ed.; John Wiley & Sons: New York, NY, USA, 2006; ISBN 0471241954.
36. Havrda, J.; Charvát, F. Quantification method of classification processes. Concept of structural α-entropy. *Kybernetika* **1967**, *3*, 30.
37. Tsallis, C. Possible generalization of Boltzmann-Gibbs statistics. *J. Stat. Phys.* **1988**, *52*, 479. [CrossRef]
38. Rényi, A. *Valószínűségszámítás*; Tankönyvkiadó: Budapest, Hungary, 1966. (English Translation: Probability Theory (North-Holland, Amsterdam, 1970)).
39. Tsallis, C. Generalized entropy-based criterion for consistent testing. *Phys. Rev. E* **1998**, *58*, 1442. [CrossRef]
40. Furuichi, S.; Yanagi, K.; Kuriyama, K. Fundamental properties of Tsallis relative entropy. *J. Math. Phys.* **2004**, *45*, 4868. [CrossRef]
41. Van Erven, T.; Harremoës, P. Rényi Divergence and Kullback-Leibler Divergence. *IEEE Trans. Inf. Theory* **2014**, *60*, 3797. [CrossRef]
42. Deutsch, D. Uncertainty in Quantum Measurements. *Phys. Rev. Lett.* **1983**, *50*, 631. [CrossRef]
43. Maassen, H.; Uffink, J.B.M. Generalized entropic uncertainty relations. *Phys. Rev. Lett.* **1988**, *60*, 1103. [CrossRef] [PubMed]
44. Durt, T.; Englert, B.-G.; Bengtsson, I.; Życzkowski, K. On mutually unbiased bases. *Int. J. Quant. Inf.* **2010**, *8*, 535. [CrossRef]

45. Bengtsson, I.; Bruzda, W.; Ericsson, Å.; Larsson, J.-Å.; Tadej, W.; Życzkowski, K. Mutually unbiased bases and Hadamard matrices of order six. *J. Math. Phys.* **2007**, *48*, 052106. [CrossRef]

46. Sanchez-Ruiz, J. Improved bounds in the entropic uncertainty and certainty relations for complementary observables. *Phys. Lett.* **1995**, *201*, 125. [CrossRef]

47. Wu, S.; Yu, S.; Mølner, K. Entropic uncertainty relation for mutually unbiased bases. *Phys. Rev. A* **2009**, *79*, 022104. [CrossRef]

48. Haapasalo, E. Robustness of incompatibility for quantum devices. *J. Phys. A* **2015**, *48*, 255303. [CrossRef]

49. Uola, R.; Luoma, K.; Moroder, T.; Heinosaari, T. Adaptive strategy for joint measurements. *Phys. Rev. A* **2016**, *94*, 022109. [CrossRef]

50. Designolle, S.; Skrzypczyk, P.; Fröwis, F.; Brunner, N. Quantifying measurement incompatibility of mutually unbiased bases. *arXiv* **2018**, arXiv:1805.09609.

51. Schwonnek, R. Additivity of entropic uncertainty relations. *Quantum* **2018**, *2*, 59. [CrossRef]

52. Rastegin, A.E. Uncertainty relations for MUBs and SIC-POVMs in terms of generalized entropies. *Eur. Phys. J. D* **2013**, *67*, 269. [CrossRef]

53. Gühne, O.; Lewenstein, M. Entropic uncertainty relations and entanglement. *Phys. Rev. A* **2004**, *70*, 022316. [CrossRef]

54. Furuichi, S. Information theoretical properties of Tsallis entropies. *J. Math. Phys.* **2006**, *47*, 023302. [CrossRef]

55. Fehr, S.; Berens, S. On the Conditional Rényi Entropy. *IEEE Trans. Inf. Theory* **2014**, *60*, 6801. [CrossRef]

56. Huang, Y. Entanglement criteria via concave-function uncertainty relations. *Phys. Rev. A* **2010**, *82*, 012335. [CrossRef]

57. Gühne, O. Characterizing Entanglement via Uncertainty Relations. *Phys. Rev. Lett.* **2004**, *92*, 117903. [CrossRef] [PubMed]

58. Zhen, Y.-Z.; Zheng, Y.-L.; Cao, W.-F.; Li, L.; Chen, Z.-B.; Liu, N.-L.; Chen, K. Certifying Einstein-Podolsky-Rosen steering via the local uncertainty principle. *Phys. Rev. A* **2016**, *93*, 012108. [CrossRef]

59. Cavalcanti, E.G.; Jones, S.J.; Wiseman, H.M.; Reid, M.D. Experimental criteria for steering and the Einstein-Podolsky-Rosen paradox. *Phys. Rev. A* **2009**, *80*, 032112. [CrossRef]

60. Horodecki, M.; Horodecki, P. Reduction criterion of separability and limits for a class of distillation protocols. *Phys. Rev. A* **1999**, *59*, 4206. [CrossRef]

61. Werner, R.F. Quantum states with Einstein-Podolsky-Rosen correlations admitting a hidden-variable model. *Phys. Rev. A* **1989**, *40*, 4277. [CrossRef]

62. Skrzypczyk, P.; Cavalcanti, D. Loss-tolerant Einstein-Podolsky-Rosen steering for arbitrary-dimensional states: Joint measurability and unbounded violations under losses. *Phys. Rev. A* **2015**, *92*, 022354. [CrossRef]

63. Bavaresco, J.; Quintino, M.T.; Guerini, L.; Maciel, T.O.; Cavalcanti, D.; Cunha, M.T. Most incompatible measurements for robust steering tests. *Phys. Rev. A* **2017**, *96*, 022110. [CrossRef]

64. Costa, A.C.S.; Angelo, R.M. Quantification of Einstein-Podolski-Rosen steering for two-qubit states. *Phys. Rev. A* **2016**, *93*, 020103(R). [CrossRef]

65. Życzkowski, K.; Penson, K.A.; Nechita, I.; Collins, B. Generating random density matrices. *J. Math. Phys.* **2011**, *52*, 062201. [CrossRef]

66. Xiao, Y.; Ye, X.-J.; Sun, K.; Xu, J.-S.; Li, C.-F.; Guo, G.-C. Demonstration of Multisetting One-Way Einstein-Podolsky-Rosen Steering in Two-Qubit Systems. *Phys. Rev. Lett.* **2017**, *118*, 140404. [CrossRef] [PubMed]

67. Peres, A. All the Bell Inequalities. *Found. Phys.* **1999**, *29*, 589. [CrossRef]

68. Skrzypczyk, P.; Navascués, M.; Cavalcanti, D. Quantifying Einstein-Podolsky-Rosen Steering. *Phys. Rev. Lett.* **2014**, *112*, 180404. [CrossRef] [PubMed]

69. Bronzan, J.B. Parametrization of SU(3). *Phys. Rev. D* **1988**, *38*, 1994. [CrossRef]

70. Schack, R.; Caves, C.M. Explicit product ensembles for separable quantum states. *J. Mod. Opt.* **2000**, *47*, 387. [CrossRef]

71. Dür, W.; Cirac, J.I. Classification of multiqubit mixed states: Separability and distillability properties. *Phys. Rev. A* **2000**, *61*, 42314. [CrossRef]

72. Gruca, J.; Laskowski, W.; Żukowski, M.; Kiesel, N.; Wieczorek, W.; Schmid, C.; Weinfurter, H. Nonclassicality thresholds for multiqubit states: Numerical analysis. *Phys. Rev. A* **2010**, *82*, 012118. [CrossRef]

73. Chen, Z.-H.; Ma, Z.-H.; Gühne, O.; Severini, S. Estimating Entanglement Monotones with a Generalization of the Wootters Formula. *Phys. Rev. Lett.* **2012**, *109*, 200503. [CrossRef] [PubMed]

74. Berta, M.; Christandl, M.; Colbeck, R.; Renes, J.M.; Renner, R. The uncertainty principle in the presence of quantum memory. *Nat. Phys.* **2010**, *6*, 659. [CrossRef]
75. Barchielli, A.; Gregoratti, M.; Toigo, A. Measurement uncertainty relations for discrete observables: Relative entropy formulation. *Commun. Math. Phys.* **2018**, *357*, 1253. [CrossRef]
76. Schneeloch, J.; Howland, G.A. Quantifying high-dimensional entanglement with Einstein-Podolsky-Rosen correlations. *Phys. Rev. A* **2018**, *97*, 042338. 10.1103/PhysRevA.97.042338. [CrossRef]

entropy

MDPI

Article

The Einstein–Podolsky–Rosen Steering and Its Certification

Yi-Zheng Zhen [1,2,3], Xin-Yu Xu [1,2], Li Li [1,2,*], Nai-Le Liu [1,2,*] and Kai Chen [1,2,*]

[1] Hefei National Laboratory for Physical Sciences at Microscale and Department of Modern Physics, University of Science and Technology of China, Hefei 230026, China; yizheng@ustc.edu.cn (Y.-Z.Z.); xxy0812@mail.ustc.edu.cn (X.-Y.X.)
[2] CAS Center for Excellence and Synergetic Innovation Center in Quantum Information and Quantum Physics, University of Science and Technology of China, Hefei 230026, China
[3] Institute for Quantum Science and Engineering, Southern University of Science and Technology (SUSTech), Shenzhen 518055, China
* Correspondence: eidos@ustc.edu.cn (L.L.); nlliu@ustc.edu.cn (N.-L.L.); kaichen@ustc.edu.cn (K.C.)

Received: 17 January 2019; Accepted: 16 April 2019; Published: 20 April 2019

Abstract: The Einstein–Podolsky–Rosen (EPR) steering is a subtle intermediate correlation between entanglement and Bell nonlocality. It not only theoretically completes the whole picture of non-local effects but also practically inspires novel quantum protocols in specific scenarios. However, a verification of EPR steering is still challenging due to difficulties in bounding unsteerable correlations. In this survey, the basic framework to study the bipartite EPR steering is discussed, and general techniques to certify EPR steering correlations are reviewed.

Keywords: EPR steering; quantum correlation; non-locality; entanglement; uncertainty relations

1. Introduction

The Einstein–Podolsky–Rosen (EPR) steering [1] depicts one of the most striking features in quantum mechanics: With local measurements, one can steer or prepare a certain state on a remote physical system without even accessing it [2,3]. This feature challenges one's intuition in a way that the set of prepared states in the EPR steering fashion cannot be produced by any local operations. Therefore, a genuine nonlocal phenomenon happens in this procedure. Whilst EPR steering requires entanglement as the basic resource to complete the remote state preparation task, the correlation implied by EPR steering is not always enough to violate any Bell inequality. In this sense, EPR steering can be seen as a subtle quantum correlation or quantum resource in between entanglement and nonlocality.

The discussion of EPR steering dated back to the emergence of quantum theory, when Einstein, Podolsky, and Rosen questioned the completeness of quantum theory in their famous 1935's paper [4]. According to their argument on local realism, quantum theory allows a curious phenomenon: the so-called "spooky action at a distance". In the next year 1936, Schrödinger firstly introduced the terminology "entanglement" and "steering" to describe such quantum "spooky action". Debates on whether quantum theory is complete and how to understand quantum entanglement lasted for the following 20 years and were finally concluded by Bohm [5] and Bell [6,7]. The celebrated Bell inequality [8] was provided in 1955 as a practical verification of such "spooky action" or equivalent "non-locality". Noteworthily, the experimental tests of nonlocality without loopholes due to the real devices have been only carried out in recent years [9–12].

Strictly speaking, Bell inequalities test nonlocal correlations of general physical theories, not necessarily the quantum theory [8]. This can be understood by that Bell inequalities are functions of general probabilities and are independent of how to realize such probabilities. Thus, it is still a question on how quantum theory realizes such "spooky action" in its own context. As proved by Werner in

1989, entanglement is a necessary resource to exhibit nonlocality but not a sufficient one [13]. Note that, in some physics research fields, e.g., condensed matters, "entanglement" is equivalently used as "nonlocality" to discuss the genuine quantum phenomenon. In Werner's paper [13], the disentangled state is termed as the "classical correlated state", when the terminology "separable state" was not often used at that time. It, thus, drives physicists to consider under what conditions can entanglement show nonlocal effects in the quantum context.

This problem was further addressed by Wiseman, Jones, and Doherty [1,14] in 2007. They showed that there exists a set of bipartite entangled states, which can exhibit EPR steering properties but are not sufficient to violate Bell inequalities. For such states, termed as "EPR steerable states", one party can remotely prepare certain quantum states on the other party, and such preparations can not be replaced by any classical or quantum local operations. It, thus, represents another form of "spooky action at a distance". This "action" is in the quantum context in the sense that the description on the other party is always quantum. Then, EPR steering stands as an intermediate between entanglement and nonlocality, and they together form a relatively complete picture. On the one hand, EPR steering can be seen as a certification of entanglement. On the other hand, EPR steering exhibits a weaker form of nonlocality in specific scenarios.

The significance of studying EPR steering follows from important applications of entanglement and nonlocality. The entanglement and nonlocality have been proved to be important resources for many quantum information tasks, from quantum communications to quantum computation. As an intermediate but subtle resource, EPR steering may help to reduce the difficulty of such tasks and helps to inspire new protocols. For instance, nonlocality offers the strongest security in quantum cryptography. Nevertheless, the realization of nonlocality is based on violating Bell inequality, which is experimentally difficult. Simultaneously, violating EPR steering inequality is relatively applicable [15–17], and the realization of EPR steering also provides a different communication security for specific tasks [18].

Numerous results have been concluded in recent years. To certify EPR steering, there have been many approaches to witness EPR steerable correlations. Besides the basic linear inequality [19,20], local uncertainty relations [21–23], entropic uncertainty relations [24], fine-grained uncertainty relations [25], the CHSH-type inequality [26], covariance matrices [27], the semidefinite programming method [28], the all-versus-nothing fashion [29,30], and other methods, have been adopted in formulating inequalities and equations to verify EPR steering. As for understanding EPR steering, the asymmetric property [31,32], the super-activation of EPR steering correlation [33,34], the quantization of EPR steering [34–37], the negativity of steerable states [38], steering in the presence of positive operator valued measure (POVMs) [39], the resource theory description [40], the multipartite case [41,42], etc. are deeply investigated. In addition, relations between EPR steering and the uncertainty principle [23,24,43,44], joint measurability [45,46], sub-channel discrimination [47], etc. have also been discussed in the literature. Experimentally, EPR steering has been tested on various physical systems and platforms [16,19,48–53].

Noteworthily, comprehensive reviews [21,22,28] have given a complete picture of EPR steering. In Reference [21], the EPR steering is introduced based on the EPR *Gedankenexperiment* [4], while proposals to realize EPR steering test are reviewed from both the theoretical and experimental perspectives. The experimental friendly criteria for certifying EPR steering is thoroughly investigated in Reference [22]. In particular, the characterization of EPR steering is reviewed through the semidefinite programming method [28], which can be explicitly used to tackle the complicated numerical problems in detecting EPR steering. Recently, the EPR steering test is further generalized to a unified framework where classical, quantum, and post-quantum steering can be investigated [54]. The black box framework in that paper is the same with the framework adopted here.

In this paper, we will mainly focus on the basic techniques to certify the bipartite EPR steering and related quantum correlations, and show how to certify EPR steerable correlation in different fashions. This survey is organized as follows. In Section 2, the basic notations and the box framework combined with trust/untrust scenarios will be introduced. After a brief discussion of entanglement and nonlocality in such a framework, EPR steering as well as other equivalent descriptions will be introduced in Section 3. In Section 4, the systematic method to formulate the criteria for certifying EPR steering will be discussed. Two types of criteria, (a) linear EPR steering inequality and (b) criterion based on uncertainty relations, will be studied in detail. Their performances on some typical states will also be given. Finally, a summary will be given in Section 5.

2. Preliminaries and Notations

In this paper, we will focus on the bipartite correlation $P\left(ab|xy\right)$ with input parameters x,y and output parameters a,b and discuss, under certain assumptions, whether the correlation can be certified as EPR steerable. Before the discussion, we firstly introduce the basic terminology and the notations that will be used throughout the paper.

2.1. The Box Framework

A typical experiment of testing a bipartite correlation can be described by the box framework, as shown in Figure 1. Suppose two parties, Alice and Bob, are in their closed labs to do the experiment. The lab is sketched as the doted rectangle, inside which there is an experimental device sketched as the solid rectangle. In each run of the experiment, Alice and Bob are distributed with a bipartite state W from a source, which may be unknown. In their own labs, combined with the subsystem they received, Alice and Bob can input x and y to the device and obtain outputs a and b, respectively. Such a run is repeated enough times so that, after the experiment, Alice and Bob can obtain the correlation $P\left(ab|xy\right)$ by announcing their input and output results. Depending on different descriptions and mechanics of the source and device, the correlation may have different structures and properties. The aim of the box framework is then to characterize the dependence of the correlation on descriptions of sources and devices.

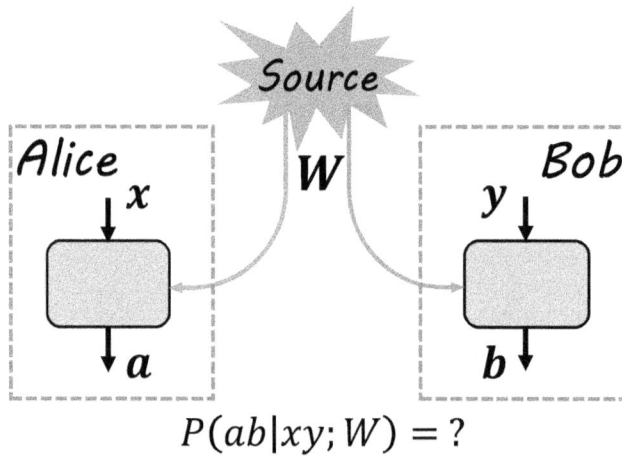

$$P(ab|xy;W) = ?$$

Figure 1. The box framework: The source distributes state W to Alice and Bob. In their own closed labs, Alice and Bob make operations on received local states. Alice's operations are labeled by inputs x, with outputs labeled by a. Bob's operations are labeled by inputs y, with outputs labeled by b. After the experiment, Alice and Bob publicize their results and the corresponding statistics are denoted by probability distribution $\{P\left(ab|xy;W\right)\}$. According to such a distribution, the local property of W can be inferred.

In general, there is no restrictions on the source, inputs, and outputs. For instance, the source W; inputs x,y; and outputs a,b can all be quantum states, with the devices being quantum instruments.

In this case, the box framework characterizes general local quantum operations on bipartite quantum states. In this paper, we will restrict the device to be the typical measurement device in labs. That is, the inputs x, y represent different measurement settings on the received subsystem and the outputs a, b represent different outcomes. Physically, x, y, a, b can be described by natural numbers $0, 1, 2, \ldots$ and corresponding sets are denoted as $\mathcal{X}, \mathcal{Y}, \mathcal{A}, \mathcal{B}$, respectively. In the scenario of steering and nonlocality, there are some common assumptions.

2.1.1. The No-Signaling Principle

Roughly speaking, the no-signaling principle describes that Alice and Bob cannot communicate with each other during the test [55,56]. In the above box framework, this principle guarantees the independence between Alice and Bob such that the correlation $P(ab|xy)$ is faithfully generated by the state W and measurements but not any other statistics shared before or during the test. Mathematically, the no-signaling principle has the following form,

$$\sum_a P(ab|xy; W) = P(b|xy; W) = P(b|y; W), \forall x \in \mathcal{X}, \tag{1}$$

$$\sum_b P(ab|xy; W) = P(a|xy; W) = P(a|x; W), \forall y \in \mathcal{Y}. \tag{2}$$

Therefore, the no-signaling principle denies the possibility that Alice and Bob can guess each other's measurement setting y or x based on their local statistics $P(a|x; W)$ or $P(b|y; W)$, respectively.

Experimentally, this principle is guaranteed by Alice and Bob being separated far away (space-like separation) and by both of them choosing measurement settings independently and randomly. The no-signaling principle is then guaranteed by two hypotheses. Firstly, two parties in the space-like separation cannot communicate with each other. Secondly, the random number generators [57] in Alice's and Bob's labs should be truly independent and random.

In the test of nonlocality and EPR steering, we suppose that the no-signaling principle has been guaranteed.

2.1.2. Trust and Untrust

If the description of boxes is restricted as quantum or classical, we can further define if a device is trusted or not for the sake of practice. A device is said to be trusted if it is believed that the function of the device is exactly what we expect. This definition comes from the sense that, without the assistance of other resources, it is, in principle, impossible to verify how an unknown device really functions based solely on statistics of measurement results. Particularly, in the rest of the paper, the device is trusted if it is a quantum device and the accurate quantum mechanical description is known.

Therefore, if we say some devices are trusted, we actually make additional assumptions. For instance, we say a measurement device is trusted if its measurement can be exactly described by a known set of POVMs $\left\{ E_b^y \right\}$, where y is the measurement settings and b is the measurement outcome. On the contrary, we say a measurement device is untrusted if we can, at most, describe the measurement results by a probability distribution $P(b|y)$.

The scenario is device-independent if all devices and the source are untrusted. Particularly, the scenario is measurement-device-independent if all measurement devices are untrusted. If some but not all measurement devices are untrusted, we say the corresponding scenario as semi-measurement-device-independent.

2.2. Entanglement and Nonlocality

In the box framework, we can discuss entanglement and nonlocality in an operational manner. Let λ label different hidden states in W and p_λ be its probability such that $\int d\lambda p_\lambda = 1$. The correlation can be written as

$$P\left(ab|xy;W\right) = \int d\lambda p_\lambda P\left(ab|xy,\lambda\right). \tag{3}$$

The local realism argues that, for any hidden variable λ, $P\left(ab|xy,\lambda\right)$ can be localized such that $P\left(ab|xy,\lambda\right) = P\left(a|x,\lambda\right)P\left(b|y,\lambda\right)$. We say the correlation $P\left(ab|xy;W\right)$ is a local correlation if all hidden states in Equation (3) can be localized.

The nonlocality is defined as the failure of local realism, usually modeled by local hidden variable (LHV) models. The main property of LHV models is that, if two parties are no longer interacting (guaranteed by space-like separation), their measurements should be local, i.e., a should be independent on y and b (similarly for b). Thus, for each hidden variable λ, the LHV models produce a localized correlation $P\left(ab|xy,\lambda\right) = P\left(a|x,\lambda\right)P\left(b|y,\lambda\right)$. The nonlocal correlation is defined as correlations that cannot explained by the local correlation

$$P_{LHV}\left(ab|xy;W\right) = \int d\lambda p_\lambda P\left(a|x,\lambda\right)P\left(b|y,\lambda\right), \tag{4}$$

where $P\left(a|x,\lambda\right)$ and $P\left(b|y,\lambda\right)$ are arbitrary probabilities. If the statistic of the experimental results cannot be explained by Equation (4), then the correlation is nonlocal and we say the source W is nonlocal.

The Bell inequality is indeed a linear constraint on all local correlations. This is based on the fact that all local correlations from Equation (4) form a convex subset. There are some correlations produced by quantum mechanics outside this subset. Precisely, in the probability space, points of local correlations form a polytope, while all probabilities produced by quantum mechanics form a superset of the polytope [8]. Thus, one can distinguish a specific nonlocal correlation from all local correlations by a linear equation. Additionally, since Alice's and Bob's measurement results are described by general probabilities, the problem of nonlocality corresponds to the device-independent scenario.

The entanglement is defined as the failure of description in the form of separable states. The separable states have a clear definition that ρ_{SEP} is separable if $\rho_{SEP} = \sum_k p_k \rho_k^A \otimes \rho_k^B$ with ρ_k^A and ρ_k^B being some local quantum states and $\sum_k p_k = 1$. Usually, the decomposition of a separable state is not unified and the verification of a separable is not an easy task. However, if the source W distributes separable states in the box framework, then the correlation is in the form of

$$P_{SEP}\left(ab|xy;W\right) = \int d\lambda p_\lambda P_Q\left(a|x,\lambda\right)P_Q\left(b|y,\lambda\right), \tag{5}$$

where $P_Q\left(a|x,\lambda\right) = \mathrm{tr}\left[E_a^x \rho_\lambda^A\right]$ and $P_Q\left(b|y,\lambda\right) = \mathrm{tr}\left[F_b^y \rho_\lambda^B\right]$ are probabilities yielded by quantum measurements. Here ρ_λ^A and ρ_λ^B are local hidden quantum states which may be unknown to Alice and Bob, while E_a^x and F_b^y are POVMs that Alice and Bob know well. If the statistic of experimental results cannot be explained by Equation (5), then the correlation is non-separable, i.e., entangled, and we say the source W is entangled.

Like the Bell inequality, one can use a linear constraint, the so-called entanglement witness, to bound all separable correlations to certify an entangled correlation. Similar to the case of local correlations, correlations produced by all separable states also form a convex set. Since all devices are assumed to be quantum, here, the entanglement corresponds to the scenario where all measurement devices are trusted.

3. The EPR Steering

3.1. Definition

From the above introduction, it is easy to see that definitions of nonlocality and entanglement have two similarities. Firstly, both of them are defined by the failure of corresponding local models in their own contexts, i.e., LHV models and separable quantum states, respectively. Secondly, as for the two local models, the descriptions on Alice's and Bob's systems are symmetric, i.e., general probabilities $P(a|x,\lambda)$ and $P(b|y,\lambda)$ in LHV models and quantum probabilities $P_Q(a|x,\lambda)$ and $P_Q(b|y,\lambda)$ in separable states. The only difference between the two definitions is whether the local descriptions are both quantum. A natural equation would be "What if the local descriptions are asymmetric?" and "Can this asymmetric property lead to novel correlations?". The answer is yes. The corresponding local model is called the local hidden state (LHS) model and its failure implies the main objective of this paper, the correlation of EPR steering [1].

Definition 1 (EPR steering). *In a box frame test, the experimental result statistics exhibits EPR steering property, if it cannot be explained by the correlation of LHS models, i.e., the correlation cannot be written as*

$$P_{LHS}(ab|xy;W) = \int d\lambda p_\lambda P(a|x,\lambda) P_Q(b|y,\lambda),\tag{6}$$

where p_λ is a probability distribution satisfying $\int d\lambda p_\lambda = 1$, $P(a|x,\lambda)$ is an arbitrary probability distribution, and $P_Q(b|y,\lambda) = \mathrm{tr}\left[F_b^y \sigma_\lambda\right]$ is a probability distribution generated by POVM F_b^y on quantum state σ_λ.

It is said that the corresponding quantum state is EPR steerable if Equation (6) is violated.

The relationship among EPR steerable states, entangled states, and nonlocal states are sketched out in Figure 2.

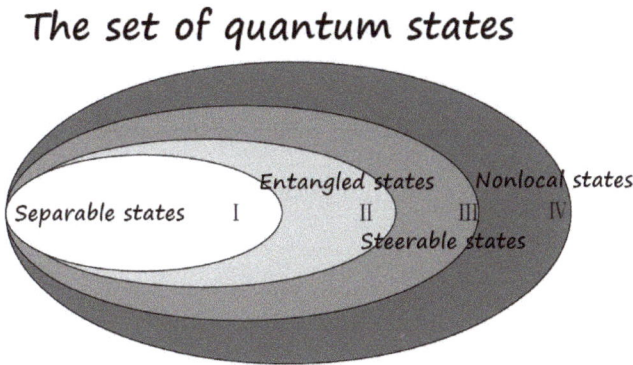

Figure 2. The set of quantum states: All quantum states form a convex set, with the boundary being the pure state. The region I represents the convex subset of separable states. The complement set, i.e., regions II, III, and IV, represent entangled states. Particularly, regions III and IV represent Einstein–Podolsky–Rosen (EPR) steerable states, and the region IV represents nonlocal states. Region II are entangled states which is neither EPR steerable nor nonlocal.

3.2. One-Sided Measurement Device Independence

The understanding of EPR steering can be more clear if we discuss it in the trust and untrust scenarios. As has been discussed before, nonlocality defies a local correlation in the device-independent scenario, while entanglement defies local correlations in the measurement-dependent scenario.

Since EPR steering is defined as the failure of LHS models, where only one party is assumed to be quantum, we have the following claim.

Remark 1. *EPR steering defies all local correlations in the one-sided measurement-device-independent scenario.*
   This scenario corresponds to the real situation when users in the communication task need different levels of security. For instance, in the communication task between banks and individuals, obviously it is easier for banks to prepare their devices to be trustworthy. For individuals, however, due to limits of costs and environments, their devices are hard to be guaranteed as trustworthy ones. In this case, let individuals be Alice and banks be Bob, such that if EPR steering correlation is certified by the violation steering inequality, then the secure quantum communications can be achieved [18].
   Different scenarios corresponding to nonlocality, entanglement, and EPR steering are shown in Figure 3.

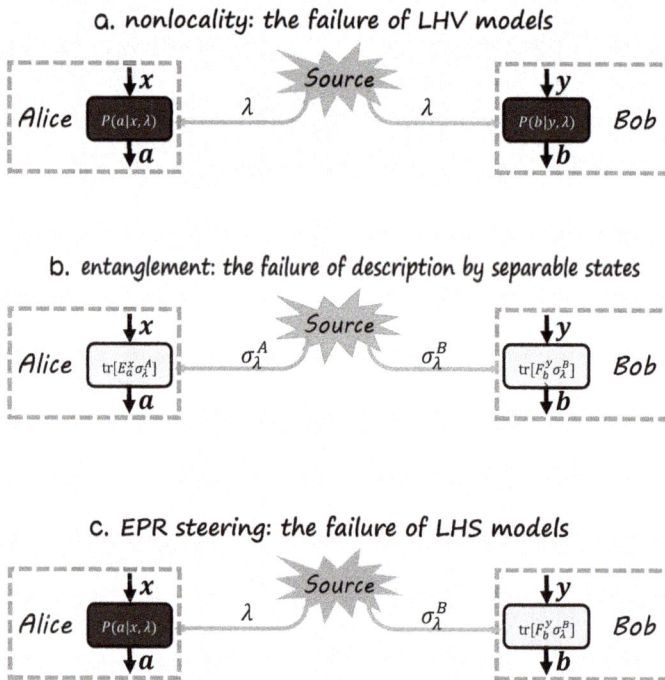

Figure 3. The box framework for nonlocality, entanglement, and EPR steering. The color black represents untrusted, gray represents unknown, and white represents trusted. (**a**) The nonlocality scenario, where the source is unknown and measurement devices are untrusted. (**b**) The entanglement scenario, where source is unknown and measurement devices are trusted. (**c**) The EPR steering scenario, where source is unknown and Alice's measurement devices are untrusted while Bob's are trusted.

3.3. Schrödinger's Steering Theorem

As an equivalent definition, one can consider the assemblage. The assemblage is defined as the collection of ensembles, denoted by $\left\{\tilde{\rho}_{a|x}\right\}_{a,x}$, where $\tilde{\rho}_{a|x}$ are unnormalized quantum states satisfying $\sum_a \tilde{\rho}_{a|x} = \sigma, \forall x$. The definition of EPR steering can be applied on the assemblage $\left\{\tilde{\rho}_{a|x}\right\}_{a,x}$ instead of correlations $P(ab|xy)$. This equivalence is guaranteed by the Schrödinger's steering theorem [2,3].

Theorem 1 (Schroödinger's steering theorem). *The following two statements hold:*

1. *For any quantum state ρ_{AB}, let $\{E_a^x\}_a$ be a complete set of POVMs satisfying $\sum_a E_a^x = \mathbb{I}, \forall x$. Then, the conditional states $\tilde{\rho}_B^{a|x} = \mathrm{tr}_A\left[E_a^x \otimes \mathbb{I}\rho_{AB}\right]$ for all x and a form an assemblage.*

2. *For any assemblage $\left\{\tilde{\rho}_{a|x}\right\}_{a,x}$ with $\sum_a \tilde{\rho}_{a|x} = \sigma$, there always exist a pure quantum state $|\psi\rangle_{AB}$ satisfying $\mathrm{tr}_A\left[|\psi\rangle_{AB}\langle\psi|\right] = \sigma$ and complete sets of POVMs $\{E_a^x\}$ satisfying $\sum_a E_a^x = \mathbb{I}$ for all x, such that $\tilde{\rho}_{a|x}$ can be produced, i.e., $\tilde{\rho}_{a|x} = \mathrm{tr}_A\left[E_a^x \otimes \mathbb{I}|\psi\rangle_{AB}\langle\psi|\right]$.*

Proof. For the first statement, it is straightforward to verity that, for all x,

$$\sum_a \tilde{\rho}_B^{a|x} = \sum_a \mathrm{tr}_A\left[E_a^x \otimes \mathbb{I}\rho_{AB}\right] = \mathrm{tr}_A\left[\sum_a E_a^x \otimes \mathbb{I}\rho_{AB}\right] = \mathrm{tr}_A\left[\rho_{AB}\right] = \rho_B.$$

For the second statement, write ρ_B in its diagonal form $\rho_B = \sum_i \lambda_i |i\rangle\langle i|$ with $\lambda_i > 0$ and let D be the diagonal matrix $D = \mathrm{diag}\left[\sqrt{\lambda_1}, \ldots, \sqrt{\lambda_d}\right]$. Denote the generalized invertible matrix of D as D^{-1}. It can be verified that $E_a^x = D^{-1}\tilde{\rho}_{a|x}^T D^{-1}$ and $|\psi\rangle_{AB} = \sum_i \sqrt{\lambda_i}|ii\rangle$ are required POVMs and the quantum state, respectively. $\square$

Then, an assemblage $\left\{\rho_{a|x}\right\}_{a,x}$ is said to be unsteerable if it can be produced by rearrangement on an LHS model $\{p_\lambda\sigma_\lambda\}$, i.e., $\rho_{a|x} = \sum_\lambda p_{x,\lambda}(a)\, p_\lambda\sigma_\lambda$ with $\sum_a p_{x,\lambda}(a) = 1$ for all x and λ. Particularly, for two-qubit states, the steered states $\tilde{\rho}_B^{a|x}$ form an ellipsoid in the Bloch sphere on Bob's side [58]. The volume of such ellipsoid indicates the steerability of the bipartite state. If the assemblage cannot be written in this manner, it is said to be EPR steerable. This EPR steering definition is equivalent to Definition 1 on the condition that Bob is allowed to do the state tomography for each conditional state $\rho_{a|x}$. Furthermore, in Reference [54] post-quantum steering is well-studied using no-signaling assemblages. If Bob's measurements are not sufficient to do the tomography, then it is hard for him to obtain each $\rho_{a|x}$ yet to verify the EPR steerability. In this case, however, the statistic of measurement results $P(ab|xy)$ is still useful. In the following discussion, Definition 1 will be mainly considered.

There is an interesting analog of the assemblage [59], from the perspective of the state-channel duality [60]. If the set of local hidden state σ_λ is replaced with a set of POVMs $\{G_\lambda\}$, then the assemblage of $\{G_\lambda\}$ can be defined as jointly measurable observables. That is, a set of POVMs $\{E_a^x\}$ is jointly measurable if $E_a^x = \sum_\lambda p_{x,\lambda}(a)\, p_\lambda G_\lambda$, with $p_{x,\lambda}(a)$ being probabilities. It has been proved that a given assemblage $\left\{\rho_{a|x}\right\}_{a,x}$ is unsteerable if and only if Alice's measurements $\{E_a^x\}_{a,x}$ is jointly measurable [45,46], which can be checked from the Proof of Theorem 1.

4. Criteria of EPR Steering

A natural question arises on how to certify the EPR steering correlation. It can be shown that unsteerable correlations, i.e., correlations produced by LHS models, form a convex subset. According to the hyperplane separate theorem, there always exists a linear constraint of all unsteerable correlations, such that steerable ones can be witnessed [22].

Suppose that the box framework is fixed, i.e., $\mathcal{X}, \mathcal{Y}, \mathcal{A}$, and $\mathcal{B}$ are all fixed. Then, the set of probability distributions $\{P(ab|xy)\,|x \in \mathcal{X}, y \in \mathcal{Y}, a \in \mathcal{A}, b \in \mathcal{B}\}$ can be seen as a point in the probability space. All correlations yielded by LHS models in Equation (6) $\{p_\lambda\sigma_\lambda\}$ form a subset $\{P_{LHS}(ab|xy)\,|x \in \mathcal{X}, y \in \mathcal{Y}, a \in \mathcal{A}, b \in \mathcal{B}\}$. This subset of usteerable correlations is convex.

Lemma 1. *The unsteerable correlations $\{P_{LHS}(ab|xy)\}$ form a convex subset.*

Proof. For any two LHS models $\left\{ p_\lambda^{(1)} \sigma_\lambda^{(1)} \right\}$ and $\left\{ p_\mu^{(2)} \sigma_\mu^{(2)} \right\}$, the correlation yielded by them are

$$P_{LHS}^{(1)}(ab|xy) = \int dp_\lambda^{(1)} P^{(1)}(a|x,\lambda) \operatorname{tr}\left[F_b^y \sigma_\lambda^{(1)} \right], \tag{7}$$

$$P_{LHS}^{(2)}(ab|xy) = \int dp_\mu^{(2)} P^{(2)}(a|x,\mu) \operatorname{tr}\left[F_b^y \sigma_\mu^{(2)} \right], \tag{8}$$

respectively. Then, any linear combination of these two, i.e., $tP_{LHS}^{(1)}(ab|xy) + (1-t) P_{LHS}^{(2)}(ab|xy)$ with $0 \leqslant t \leqslant 1$, can always be written as the correlation yielded by another LHS model $\{q_\nu \tau_\nu\}$, where

$$q_\nu = t p_\lambda^{(1)} \delta_{\nu\lambda} + (1-t) p_\mu^{(2)} \delta_{\nu\mu}, \tag{9}$$

$$\tau_\nu = \sigma_\lambda^{(1)} \delta_{\nu\lambda} + \sigma_\mu^{(2)} \delta_{\nu\mu}. \tag{10}$$

It is easy to verify that

$$P_{LHS}^{(3)}(ab|xy) = \int dq_\nu P(a|x,\nu) \operatorname{tr}\left[F_b^y \tau_\nu \right] \tag{11}$$

$$= t \int dp_\lambda^{(1)} P^{(1)}(a|x,\lambda) \operatorname{tr}\left[F_b^y \sigma_\lambda^{(1)} \right] \tag{12}$$

$$+ (1-t) \int dp_\mu^{(2)} P^{(2)}(a|x,\mu) \operatorname{tr}\left[F_b^y \sigma_\mu^{(2)} \right] \tag{13}$$

$$= t P_{LHS}^{(1)}(ab|xy) + (1-t) P_{LHS}^{(2)}(ab|xy). \tag{14}$$

Therefore, the subset of all unsteerable correlation is convex. □

Any convex subset can be bounded by a linear equation, which is guaranteed by the hyperplane separation theorem [61].

Lemma 2. *(Hyperplane separation theorem) Let A and B be two disjoint nonempty convex subsets of $\mathbb{R}_n$. Then, there exists a nonzero vector v and a real number c such that*

$$\langle x, v \rangle \geq c \qquad and \qquad \langle y, v \rangle \leq c$$

for all x in A and y in B, i.e., the hyperplane $\langle \cdot, v \rangle = c$ and v the normal vector, separates A and B.

The proof can be found in many Linear Algebra textbooks (like Reference [61]) and is skipped here. Based on these two lemmas, one can certify EPR correlations by linear inequalities [22].

Theorem 2. *Any EPR steerable correlation can be verified by an inequality.*

Proof. According to Lemma 2, let the set A be the set of all unsteerable correlations, which is proved by Lemma 1. For any EPR steerable correlation $P_{STE}(ab|xy)$, let B be a sufficient open ball containing $P_{STE}(ab|xy)$, such that the open ball is disjoint with the subset A. Then, there always exists a hyperplane $v(P(ab|xy)) = \sum_{abxy} v_{ab}^{xy} P(ab|xy) = c$, such that $v(P_{LHS}(ab|xy)) \geqslant c$ holds for all unsteerable correlations $P_{LHS}(ab|xy)$ while $v(P_{STE}(ab|xy)) < c$ holds for the certain EPR steerable correlation $P_{STE}(ab|xy)$. □

4.1. Linear EPR Steering Inequality

Perhaps the most straightforward criteria to verify EPR steering is the linear steering inequality. The linear steering inequality to certify EPR steering is like the Bell inequality to nonlocality and the entanglement witness to entanglement. From the Proof of Theorem 2, the linear steering inequality has a general from, i.e., for all unsteerable correlations, the following inequality holds:

$$I(P) = \sum_{a,b,x,y} V_{ab}^{xy} P(ab|xy) \leqslant B_{LHS}, \tag{15}$$

$$B_{LHS} = \max_{P_{LHS}} \sum_{a,b,x,y} V_{ab}^{xy} P_{LHS}(ab|xy), \tag{16}$$

where $P = \{P(ab|xy)\}$ denotes the correlation $\{P(ab|xy) | a \in \mathcal{A}, b \in \mathcal{B}, x \in \mathcal{X}, y \in \mathcal{Y}\}$, $V_{ab}^{xy} \in \mathbb{R}$ are some coefficients, and B_{LHS} is the bound of all unsteerable correlations.

Then, if for a certain correlation $Q = \{Q(ab|xy)\}$ satisfies $I(Q) > B_{LHS}$, i.e., the linear steering inequality in Equation (15) is violated, then it can be conclude that Q cannot be explained by any LHS correlations, i.e., Q is EPR steerable.

In practice, the expectation value of the measurement results is usually considered for convenience and clarity. Combined with the scenario of EPR steering where Alice's and Bob's measurement devices are untrusted and trusted, respectively, denote $A_x = \{a_x \in \mathbb{R}\}$ as the random variable corresponding to Alice's measurements and $B_y = \sum_b b_y F_b^y$ as the general quantum measurement for Bob's measurements, with F_b^y being the POVM corresponding to the result b_y.

Suppose that, in an EPR steering test experiment, Alice and Bob randomly and independently choose n pairs of measurements A_k and B_k, respectively, labeled by $k = 1, 2, \ldots, n$. After the experiment, the value of each pair of measurements is

$$\langle A_k B_k \rangle = \sum_{a_k, b_k} a_k b_k P(ab|A_k B_k). \tag{17}$$

Then, the following linear steering inequality holds for all unsteerable correlations [19,20].

Theorem 3 (The linear EPR steering inequality). *If the result of an EPR steering test violates the following inequality*

$$S_n = \frac{1}{n} \sum_{k=1}^{n} g_k \langle A_k B_k \rangle \leqslant C_n, \tag{18}$$

where g_k are real numbers and C_n satisfies

$$C_n = \max_{a_k \in A_k} \left\{ \lambda_{max} \left(\frac{1}{n} \sum_{n=1}^{n} g_k a_k B_k \right) \right\}, \tag{19}$$

with $\lambda_{max}(\cdot)$ the maximal eigenvalue of the matrix, then the correlation of the test shows EPR steering. The corresponding quantum state ρ_{AB} is EPR steerable, and more precisely, Alice can steer Bob.

Proof. By definition, $S_n \leqslant C_n$ is an EPR steering inequality when it holds for all unsteerable correlation P_{LHS}. P_{LHS} has a general form as defined by Equation (6), i.e.,

$$P_{LHS}(ab|xy) = \int dp_\lambda P(a|x,\lambda) \, \text{tr} \left[F_b^y \sigma_\lambda \right].$$

It is straightforward to verify that

$$S_n \left(\mathbf{P}_{LHS} \right) = \frac{1}{n} \sum_{k=1}^{n} g_k \sum_{a_k} a_k \int dp_\lambda P \left(a_k | A_k, \lambda \right) \text{tr} \left[B_k \sigma_\lambda \right] \tag{20}$$

$$\leqslant \max_{a_k \in A_k} \frac{1}{n} \sum_{k=1}^{n} g_k a_k \int dp_\lambda \text{tr} \left[B_k \sigma_\lambda \right] \tag{21}$$

$$\leqslant \max_{a_k \in A_k} \lambda_{\max} \left(\frac{1}{n} \sum_{k=1}^{n} g_k a_k B_k \right) \tag{22}$$

$$= C_n. \tag{23}$$

Here, the second line comes from $\sum_{a_k} a_k P \left(a_k | A_k, \lambda \right) \text{tr} \left[B_k \sigma_\lambda \right] \leq \max_{a_k \in A_k} a_k \text{tr} \left[B_k \sigma_\lambda \right]$, and the third line comes from

$$\frac{1}{n} \sum_{k=1}^{n} g_k a_k \int dp_\lambda \text{tr} \left[B_k \sigma_\lambda \right] = \text{tr} \left[\left(\frac{1}{n} \sum_{k=1}^{n} g_k a_k B_k \right) \int dp_\lambda \sigma_\lambda \right] \leqslant \lambda_{\max} \left(\frac{1}{n} \sum_{k=1}^{n} g_k a_k B_k \right). \tag{24}$$

□

Here, g_k are flexible coefficients to help to form efficient inequalities.

Example 1. *The 2-qubit Werner state [13].*

As a simple example, one can consider the 2-qubit Werner state, which is an often-used bipartite quantum states in quantum information processes. It can be constructed as the mixture of the maximally entangled state $|\Psi^-\rangle = \left(|01\rangle - |10\rangle \right) / \sqrt{2}$ and the white noise $\mathbb{I}/4$, i.e.,

$$W_\mu = \mu \left| \Psi^- \right\rangle \left\langle \Psi^- \right| + (1 - \mu) \frac{\mathbb{I}}{4}, \tag{25}$$

where $\mu \in [0, 1]$. It can be theoretically proved that W_μ is entangled when $\mu > 1/3$ and is separable when $\mu \leq 1/3$ [13]. When $\mu > 1/\sqrt{2}$, there exists certain observables such that the CHSH inequality is violated [62], i.e., W_μ is nonlocal when $\mu > 1/\sqrt{2}$. When $\mu \lesssim 0.66$, any measurement results of W_μ can be explained by some LHV models, i.e., W_μ never exhibits a nonlocality when $\mu \lesssim 0.66$ [63]. It is an open question of whether W_μ is nonlocal when $0.66 \lesssim \mu \leq 1/\sqrt{2}$.

It has been proved that $\mu > \frac{1}{2}$ is the critical bound for the EPR steerability of W_μ [1], i.e., any measurement results of W_μ can be explained by LHS models when $\mu \leq \frac{1}{2}$.

It is easy to see that the performance of linear EPR steering inequality (Theorem 3) depends on the number of Alice and Bob's measurement pairs and Bob's observables. Furthermore, from the symmetric property of the 2-qubit Werner state, when Bob's k'th observable is $B_k = \mathbf{n}_k \cdot \boldsymbol{\sigma}$, where $\mathbf{n}_k = \left(n_x^{(k)}, n_y^{(k)}, n_z^{(k)} \right)$ is a unit vector and $\boldsymbol{\sigma} = (\sigma_x, \sigma_y, \sigma_z)$ is the set of Pauli matrices, i.e.,

$$\sigma_x = \begin{pmatrix} 0 & 1 \\ 1 & 0 \end{pmatrix}, \sigma_y = \begin{pmatrix} 0 & -i \\ i & 0 \end{pmatrix}, \sigma_z = \begin{pmatrix} 1 & 0 \\ 0 & -1 \end{pmatrix}, \tag{26}$$

Alice can always choose her observable as $A_k = -\mathbf{n}_k \cdot \boldsymbol{\sigma}$, such that the expectation value of the measurement pair $\text{tr} \left[A_k \otimes B_k W_\mu \right] = \text{tr} \left[-\mathbf{n}_k \cdot \boldsymbol{\sigma} \otimes \mathbf{n}_k \cdot \boldsymbol{\sigma} W_\mu \right] = \mu$. If we further let $g_k = 1$, $S_n = \mu$ always holds independent of the number of measurements.

The bound C_n, however, depends on n and the form of B_k. More precisely, when $n = 2$, let $B_1 = \sigma_x$ and $B_2 = \sigma_y$. The corresponding $C_2 = 1/\sqrt{2}$ and, thus, W_μ is steerable when $\mu > 1/\sqrt{2} \approx 0.707$. When $n = 3$, let $B_1 = \sigma_x$, $B_2 = \sigma_y$, and $B_3 = \sigma_z$. The corresponding $C_3 = 1/\sqrt{3}$ and, thus, W_μ is steerable when $\mu > 1/\sqrt{3} \approx 0.577$. It can be proved that, for $n = 2, 3$, the above Bob's observables are optimal [15,19]. When $n = 4$, it is a little complicated, but one can let $B_1 = \sigma_x$, $B_2 = \sigma_y$, $B_3 = \left(\sigma_y + \sqrt{3}\sigma_z \right)/2$, and $B_4 = \left(\sigma_y - \sqrt{3}\sigma_z \right)/2$. The corresponding $C_4 = \sqrt{5}/4$ and W_μ is steerable when $\mu > \sqrt{5}/4 \approx 0.559$. In this case, the observables $\{B_1, B_2, B_3, B_4\}$ may not be optimal. It can be concluded that the larger the number of measurement pairs, the lower bound of μ can be detected by the linear inequality. In principle, when $n \to \infty$, which can be understood as the state tomography, one can image that the critical bound for the EPR steerability can be finally found, i.e., $\mu > 1/2$ [15,19].

This example shows the application of the linear EPR steering inequality, as well as its limitations. Firstly, the linear inequality (Equation (18)) may not give the critical bound of the EPR steerability when testing some kinds of quantum states. This makes sense as the linear inequality represents only one hyperplane in the probability space, while the sufficient and necessary condition for the EPR steerability usually requires numerous such hyperplanes. Secondly, the linear inequality (Equation (18)) closely relies on observables that would be chosen. Thus, in practice, a natural question is how to choose Alice's and Bob's observables such that the detection of EPR steering is efficient. Thirdly, as seen from the example, the more measurements, the better the performance of the linear inequality. However, the complexity to compute C_n is also increasing when n becomes large. In fact, the method in Equation (19) to calculate C_n needs to maximize all $a_k \in A_k$ for all k, which leads the complexity of C_n exponentially increasing with n. Therefore, it is motivated to specify systematic techniques of choosing proper observables and obtaining C_n more efficiently.

4.1.1. Optimal Observables for Alice

Usually, Bob's observables $\{B_k\}$ are fixed due to the measurement devices are trusted in his lab. Here, the problem of how Alice chooses proper measurement settings according to Bob's observables is discussed. The main idea is that, to violate the linear inequality (Equation (18)) more obviously, Alice should choose observables such that $\langle A_k B_k \rangle$ is larger when $g_k > 0$ and $\langle A_k B_k \rangle$ is smaller when $g_k < 0$. In this sense, the value of S_n can be made as large as possible so as to violate the unsteerable bound. This technique can be formulated based on the following lemma [64].

Lemma 3. *For any two $n \times n$-dimensional Hermite matrices A and B, the following equation holds,*

$$\max_{U} \operatorname{tr}\left[AU^\dagger BU\right] = \sum_{i=1}^{n} \alpha_i \beta_i, \tag{27}$$

where U is an arbitrary unitary matrix and $\alpha_1 \geq \alpha_2 \geq \cdots \geq \alpha_n$ and $\beta_1 \geq \beta_2 \geq \cdots \geq \beta_n$ are the eigenvalues of A and B, respectively.

Proof. Write $A = \sum \alpha_i e_i$ and $B = \sum \beta_j f_j$ in the diagonal form, where $\{e_i\}$ and $\{f_j\}$ are specific bases of the operator space, respectively satisfying $\operatorname{tr}\left[e_i e_j^\dagger\right] = \delta_{ij} = \operatorname{tr}\left[f_i f_j^\dagger\right]$ and $\sum_i e_i = \mathbb{I} = \sum_j f_j$. Then,

$$\max_{U} \operatorname{tr}\left[AU^\dagger BU\right] = \max_{U} \sum_{ij} \alpha_i \beta_j \operatorname{tr}\left[U e_i U^\dagger e_j\right] \tag{28}$$

$$= \max_{U} \sum_{ij} \alpha_i \beta_j \operatorname{tr}\left[\tilde{e}_i e_j\right] = \max_{D} \sum_{ij} \alpha_i \beta_j D_{ij}. \tag{29}$$

Here, $\{\tilde{e}_i = U e_i U^\dagger\}$ is another bases of the operator space, and it is straightforward to verify that the transition matrix $D_{ij} = \operatorname{tr}\left[\tilde{e}_i e_j\right]$ is a doubly stochastic matrix, i.e., $\sum_i D_{ij} = 1$ and $\sum_j D_{ij} = 1$. As the doubly stochastic matrix can always been written as the convex combination of permutation matrices [61], the following equation holds:

$$\max_{D} \sum_{ij} \alpha_i \beta_j D_{ij} = \max_{\sigma} \sum_i \alpha_i \beta_{\sigma(i)} = \sum_i \alpha_i \beta_i,$$

where σ is a certain permutation. $\square$

Then, the following technique to choose Alice's observables $\{A_k\}$ can be specified [20].

Theorem 4. *When the quantum state ρ_{AB} is to be tested and Bob's observables are fixed as $\{B_k\}$, $\langle A_k \otimes B_k \rangle$ is maximal if Alice's observables satisfy the following conditions.*

1. A_k *and* $\tilde{\rho}_k = \mathrm{tr_B}\left[(\mathbb{I}_A \otimes B_k)\,\rho_{AB}\right]$ *are diagonalized in the same bases* $\{e_i^A\}$.
2. *Eigenvalues of* A_k *and eigenvalues of* $\tilde{\rho}_k = \mathrm{tr_B}\left[(\mathbb{I}_A \otimes B_k)\,\rho_{AB}\right]$ *have the same order.*

Then,

$$\langle A_k \otimes B_k \rangle = \sum_i a_k^{(i)} \beta_k^{(i)},$$

where $\alpha_k^{(1)} \geq \alpha_k^{(2)} \geq \cdots \geq \alpha_k^{(n)}$ *and* $\beta_k^{(1)} \geq \beta_k^{(2)} \geq \cdots \geq \beta_k^{(n)}$ *are eigenvalues of* A_k *and* $\tilde{\rho}_k$, *respectively.*

Proof. For any observables A_k and B_k on a quantum state ρ_{AB}, the expectation value of $A_k \otimes B_k$ is

$$\langle A_k \otimes B_k \rangle = \mathrm{tr}\left[A_k \otimes B_k \rho_{AB}\right] = \mathrm{tr_A}\left\{A_k \mathrm{tr_B}\left[(\mathbb{I}_A \otimes B_k)\,\rho_{AB}\right]\right\} = \mathrm{tr}\left[A_k \tilde{\rho}_k\right] \tag{30}$$

$$= \mathrm{tr}\left[U_k D_k U_k^\dagger \tilde{\rho}_k\right] = \mathrm{tr}\left[D_k U_k^\dagger \tilde{\rho}_k U_k\right], \tag{31}$$

where U_k is a unitary matrix, D_k is a diagonal matrix, and $A_k = U_k D_k U_k^\dagger$ holds. From Lemma 3, $\mathrm{tr}\left[U_k D_k U_k^\dagger \tilde{\rho}_k\right]$ is maximized when U_k can diagonalize $\tilde{\rho}_k$ simultaneously, i.e., $U_k^\dagger \tilde{\rho}_k U_k$ is a diagonal matrix, and D_k has the same order of diagonal values with $U_k^\dagger \tilde{\rho}_k U_k$. In this case, $\langle A_k \otimes B_k \rangle = \sum_i a_k^{(i)} \beta_k^{(i)}$ is the maximal over all Alice's observables, where $\alpha_k^{(1)} \geq \alpha_k^{(2)} \geq \cdots \geq \alpha_k^{(n)}$ and $\beta_k^{(1)} \geq \beta_k^{(2)} \geq \cdots \geq \beta_k^{(n)}$ are eigenvalues of A_k and $\tilde{\rho}_k$, respectively. □

Note that, when ρ_k contains degenerate eigenvalues, the optimal A_k by this method are not unique. As an example, we consider the 3×3-dimensional isotropic state [23].

Example 2. *The 3×3-dimensional isotropic state.*
    The 3×3-dimensional isotropic state has the following form

$$\rho_\eta = \eta \, |\phi^+\rangle \langle \phi^+| + (1 - \eta)\,\frac{\mathbb{I}}{9}, \tag{32}$$

where $|\phi^+\rangle = (|00\rangle + |11\rangle + |22\rangle)/\sqrt{3}$. *From the partial transpose criterion [65], ρ_η can be certified entangled if $\eta > 1/4$. To detect its steerability, let Bob's observables be the Gell–Mann matrices:*

$$G_1 = \tfrac{1}{\sqrt{2}}\begin{pmatrix} 0 & 1 & 0 \\ 1 & 0 & 0 \\ 0 & 0 & 0 \end{pmatrix}, \quad G_2 = \tfrac{1}{\sqrt{2}}\begin{pmatrix} 0 & 0 & 1 \\ 0 & 0 & 0 \\ 1 & 0 & 0 \end{pmatrix}, \quad G_3 = \tfrac{1}{\sqrt{2}}\begin{pmatrix} 0 & 0 & 0 \\ 0 & 0 & 1 \\ 0 & 1 & 0 \end{pmatrix},$$

$$G_4 = \tfrac{1}{\sqrt{2}}\begin{pmatrix} 0 & -i & 0 \\ i & 0 & 0 \\ 0 & 0 & 0 \end{pmatrix}, \quad G_5 = \tfrac{1}{\sqrt{2}}\begin{pmatrix} 0 & 0 & -i \\ 0 & 0 & 0 \\ i & 0 & 0 \end{pmatrix}, \quad G_6 = \tfrac{1}{\sqrt{2}}\begin{pmatrix} 0 & 0 & 0 \\ 0 & 0 & -i \\ 0 & i & 0 \end{pmatrix}, \tag{33}$$

$$G_7 = \tfrac{1}{\sqrt{2}}\begin{pmatrix} 1 & 0 & 0 \\ 0 & -1 & 0 \\ 0 & 0 & 0 \end{pmatrix}, \quad G_8 = \tfrac{1}{\sqrt{6}}\begin{pmatrix} 1 & 0 & 0 \\ 0 & 1 & 0 \\ 0 & 0 & -2 \end{pmatrix}.$$

Then, from Theorem 4, Alice's observables can be chosen as $G_k^A = \left(G_k^B\right)^T$, such that $\langle G_k^A \otimes G_k^B \rangle = \eta/3$ obtains its maximal value and $S_n = \eta/3$.
    For the LHS bound C_n, we have the following results. When $n = 3$ and Bob chooses G_1, G_2, and G_3, the state is steerable if $\eta > 0.8660$. When $n = 4$ and Bob chooses G_1, G_2, G_4, and G_8, the state is steerable if $\eta > 0.7318$. When $n = 5$ and Bob chooses G_3, G_4, G_5, G_6, and G_7, the state is steerable if $\eta > 0.6708$. When $n = 6$ and Bob chooses G_1, G_2, G_3, G_4, G_5, and G_8, the state is steerable if $\eta > 0.6424$. When $n = 7$ and

Bob chooses observables from G_1 to G_7, the state is steerable if $\eta > 0.6204$. Finally, when $n = 8$ and Bob chooses all Gell–Mann matrices, the state is steerable if $\eta > 0.5748$. Note that, in this case, when Bob chooses only two observables from Gell–Mann matrices, the corresponding linear inequality will not detect any steerability of the state.

4.1.2. A Flexible Bound on Unsteerable Correlations

As discussed above, the unsteerable bound C_n in the linear inequality from Equation (19) contains a maximization over all Alice's measurement results. The complexity to compute C_n is exponentially increasing with the number of n. This property can also be concluded from the above two examples. Therefore, when the number of measurements are large, a simpler bound is needed [66].

Theorem 5. *If the result of an EPR steering test violates the following inequality*

$$S_n = \frac{1}{n} \sum_{k=1}^{n} g_k \langle A_k B_k \rangle \le C'_n = \Lambda_A^{1/2} \Lambda_B^{1/2}, \tag{34}$$

where g_k are some real numbers and $\Lambda_A \Lambda_B$ satisfies

$$\Lambda_A = \sum_{k=1}^{n} g_k^2 \bar{A}_k^2, \Lambda_B = \max_{\rho} \left\{ \sum_{k=1}^{n} \langle B_k \rangle_\rho^2 \right\}, \tag{35}$$

with $\bar{A}_k^2 = \sum_{a_k} a_k^2 p(a_k)$, then the correlation of the test is EPR steering. The corresponding quantum state ρ_{AB} is EPR steerable, and more precisely, Alice can steer Bob.

Proof. Take in the definition of unsteerable correlation (Equation (6)),

$$P_{LHS}(ab|xy) = \int dp_\lambda P(a|x,\lambda) \, \text{tr} \left[F_b^y \sigma_\lambda \right].$$

Then

$$S_n = \sum_{k=1}^{n} g_k \langle A_k \otimes B_k \rangle = \int dp_\lambda \sum_{k=1}^{n} g_k \sum_{a_k} a_k P(a_k|A_k,\lambda) \, \text{tr}[B_k \sigma_\lambda] \tag{36}$$

$$= \int dp_\lambda \sum_{k=1}^{n} g_k (\bar{A}_k)_\lambda \langle B_k \rangle_\lambda \tag{37}$$

$$\le \int dp_\lambda \left[\sum_{k=1}^{n} g_k^2 (\bar{A}_k)_\lambda^2 \right]^{1/2} \left(\sum_{k=1}^{n} \langle B_k \rangle_{\sigma_\lambda}^2 \right)^{1/2} \tag{38}$$

$$\le \int dp_\lambda \left[\sum_{k=1}^{n} g_k^2 (\bar{A}_k^2)_\lambda \right]^{1/2} \max_{\rho} \left(\sum_{k=1}^{n} \langle B_k \rangle_\rho^2 \right)^{1/2} \tag{39}$$

$$\le \left[\sum_{k=1}^{n} g_k^2 \int dp_\lambda (\bar{A}_k^2)_\lambda \right]^{1/2} \Lambda_B^{1/2} \tag{40}$$

$$= \left[\sum_{k=1}^{n} g_k^2 (\bar{A}_k^2) \right]^{1/2} \Lambda_B^{1/2} = \Lambda_A^{1/2} \Lambda_B^{1/2}. \tag{41}$$

Here, $(\bar{A}_k)_\lambda = \sum a_k P(a_k|A_k,\lambda)$ is the expectation value of A_k under the probability distribution $P(a_k|A_k,\lambda)$ and $(\bar{A}_k^2)_\lambda$ is the expectation value of A_k^2 under the probability distribution $P(a_k|A_k,\lambda)$. The third line is based on the Cauchy–Schwarz inequality $u \cdot v \le |u||v|$, where we let $u = (\ldots g_k(\bar{A}_k)_\lambda \ldots)$ and $v = (\ldots (B_k)_\lambda \ldots)$. The fourth line comes from $(\bar{A}_k)_\lambda^2 \le (A_k^2)_\lambda$ and $\sum_{k=1}^{n} \langle B_k \rangle_{\sigma_\lambda}^2 \le \max_\rho \sum_{k=1}^{n} \langle B_k \rangle_\rho^2$. The fifth line is due to the concavity of the function $y = x^{1/2}$. □

Compared with the bound (Equation (19)) in the linear EPR steering inequality (Equation (18)), here, the unsteerable bound C'_n is simpler to compute and the complexity to obtain Λ_A and Λ_B increases linearly with n. However, C'_n may not as tight as C_n, i.e., some steerable states may be detectable by bound C_n but not with bound C'_n.

4.2. EPR Steering Inequality Based on Local Uncertainty Relations

For a random variable $X = \{x_i\}$, the variance is defined as $\delta^2 (X) = \overline{X^2} - \overline{X}^2$, where $\overline{X^2} = \sum_i p(x_i) x_i^2$ is the mean of the square of X and $\overline{X}^2 = (\sum_i p(x_i) x_i)^2$ is the square of the mean of X. For any random variable X, $\delta^2 (X) \geq 0$ always holds. In quantum mechanics, the variance describes the uncertainty of measurement results. For instance, consider the projective measurement $M = \sum_k m_k \Pi_k$, where Π_k are projectors and m_k are the corresponding outcome. The variance of measurement results $\{m_i\}$ on a quantum state ρ is in the form of $\delta^2 (M)_\rho = \langle M^2 \rangle_\rho - \langle M \rangle_\rho^2$, where $\langle M \rangle_\rho = \mathrm{tr}\,[M\rho]$ is the expectation value of measurement M on ρ and $\langle M^2 \rangle_\rho = \mathrm{tr}\,[M^2\rho]$ is the expectation value of the square of measurement M on ρ. In the following, the subscript ρ is omitted for simplicity. The uncertainty relation can be described as, for a set of measurements $\{M_i | i = 1, \ldots, n\}$, the sum of variances is larger than a certain value, i.e., $\sum_i \delta^2 (M_i) \geq C_M$ with $C_M = \min_\rho \sum_i \delta^2 (M_i)_\rho$. In a nontrivial case, where $\{M_i\}$ has no common eigenvectors, C_M is positive, i.e., $C_M > 0$ [67–69].

In the EPR steering test, only Bob's measurements are assumed to be quantum. Then, the local uncertainty relations (LUR) on Bob's side can help to certify EPR steering correlation [23].

Theorem 6 (Steering inequality based on LUR). *If the result of an EPR steering test violates the following inequality*

$$\sum_{k=1}^n \delta^2 (\alpha_i A_i + B_i) \geq C_B, \tag{42}$$

where α_i are some real numbers and $C_B = \min_\rho \sum_i \delta^2 (B_i)_\rho$, then the correlation of the test is EPR steering. The corresponding quantum state ρ_{AB} is EPR steerable, and more precisely, Alice can steer Bob.

Proof. Generally, for any two random variables X and Y, let $p(xy)$ be the joint probability distribution and $p(y|x) = p(xy) / p(x)$ be the conditioned probability distribution. Then, the variance of Y satisfies

$$\delta^2 (Y) = \sum_y p(y) y^2 - \left[\sum_y p(y) y \right]^2 \tag{43}$$

$$= \sum_{y,x} p(x) p(y|x) y^2 - \left[\sum_{y,x} p(x) p(y|x) y \right]^2 \tag{44}$$

$$\geq \sum_x p(x) \left[\sum_y p(y|x) y^2 - \left(\sum_y p(y|x) y \right)^2 \right] \tag{45}$$

$$= \sum_x p(x) \delta^2 (y)_x, \tag{46}$$

where the third line comes from the concavity of function $f(t) = t^2$ and $\delta^2 (y)_x$ is the variance of Y under the distribution $\{p(y|x)\}$. Now, consider the definition of unsteerable correlation

$$P_{LHS} (ab|xy) = \int d\lambda p_\lambda P (a|x, \lambda)\, \mathrm{tr}\, \left[F_b^y \sigma_\lambda \right].$$

One has

$$\sum_i \delta^2 (\alpha_i A_i + B_i) \geq \sum_i \int d\lambda p_\lambda \delta^2 (\alpha_i A_i + B_i)_\lambda \tag{47}$$

$$= \sum_i \int d\lambda p_\lambda \left[(\alpha_i A_i + B_i)_\lambda^2 - (\alpha_i \overline{A}_i + \overline{B}_i)_\lambda^2 \right] \tag{48}$$

$$= \sum_i \int d\lambda p_\lambda \left[\alpha_i^2 \left(\overline{A_i^2} - \overline{A}_i^2 \right) + \left(\overline{B_i^2} - \overline{B}_i^2 \right) \right]_\lambda \tag{49}$$

$$= \sum_i \int d\lambda p_\lambda \left[\alpha_i^2 \delta^2 (A_i)_\lambda + \delta^2 (B_i)_{\sigma_\lambda} \right] \tag{50}$$

$$\geq \sum_i \int d\lambda p_\lambda \left[0 + \sum_i \delta^2 (B_i)_{\sigma_\lambda} \right] \geq C_B, \tag{51}$$

where trivial results $\delta^2 (A_i)_\lambda \geqslant 0$ is used. □

Here, $\{\alpha_i\}$ are some flexible real variables. For a certain probability distribution $\{P(a_k b_k | A_k B_k)\}$ generated from an EPR steering test, the optimal $\{\alpha_i\}$ can be calculated such that the inequality from Equation (42) is maximally violated. For each term in Equation (42), $\delta^2 (\alpha_i A_i + B_i) = \alpha_i^2 \delta^2 (A_i) + 2\alpha_i C(A_i, B_i) + \delta^2 (B_i)$ holds where $C(A_i, B_i) = \langle A_i B_i \rangle - \langle A_i \rangle \langle B_i \rangle$ is the covariance. Therefore, $\delta^2 (\alpha_i A_i + B_i)$ can be seen as a quadratic polynomial of α_i, from which the optimal α_i can be obtained, i.e.,

$$\alpha_i = \begin{cases} -C(A_i, B_i)/\delta^2 (A_i), & \text{if } \delta^2 (A_i) \neq 0; \\ -\delta^2 (B_i)/2C(A_i, B_i), & \text{if } \delta^2 (A_i) = 0, C(A_i, B_i) \neq 0; \\ 0, & \text{if } \delta^2 (A_i) \neq 0, C(A_i, B_i) = 0. \end{cases}$$

It is noteworthy that, here, like the case in the linear inequality of Equation (34), the complexity to compute unsteerable bound C_B also increases linearly with the number of measurements n, better than the case in inequality (Equation (18)), where the complexity increases exponentially with n.

Remark 2. *The use of LUR in quantum correlations.*

In the case of EPR steering, the inequality from Equation (42) shows that, for unsteerable correlations, the uncertainty of the total system AB is always larger than that of one subsystem B. This conclusion is consistent with the definition of LHS models, where only Bob has the quantum description. One property of EPR steering is, thus, that the uncertainty of the correlated measurement results can be less than the uncertainty of one subsystem. In this sense, the violation of LUR indicates the amount of quantum correlations.

Furthermore, if quantum entanglement is considered in this fashion, for any separable states $\sigma_{AB}^{SEP} = \sum_k p_k \sigma_k^A \otimes \sigma_k^B$, it has been proved that

$$\sum_i \delta^2 (A_i + B_i)_{SEP} \geq C_A + C_B, \tag{52}$$

where $C_A = \min_\rho \sum_i \delta^2 (A_i)_\rho$ [70]. That is, in the case of quantum separable states, where both Alice and Bob can be described as quantum but classically correlated, the uncertainty of the total system is always larger than the sum of the local uncertainty relations of all subsystems.

However, for the nonlocality, the probability distribution of LHV models always satisfies

$$\sum_i \delta^2 (A_i + B_i) \geq 0, \tag{53}$$

which is a trivial result, and no violation can be detected. In fact, formulating a nonlinear form of Bell inequalities is a difficult problem.

It is noteworthy that in Reference [43], the violation of the CHSH inequality [62] can be restricted by the so-called fine-grained uncertainty relations combined by a properly-defined steerability. Such a restriction holds only when a specific form of the Bell inequalities are selected [44]. Different from the variance-based uncertainties discussed here or entropies [24], the fine-grained uncertainty relation are described in a linear form of the set of measurement observables, which can also be used as the certification of EPR steering [25] .

Example 3. *Bell diagonal states*

Bell diagonal states has the following simple form,

$$\rho_c = \frac{1}{4} \left[\mathbb{I} + \sum_j c_j \sigma_j \otimes \sigma_j \right], \tag{54}$$

where $\{\sigma_j, j = x, y, z\}$ is the set of Pauli matrices. In another form, ρ_c can be written in the diagonal form

$$\rho_c = t_1 \left| \psi^+ \right\rangle \left\langle \psi^+ \right| + t_2 \left| \psi^- \right\rangle \left\langle \psi^- \right| + t_3 \left| \phi^+ \right\rangle \left\langle \phi^+ \right| + t_4 \left| \phi^- \right\rangle \left\langle \phi^- \right|, \tag{55}$$

where $\left| \psi^\pm \right\rangle = (\left| 00 \right\rangle \pm \left| 11 \right\rangle) / \sqrt{2}$ and $\left| \phi^\pm \right\rangle = (\left| 01 \right\rangle \pm \left| 10 \right\rangle) / \sqrt{2}$ are four Bell states and $\sum_i t_i = 1$.

If three Pauli matrices are selected as the observables, the linear EPR steering inequality (18) can be simplified as $\left| \sum_i \omega_i \left\langle \sigma_i^A \otimes \sigma_i^B \right\rangle \right| < \sqrt{3}$ with $\omega_i \in \{\pm 1\}$. Here, the absolute value and binary ω_i suggest that there are a set of linear inequalities. The violation implies that ρ_c is steerable if $\left| c_x \pm c_y \pm c_z \right| > \sqrt{3}$. Nevertheless, the EPR steering inequalities (42) based on LUR can be optimized as $\sum_i \delta^2 \left(\sigma_i^B \right) - C^2 \left(\sigma_i^A, \sigma_i^B \right) / \delta^2 \left(\sigma_i^A \right) \geqslant 2$, the violation of which implies $\sum_i c_i^2 > 1$. As a comparison, it can be verified that, in this example, the inequality based on LUR certifies a larger steerable region of Bell diagonal states than the linear inequality [23].

4.3. Realignment Method

From the EPR steering inequality based on LUR, the realignment method for certifying entanglement also works for the EPR steering case. Generally, the realignment criterion [71] or the computable cross-norm criterion [72] are important techniques to certify bound quantum entanglement, i.e., entangled states with a positive partial transpose. Mathematically, the realignment is a map on a quantum state ρ_{AB} such that $\mathcal{R}(\rho_{AB}) : \rho_{AB} \mapsto \left\langle m \right| \left\langle \mu \right| \mathcal{R}(\rho_{AB}) \left| n \right\rangle \left| v \right\rangle = \left\langle m \right| \left\langle n \right| \rho_{AB} \left| v \right\rangle \left| \mu \right\rangle$. If ρ_{AB} is separable, then the trace norm of the matrix $\mathcal{R}(\rho)$ is not larger than 1.

To obtain the norm of $\mathcal{R}(\rho)$, one can seek for the complete set of local orthogonal observables (LOOs). A complete set of LOOs is a collection of observables $\{G_k\}$ satisfying $G_k^\dagger = G_k$, $\mathrm{tr}\,[G_k G_l] = \delta_{kl}$, and $\sum_k G_k^2 = \mathbb{I}$. Indeed, $\{G_k\}$ forms a complete set of orthonormal bases for the corresponding operator space. Then, a state ρ can be written as $\rho = \sum_k \mu_k G_k$, where $\mu_k = \mathrm{tr}\,[\rho G_k]$. For example, in the case of qubits, the identity matrix and three Pauli matrices form a complete set of LOOs, and in the case of qutrits, the identity matrix and eight Gell-Mann matrices form a complete set of LOOs.

For any bipartite quantum state ρ_{AB}, suppose that the maximal dimension of Alice's Hilbert space and Bob's Hilbert space is d. Let the complete sets of LOOs for Alice's operator space and Bob's operator space be $\{\tilde{G}_k^A\}$ and $\{\tilde{G}_k^B\}$, respectively. Then, ρ_{AB} can always be written as $\rho_{AB} = \sum_{kl} \mu_{kl} \tilde{G}_k^A \otimes \tilde{G}_k^B$, where $\mu_{kl} = \mathrm{tr}\,[\rho_{AB} \tilde{G}_k^A \otimes \tilde{G}_k^B]$. The singular value decomposition on the matrix $\mu = (\mu_{kl})$ yields $\mu = S \lambda T^T$, where $\lambda = \mathrm{diag}\,\{\lambda_1, \ldots, \lambda_{d2}\}$ is the diagonal matrix with $\lambda_k \geq 0$, $S = (s_{ij})$ and $T = (t_{ij})$ are two orthogonal matrices, i.e., $SS^T = TT^T = \mathbb{I}$. Take $\mu = S \lambda T^T$ into the expression of ρ_{AB}, and finally, the Hilbert–Schmidt decomposition of ρ_{AB} can be obtained:

$$\rho_{AB} = \sum_k \lambda_k G_k^A \otimes G_k^B, \tag{56}$$

where $G_k^A = \sum_m s_{mk} \tilde{G}_m^A$ and $G_k^B = \sum_m t_{mk} \tilde{G}_m^B$. It can be verified that $\{G_k^A\}$ and $\{G_k^B\}$ are another two complete sets of LOOs, and $\lambda_k = \mathrm{tr}\,[\rho_{AB} G_k^A \otimes G_k^B]$.

In a certifying entanglement, if ρ_{AB} is separable, then the realignment [71,72] method guarantees that

$$\sum_k \lambda_k \leqslant 1. \tag{57}$$

In certifying EPR steering, a similar result can be concluded [23].

Theorem 7 (Realignment for EPR steering). *If $\rho_{AB} = \sum_k \lambda_k G_k^A \otimes G_k^B$ satisfies*

$$\sum_k \lambda_k > \sqrt{d}, \tag{58}$$

then ρ_{AB} is EPR steerable. In this case, Alice can steer Bob and Bob can also steer Alice.

Proof. From the EPR steering inequality based on LUR, for a bipartite quantum state $\rho_{AB} = \sum_k \lambda_k G_k^A \otimes G_k^B$, let Alice's and Bob's observables be $\{G_k^A\}$ and $\{G_k^B\}$ and $g_k = -g$. The violation of Equation (42) implies $g^2 d + d - 2g \sum_k \lambda_k - \sum_k \left(g \left\langle G_k^A \right\rangle - \left\langle G_k^B \right\rangle \right)^2 < d - 1$. A sufficient condition of this inequality is omitting the quadratic term, i.e., $g^2 d + d - 2g \sum_k \lambda_k < d - 1$. Finally, let $g = \sum_k \lambda_k / d$, and the inequality (58) is concluded. $\square$

Different from the linear inequality and the inequality based on LUR, the realignment method does not require an EPR steering test. For any quantum state ρ_{AB}, there is a possibility that one can know whether this state is EPR steerable or not, regardless of how to certify it in the test. A limitation is that, as a corollary of the inequality, the realignment method will not perform better than the inequality.

In the entanglement case, where the state is entangled if the value $\sum_k \lambda_k$ is larger than 1. Here, this quantity should be larger than $\sqrt{d}$ to certify the EPR steerability. Although the realignment method can certify positive partial transpose (PPT) entanglement, it remains an open question if it can certify PPT EPR steering, i.e., EPR steerable states with PPT. Note that there have been numerical results proving the existence of such states [73,74].

5. Summary

In this survey, the basic technique to discuss and certify EPR steering is discussed. Particularly, the box framework and trust-untrust scenario is adopted. The linear criterion and local-uncertainty-relation-based criterion are summarized. Both criteria are constructed in an experimentally friendly manner, i.e., they can be directly applied in real experiments for arbitrary measurement settings and arbitrary outcomes, with a reduced complexity to obtain the unsteerable bound. Moreover, an analytical method for the optimization of EPR steering detection is also maintained. Furthermore, from these criteria, LUR are shown to play an important role in the correlation exhibition of quantum bipartite systems.

There have also been other useful criteria, as has been listed in Section 1. Most of them are formulated in the same fashion as introduced in this survey. Therefore, the discussed techniques to find a computable unsteerable bound and optimal observables can be directly applied. There still remains an open problem of how much entanglement is sufficient for EPR steering and how much EPR steering is sufficient for nonlocality. Solving this problem would technically advance the realization of nonlocality-based quantum protocols and finally contributes to the application of quantum information technologies.

Author Contributions: Conceptualization, K.C.; investigation, Y.-Z.Z., X.-Y.X., L.L., N.-L.L., and K.C.; project administration, K.C.; supervision, L.L., N.-L.L., and K.C.; writing—original draft, Y.-Z.Z. and X.-Y.X.; writing—review and editing, L.L., N.-L.L., and K.C.

Funding: This work has been supported by the Chinese Academy of Science, the National Fundamental Research Program, and the National Natural Science Foundation of China (Grants No. 11575174, No. 11374287, No. 11574297, and No. 61771443), as well as Anhui Initiative in Quantum Information Technologies

Acknowledgments: We thank Wen-Ge Wang, Yu-Lin Zheng, Wen-Fei Cao, and Shuai Zhao for the valuable and enlightening discussions. Y.-Z.Z. specially thank Yingqiu Mao for her love and support.

Conflicts of Interest: The authors declare no conflict of interest.

Abbreviations

The following abbreviations are used in this manuscript:

EPR	Einstein–Podolsky–Rosen
LHV	local hidden variable
LHS	local hidden state
LOO	local orthogonal observable
LUR	local uncertainty relations
POVM	positive operator valued measure
PPT	positive partial transpose

References

1. Wiseman, H.M.; Jones, S.J.; Doherty, A.C. Steering, Entanglement, Nonlocality, and the Einstein-Podolsky-Rosen Paradox. *Phys. Rev. Lett.* **2007**, *98*, 140402. [CrossRef] [PubMed]
2. Schrödinger, E. Discussion of Probability Relations between Separated Systems. *Math. Proc. Camb.* **1935**, *31*, 555–563. [CrossRef]
3. Schrödinger, E. Probability relations between separated systems. *Math. Proc. Camb.* **1936**, *32*, 446–452. [CrossRef]
4. Einstein, A.; Podolsky, B.; Rosen, N. Can Quantum-Mechanical Description of Physical Reality Be Considered Complete? *Phys. Rev.* **1935**, *47*, 777–780. [CrossRef]
5. Bohm, D. *Quantum Theory*; Dover Publishications: Mineola, NY, USA, 1951.
6. Bell, J.S. On the Einstein-Podolsky-Rosen Paradox. *Physics* **1964**, *1*, 195–200. [CrossRef]
7. Bell, J.S. *Speakable and Unspeakable in Quantum Mechanics: Collected Papers on Quantum Philosophy*, 2nd ed.; Cambridge University Press: Cambridge, UK, 2004. [CrossRef]
8. Brunner, N.; Cavalcanti, D.; Pironio, S.; Scarani, V.; Wehner, S. Bell nonlocality. *Rev. Mod. Phys.* **2014**, *86*, 419–478. [CrossRef]
9. Hensen, B.; Bernien, H.; Dreau, A.E.; Reiserer, A.; Kalb, N.; Blok, M.S.; Ruitenberg, J.; Vermeulen, R.F.L.; Schouten, R.N.; Abellan, C.; et al. Detection-Loophole-Free Test of Quantum Nonlocality, and Applications. *Nature* **2015**, *526*, 682–686. [CrossRef] [PubMed]
10. Giustina, M.; Versteegh, M.A.M.; Wengerowsky, S.; Handsteiner, J.; Hochrainer, A.; Phelan, K.; Steinlechner, F.; Kofler, J.; Larsson, J.; Abellán, C.; et al. Significant-Loophole-Free Test of Bell's Theorem with Entangled Photons. *Phys. Rev. Lett.* **2015**, *115*, 250401. [CrossRef]
11. Shalm, L.K.; Meyer-Scott, E.; Christensen, B.G.; Bierhorst, P.; Wayne, M.A.; Stevens, M.J.; Gerrits, T.; Glancy, S.; Hamel, D.R.; Allman, M.S.; et al. Strong Loophole-Free Test of Local Realism. *Phys. Rev. Lett.* **2015**, *115*, 250402. [CrossRef] [PubMed]
12. Liu, Y.; Zhao, Q.; Li, M.H.; Guan, J.Y.; Zhang, Y.; Bai, B.; Zhang, W.; Liu, W.Z.; Wu, C.; Yuan, X.; et al. Device-independent quantum random-number generation. *Nature* **2018**, *562*, 548–551. [CrossRef] [PubMed]
13. Werner, R.F. Quantum states with Einstein-Podolsky-Rosen correlations admitting a hidden-variable model. *Phys. Rev. A* **1989**, *40*, 4277–4281. [CrossRef]
14. Jones, S.J.; Wiseman, H.M.; Doherty, A.C. Entanglement, Einstein-Podolsky-Rosen correlations, Bell nonlocality, and steering. *Phys. Rev. A* **2007**, *76*, 052116. [CrossRef]
15. Bennet, A.J.; Evans, D.A.; Saunders, D.J.; Branciard, C.; Cavalcanti, E.G.; Wiseman, H.M.; Pryde, G.J. Arbitrarily Loss-Tolerant Einstein-Podolsky-Rosen Steering Allowing a Demonstration over 1 km of Optical Fiber with No Detection Loophole. *Phys. Rev. X* **2012**, *2*, 031003. [CrossRef]
16. Wittmann, B.; Ramelow, S.; Steinlechner, F.; Langford, N.K.; Brunner, N.; Wiseman, H.M.; Ursin, R.; Zeilinger, A. Loophole-free Einstein-Podolsky-Rosen experiment via quantum steering. *New J. Phys.* **2012**, *14*, 053030. [CrossRef]

17. Christensen, B.G.; McCusker, K.T.; Altepeter, J.B.; Calkins, B.; Gerrits, T.; Lita, A.E.; Miller, A.; Shalm, L.K.; Zhang, Y.; Nam, S.W.; et al. Detection-Loophole-Free Test of Quantum Nonlocality, and Applications. *Phys. Rev. Lett.* **2013**, *111*, 130406. [CrossRef]

18. Branciard, C.; Cavalcanti, E.G.; Walborn, S.P.; Scarani, V.; Wiseman, H.M. One-sided device-independent quantum key distribution: Security, feasibility, and the connection with steering. *Phys. Rev. A* **2012**, *85*, 010301. [CrossRef]

19. Saunders, D.J.; Jones, S.J.; Wiseman, H.M.; Pryde, G.J. Experimental EPR-steering using Bell-local states. *Nat. Phys.* **2010**, *6*, 845–849. [CrossRef]

20. Zheng, Y.L.; Zhen, Y.Z.; Cao, W.F.; Li, L.; Chen, Z.B.; Liu, N.L.; Chen, K. Optimized detection of steering via linear criteria for arbitrary-dimensional states. *Phys. Rev. A* **2017**, *95*, 032128. [CrossRef]

21. Reid, M.D.; Drummond, P.D.; Bowen, W.P.; Cavalcanti, E.G.; Lam, P.K.; Bachor, H.A.; Andersen, U.L.; Leuchs, G. Colloquium: The Einstein-Podolsky-Rosen paradox: From concepts to applications. *Rev. Mod. Phys.* **2009**, *81*, 1727–1751. [CrossRef]

22. Cavalcanti, E.G.; Jones, S.J.; Wiseman, H.M.; Reid, M.D. Experimental criteria for steering and the Einstein-Podolsky-Rosen paradox. *Phys. Rev. A* **2009**, *80*, 032112. [CrossRef]

23. Zhen, Y.Z.; Zheng, Y.L.; Cao, W.F.; Li, L.; Chen, Z.B.; Liu, N.L.; Chen, K. Certifying Einstein-Podolsky-Rosen steering via the local uncertainty principle. *Phys. Rev. A* **2016**, *93*, 012108. [CrossRef]

24. Schneeloch, J.; Broadbent, C.J.; Walborn, S.P.; Cavalcanti, E.G.; Howell, J.C. Einstein-Podolsky-Rosen steering inequalities from entropic uncertainty relations. *Phys. Rev. A* **2013**, *87*, 062103. [CrossRef]

25. Pramanik, T.; Kaplan, M.; Majumdar, A.S. Fine-grained Einstein-Podolsky-Rosen-steering inequalities. *Phys. Rev. A* **2014**, *90*, 050305. [CrossRef]

26. Żukowski, M.; Dutta, A.; Yin, Z. Geometric Bell-like inequalities for steering. *Phys. Rev. A* **2015**, *91*, 032107. [CrossRef]

27. Ji, S.W.; Lee, J.; Park, J.; Nha, H. Steering criteria via covariance matrices of local observables in arbitrary-dimensional quantum systems. *Phys. Rev. A* **2015**, *92*, 062130. [CrossRef]

28. Cavalcanti, D.; Skrzypczyk, P. Quantum steering: A review with focus on semidefinite programming. *Prog. Phys.* **2017**, *80*, 024001 [CrossRef]

29. Chen, J.L.; Ye, X.J.; Wu, C.; Su, H.Y.; Cabello, A.; Kwek, L.C.; Oh, C.H. All-Versus-Nothing Proof of Einstein-Podolsky-Rosen Steering. *Sci. Rep.* **2013**, *3*, 2143. [CrossRef]

30. Sun, K.; Xu, J.S.; Ye, X.J.; Wu, Y.C.; Chen, J.L.; Li, C.F.; Guo, G.C. Experimental Demonstration of the Einstein-Podolsky-Rosen Steering Game Based on the All-Versus-Nothing Proof. *Phys. Rev. Lett.* **2014**, *113*, 140402. [CrossRef]

31. Händchen, V.; Eberle, T.; Steinlechner, S.; Samblowski, A.; Franz, T.; Werner, R.F.; Schnabel, R. Observation of one-way Einstein-Podolsky-Rosen steering. *Nat. Photon* **2012**, *6*, 596–599. [CrossRef]

32. Bowles, J.; Vértesi, T.; Quintino, M.T.; Brunner, N. One-way Einstein-Podolsky-Rosen Steering. *Phys. Rev. Lett.* **2014**, *112*, 200402. [CrossRef]

33. Quintino, M.T.; Brunner, N.; Huber, M. Superactivation of quantum steering. *Phys. Rev. A* **2016**, *94*, 062123. [CrossRef]

34. Hsieh, C.Y.; Liang, Y.C.; Lee, R.K. Quantum steerability: Characterization, quantification, superactivation, and unbounded amplification. *Phys. Rev. A* **2016**, *94*, 062120. [CrossRef]

35. Skrzypczyk, P.; Navascués, M.; Cavalcanti, D. Quantifying Einstein-Podolsky-Rosen Steering. *Phys. Rev. Lett.* **2014**, *112*, 180404. [CrossRef]

36. Costa, A.C.S.; Angelo, R.M. Quantification of Einstein-Podolski-Rosen steering for two-qubit states. *Phys. Rev. A* **2016**, *93*, 020103. [CrossRef]

37. Das, D.; Datta, S.; Jebaratnam, C.; Majumdar, A.S. Cost of Einstein-Podolsky-Rosen steering in the context of extremal boxes. *Phys. Rev. A* **2018**, *97*, 022110. [CrossRef]

38. Pusey, M.F. Negativity and steering: A stronger Peres conjecture. *Phys. Rev. A* **2013**, *88*, 032313. [CrossRef]

39. Nguyen, H.C.; Milne, A.; Vu, T.; Jevtic, S. Quantum steering with positive operator valued measures. *J. Phys. A Math. Theor.* **2018**, *51*, 355302. [CrossRef]

40. Gallego, R.; Aolita, L. Resource Theory of Steering. *Phys. Rev. X* **2015**, *5*, 041008. [CrossRef]

41. He, Q.Y.; Reid, M.D. Genuine Multipartite Einstein-Podolsky-Rosen Steering. *Phys. Rev. Lett.* **2013**, *111*, 250403. [CrossRef]

42. Li, C.M.; Chen, K.; Chen, Y.N.; Zhang, Q.; Chen, Y.A.; Pan, J.W. Genuine High-Order Einstein-Podolsky-Rosen Steering. *Phys. Rev. Lett.* **2015**, *115*, 010402. [CrossRef]

43. Oppenheim, J.; Wehner, S. The Uncertainty Principle Determines the Nonlocality of Quantum Mechanics. *Science* **2010**, *330*, 1072–1074. [CrossRef] [PubMed]

44. Zhen, Y.Z.; Goh, K.T.; Zheng, Y.L.; Cao, W.F.; Wu, X.; Chen, K.; Scarani, V. Nonlocal games and optimal steering at the boundary of the quantum set. *Phys. Rev. A* **2016**, *94*, 022116. [CrossRef]

45. Quintino, M.T.; Vértesi, T.; Brunner, N. Joint Measurability, Einstein-Podolsky-Rosen Steering, and Bell Nonlocality. *Phys. Rev. Lett.* **2014**, *113*, 160402. [CrossRef]

46. Uola, R.; Budroni, C.; Gühne, O.; Pellonpää, J.P. One-to-One Mapping between Steering and Joint Measurability Problems. *Phys. Rev. Lett.* **2015**, *115*, 230402. [CrossRef] [PubMed]

47. Piani, M.; Watrous, J. Necessary and Sufficient Quantum Information Characterization of Einstein-Podolsky-Rosen Steering. *Phys. Rev. Lett.* **2015**, *114*, 060404. [CrossRef] [PubMed]

48. Armstrong, S.; Wang, M.; Teh, R.Y.; Gong, Q.; He, Q.; Janousek, J.; Bachor, H.A.; Reid, M.D.; Lam, P.K. Multipartite Einstein-Podolsky-Rosen steering and genuine tripartite entanglement with optical networks. *Nat. Phys.* **2015**, *11*, 167–172. [CrossRef]

49. Wasak, T.; Chwedeńczuk, J. Bell Inequality, Einstein-Podolsky-Rosen Steering, and Quantum Metrology with Spinor Bose-Einstein Condensates. *Phys. Rev. Lett.* **2018**, *120*, 140406. [CrossRef] [PubMed]

50. Tischler, N.; Ghafari, F.; Baker, T.J.; Slussarenko, S.; Patel, R.B.; Weston, M.M.; Wollmann, S.; Shalm, L.K.; Verma, V.B.; Nam, S.W.; et al. Conclusive Experimental Demonstration of One-Way Einstein-Podolsky-Rosen Steering. *Phys. Rev. Lett.* **2018**, *121*, 100401. [CrossRef]

51. Kunkel, P.; Prüfer, M.; Strobel, H.; Linnemann, D.; Frölian, A.; Gasenzer, T.; Gärttner, M.; Oberthaler, M.K. Spatially distributed multipartite entanglement enables EPR steering of atomic clouds. *Science* **2018**, *360*, 413–416. [CrossRef]

52. Lange, K.; Peise, J.; Lücke, B.; Kruse, I.; Vitagliano, G.; Apellaniz, I.; Kleinmann, M.; Tóth, G.; Klempt, C. Entanglement between two spatially separated atomic modes. *Science* **2018**, *360*, 416–418. [CrossRef] [PubMed]

53. Fadel, M.; Zibold, T.; Décamps, B.; Treutlein, P. Spatial entanglement patterns and Einstein-Podolsky-Rosen steering in Bose-Einstein condensates. *Science* **2018**, *360*, 409–413. [CrossRef]

54. Sainz, A.B.; Aolita, L.; Piani, M.; Hoban, M.J.; Skrzypczyk, P. A formalism for steering with local quantum measurements. *New J. Phys.* **2018**, *20*, 083040. [CrossRef]

55. Masanes, L.; Acin, A.; Gisin, N. General properties of nonsignaling theories. *Phys. Rev. A* **2006**, *73*, 012112. [CrossRef]

56. Barrett, J. Information processing in generalized probabilistic theories. *Phys. Rev. A* **2007**, *75*, 032304. [CrossRef]

57. Pironio, S.; Acín, A.; Massar, S.; de la Giroday, A.B.; Matsukevich, D.N.; Maunz, P.; Olmschenk, S.; Hayes, D.; Luo, L.; Manning, T.A.; et al. Random numbers certified by Bell's theorem. *Nature* **2010**, *464*, 1021. [CrossRef]

58. Jevtic, S.; Pusey, M.; Jennings, D.; Rudolph, T. Quantum Steering Ellipsoids. *Phys. Rev. Lett.* **2014**, *113*, 020402. [CrossRef]

59. Heinosaari, T.; Reitzner, D.; Stano, P. Notes on Joint Measurability of Quantum Observables. *Found. Phys.* **2008**, *38*, 1133–1147. [CrossRef]

60. Kraus, K. *States, Effects and Operations: Fundamental Notions of Quantum Theory*; Springer: Berlin, Germany, 1983. [CrossRef]

61. Boyd, S.; Vandenberghe, L. *Convex Optimization*; Cambridge University Press: New York, NY, USA, 2004.

62. Clauser, J.F.; Horne, M.A.; Shimony, A.; Holt, R.A. Proposed Experiment to Test Local Hidden-Variable Theories. *Phys. Rev. Lett.* **1969**, *23*, 880–884. [CrossRef]

63. Acín, A.; Gisin, N.; Toner, B. Grothendieck's constant and local models for noisy entangled quantum states. *Phys. Rev. A* **2006**, *73*, 062105. [CrossRef]

64. Coope, I.; Renaud, P. Trace Inequalities with Applications to Orthogonal Regression and Matrix Nearness Problems. *J. Inequal. Pure Appl. Math.* **2009**, *10*, 92.

65. Peres, A. Separability Criterion for Density Matrices. *Phys. Rev. Lett.* **1996**, *77*, 1413–1415. [CrossRef]

66. Zheng, Y.L.; Zhen, Y.Z.; Chen, Z.B.; Liu, N.L.; Chen, K.; Pan, J.W. Efficient linear criterion for witnessing Einstein-Podolsky-Rosen nonlocality under many-setting local measurements. *Phys. Rev. A* **2017**, *95*, 012142. [CrossRef]

67. Heisenberg, W. Über den anschaulichen Inhalt der quantentheoretischen Kinematik und Mechanik. *Zeitschrift Phys.* **1927**, *43*, 172–198. [CrossRef]

68. Robertson, H.P. The Uncertainty Principle. *Phys. Rev.* **1929**, *34*, 163–164. [CrossRef]

69. Deutsch, D. Uncertainty in Quantum Measurements. *Phys. Rev. Lett.* **1983**, *50*, 631–633. [CrossRef]

70. Hofmann, H.F.; Takeuchi, S. Violation of local uncertainty relations as a signature of entanglement. *Phys. Rev. A* **2003**, *68*, 032103. [CrossRef]

71. Chen, K.; Wu, L.A. A matrix realignment method for recognizing entanglement. *Quant. Inf. Comput.* **2003**, *3*, 193.

72. Rudolph, O. On the cross norm criterion for separability. *J. Phys. A Math. Gen.* **2003**, *36*, 5825. [CrossRef]

73. Moroder, T.; Gittsovich, O.; Huber, M.; Gühne, O. Steering Bound Entangled States: A Counterexample to the Stronger Peres Conjecture. *Phys. Rev. Lett.* **2014**, *113*, 050404. [CrossRef]

74. Vértesi, T.; Brunner, N. Disproving the Peres conjecture by showing Bell nonlocality from bound entanglement. *Nat. Commun.* **2014**, *5*, 5297. [CrossRef]

entropy

MDPI

Article
Discrimination of Non-Local Correlations

Alberto Montina * and Stefan Wolf *

Facoltà di Informatica, Università della Svizzera italiana, 6900 Lugano, Switzerland
* Correspondence: montia@usi.ch (A.M.); wolfs@usi.ch (S.W.)

Received: 15 December 2018; Accepted: 17 January 2019; Published: 23 January 2019

Abstract: In view of the importance of quantum non-locality in cryptography, quantum computation, and communication complexity, it is crucial to decide whether a given correlation exhibits non-locality or not. As proved by Pitowski, this problem is NP-complete, and is thus computationally intractable unless NP is equal to P. In this paper, we first prove that the Euclidean distance of given correlations from the local polytope can be computed in polynomial time with arbitrary fixed error, granted the access to a certain oracle; namely, given a fixed error, we derive two upper bounds on the running time. The first bound is linear in the number of measurements. The second bound scales with the number of measurements to the sixth power. The former holds only for a very high number of measurements, and is never observed in the performed numerical tests. We, then, introduce a simple algorithm for simulating the oracle. In all of the considered numerical tests, the simulation of the oracle contributes with a multiplicative factor to the overall running time and, thus, does not affect the sixth-power law of the oracle-assisted algorithm.

Keywords: local polytope; quantum nonlocality; communication complexity; optimization

---

1. Introduction

Non-local correlations, displayed by certain entangled quantum systems, mark a clear departure from the classical framework made up of well-defined, locally interacting quantities [1]. Besides their importance in foundation of quantum theory, non-local correlations have gained interest as information-processing resources in cryptography [2–8], randomness amplification [9,10], quantum computation, and communication complexity [11]. In view of their importance, a relevant problem— hereafter called the *non-locality problem*—is to find a criterion for deciding if observed correlations are actually non-local. Such a criterion is, for example, provided by the *Bell inequalities* [12]. However, a result by Pitowski [12] suggests that the problem of discriminating between local and non-local correlations is generally intractable. Pitowski proved that deciding membership to the correlation polytope is NP-complete, and is therefore intractable unless NP is equal to P. This result also implies that the opposite problem, deciding whether given correlations are outside the polytope, is not even in NP, unless NP=co-NP—which is believed to be false.

In this paper, we present an algorithm whose numerical tests suggest a polynomial running time for all the considered quantum-correlation problems. More precisely, the algorithm computes the distance from the local polytope. First, we prove that the time cost of computing the distance with an arbitrary fixed error grows polynomially in the size of the problem input (number of measurements and outcomes), granted the access to a certain oracle. Namely, given a fixed error, we derive two upper bounds on the running time. The first bound is linear in the number of measurements. The second bound scales with the number of measurements to the sixth power. The former holds only for a very high number of measurements, and is never observed in the performed numerical tests. Thus, the problem of computing the distance is reduced to determining an efficient simulation of the oracle. Then, we introduce a simple algorithm that simulates the oracle. The algorithm is probabilistic and provides the right answer in a subset of randomized inputs. Thus, to have a correct answer with sufficiently high probability, the simulation of the oracle has to be performed with a suitably high number of initial random inputs. In the numerical tests, the number of random initial trials has pragmatically been chosen such that the simulation of the oracle contributes to the overall running time with a multiplicative factor and, thus, does not affect the sixth-power law of the oracle-assisted algorithm. In all of the performed numerical tests, the overall algorithm always computes the distance within the desired accuracy. The scaling of the running time observed in the tests is compatible with the sixth-power law, derived theoretically.

Similar results have independently been published in [13], almost simultaneously to a first version of this paper [14]. The algorithm in [13] is a modification of Gilbert's algorithm for minimizing quadratic forms in a convex set. In its original form, the algorithm uses the following strategy for generating a sequence of points, which converge to the minimizer: Given a point P_n of the sequence, a procedure of linear optimization generates another point Q_n, such that the next point P_{n+1} of the sequence is computed as a convex combination of P_n and Q_n. If the convex set is a polytope, the points $Q_1, \ldots$ turn out to be vertices of the polytope. The modified algorithm, introduced in [13], keeps track of the previous vertices $Q_{n-m}, Q_{n-m+1}, \ldots, Q_n$, m being some fixed parameter, and computes the next point P_{n+1} as convex combination of these points and P_n. In our algorithm, we compute P_{n+1} as a convex combination of a suitable set of previously computed vertices, without using the point P_n (Section 5). This difference does not result in substantial computational differences. However, our approach has the advantage of keeping track of the minimal number of vertices required for a convex representation of the optimizer. In particular, in the case of local correlations, the algorithm immediately gives a minimal convex representation of them. This representation provides a certificate, which another party can use for directly proving locality. As another minor difference, our algorithm actually computes the distance from what we will call the *local cone*. This allows us to eliminate a normalization constraint from the optimization problem.

The paper is organized as follows. In Section 2, we introduce our general scenario. For the sake of simplicity, we will discuss only the two-party case, but the results can be extended to the general case of many parties. After introducing the local polytope in Section 3, we formulate the non-locality problem as a minimization problem; namely, the problem of computing the distance from the local polytope (Section 4). In Section 5, the algorithm is introduced. The convergence and the computational cost are then discussed in Section 6. After introducing the algorithm for solving the oracle, we finally discuss the numerical results in Section 7.

2. Nonsignaling Box

In a Bell scenario, two quantum systems are prepared in an entangled state and delivered to two spatially separate parties; say, Alice and Bob. These parties each perform a measurement on their system and get an outcome. In general, Alice and Bob are allowed to choose among their respective sets of possible measurements. We assume that the sets are finite, but arbitrarily large. Let us denote the measurements performed by Alice and Bob by the indices $a \in \{1, \ldots, A\}$ and $b \in \{1, \ldots, B\}$, respectively. After the measurements, Alice gets an outcome $r \in \mathcal{R}$ and Bob an outcome $s \in \mathcal{S}$, where $\mathcal{R}$ and $\mathcal{S}$ are two sets with cardinality R and S, respectively. The overall scenario is described by the joint conditional probability $P(r, s|a, b)$ of getting (r, s), given (a, b). Since the parties are spatially separate, causality and relativity imply that this distribution satisfies the nonsignaling conditions

$$P(r|a, b) = P(r|a, \bar{b}) \quad \forall r, a, b, \bar{b}, \text{ and}$$
$$P(s|a, b) = P(s|\bar{a}, b) \quad \forall s, b, a, \bar{a}, \tag{1}$$

where $P(r|a, b) \equiv \sum_s P(r, s|a, b)$ and $P(s|a, b) \equiv \sum_r P(r, s|a, b)$ are the marginal conditional probabilities of r and s, respectively. In the following discussion, we consider a more general scenario than quantum correlations, and we just assume that $P(r, s|a, b)$ satisfies the nonsignaling conditions. The abstract machine producing the correlated variables r and s from the inputs a and b will be called the *nonsignaling box* (briefly, NS-box).

3. Local Polytope

The correlations between the outcomes r and s, associated with the measurements a and b, are *local* if and only if the conditional probability $P(r,s|a,b)$ can be written in the form

$$P(r,s|a,b) = \sum_x P^A(r|a,x)P^B(s|b,x)P^S(x),\tag{2}$$

where P^A, P^B, and P^S are suitable probability distributions. It is always possible to write the conditional probabilities P^A and P^B as convex combination of local deterministic processes, that is,

$$\begin{aligned}
P^A(r|a,x) &= \sum_{\mathbf{r}} P^A_{\det}(r|\mathbf{r},a)\rho^A(\mathbf{r}|x), \text{ and}\\
P^B(s|b,x) &= \sum_{\mathbf{s}} P^B_{\det}(s|\mathbf{s},b)\rho^B(\mathbf{s}|x),
\end{aligned}\tag{3}$$

where $\mathbf{r} \equiv (r_1,\dots,r_A)$, $\mathbf{s} \equiv (s_1,\dots,s_B)$, $P^A_{\det}(r|\mathbf{r},a) = \delta_{r_a,r}$, and $P^B_{\det}(s|\mathbf{s},b) = \delta_{s_b,s}$. Using this decomposition, Equation (2) takes the form of a convex combination of local deterministic distributions. That is,

$$\begin{aligned}
P(r,s|a,b) &= \sum_{\mathbf{r},\mathbf{s}} P^A_{\det}(r|\mathbf{r},a)P^B_{\det}(s|\mathbf{s},b)P^{AB}(\mathbf{r},\mathbf{s})\\
&= \sum_{\mathbf{r},\mathbf{s}} \delta_{r,r_a}\delta_{s,s_b} P^{AB}(\mathbf{r},\mathbf{s})\\
&= \sum_{\mathbf{r},r_a=r}\sum_{\mathbf{s},s_b=s} P^{AB}(\mathbf{r},\mathbf{s}),
\end{aligned}\tag{4}$$

where $P^{AB}(\mathbf{r},\mathbf{s}) \equiv \sum_x \rho^A(\mathbf{r}|x)\rho^B(\mathbf{s}|x)P^S(x)$ and $\delta_{i,j}$ is the Kronecker delta. Equation (5) is known as Fine's theorem [15]. Thus, a local distribution can always be written as convex combination of local deterministic distributions. Clearly, the converse is also true and a convex combination of local deterministic distributions is local. Therefore, the set of local distributions is a polytope, called a *local polytope*. As the deterministic probability distributions $P^A_{\det}(r|\mathbf{r},a)P^B_{\det}(s|\mathbf{s},b)$ are not convex combinations of other distributions, they all define the vertices of the local polytope. Thus, there are $R^A S^B$ vertices, each one specified by the sequences $\mathbf{r}$ and $\mathbf{s}$. Let us denote the map from $(\mathbf{r},\mathbf{s})$ to the associated vertex by $\vec{V}$. That is, $\vec{V}$ maps the sequences to a deterministic local distribution,

$$\vec{V}(\mathbf{r},\mathbf{s}) \equiv P_{\det} : (r,s,a,b) \mapsto \delta_{r,r_a}\delta_{s,s_b}.\tag{5}$$

Since the elements of the local polytope are normalized distributions and satisfy the nonsignaling conditions (1), the $RSAB$ parameters defining $P(r,s|a,b)$ are not independent and the polytope lives in a lower-dimensional subspace. The dimension of this subspace and, more generally, of the subspace of NS-boxes, is equal to [16]

$$d_{NS} \equiv AB(R-1)(S-1) + A(R-1) + B(S-1).\tag{6}$$

By the Minkowski–Weyl theorem, the local polytope can be represented as the intersection of finitely many half-spaces. A half-space is defined by an inequality

$$\sum_{r,s,a,b} P(r,s|a,b)B(r,s;a,b) \le L.\tag{7}$$

In the case of the local polytope, these inequalities are called *Bell inequalities*. Given the coefficients $B(r,s;a,b)$, we can choose L such that the inequality is as restrictive as possible. This is attained by imposing that at least one vertex of the local polytope is at the boundary of the half-space; that is, by taking

$$L = \max_{r,s} \sum_{a,b} B(r_a, s_b; a, b).$$ (8)

The oracle, which is central in this work, and introduced later in Section 4, returns the value L from the coefficients $B(r, s; a, b)$.

A minimal representation of a polytope is given by the set of facets of the polytope. A half-space $\sum_{r,s,a,b} P(r, s|a, b) B(r, s; a, b) \leq L$ specifies a facet if the associated hyperplane $\sum_{r,s,a,b} P(r, s|a, b) B(r, s; a, b) = L$ intersects the boundary of the polytope in a set with dimension equal to the dimension of the polytope minus one. A distribution $P(r, s|a, b)$ is local if and only if every facet inequality is not violated. Deciding whether some inequality is violated is generally believed to be intractable, due to a result by Pitowski [12], but to test the membership of a distribution to the local polytope can be done in polynomial time, once the vertices—of which the distribution is a convex combination—are known. Thus, deciding membership to the local polytope is an NP problem. Furthermore, the problem is NP-complete [12].

4. Distance from the Local Polytope

The non-locality problem can be reduced to a convex optimization problem, such as the computation of the nonlocal capacity, introduced in [17], and the distance from the local polytope, which can be reduced to a linear program if the L^1 norm is employed [18]. Here, we define the distance of a distribution $P(r, s|a, b)$ from the local polytope as the Euclidean distance between $P(r, s|a, b)$ and the closest local distribution. As mentioned in Section 3 (see Equation (5)), and stated by Fine's theorem [15], a conditional distribution $\rho(r, s|a, b)$ is local if and only if there is a non-negative function $\chi(\mathbf{r}, \mathbf{s})$ such that

$$\rho(r, s|a, b) = \sum_{\mathbf{r}, r_a = r} \sum_{\mathbf{s}, s_b = s} \chi(\mathbf{r}, \mathbf{s}).$$ (9)

That is, a conditional distribution $\rho(r, s|a, b)$ is local if it is the marginal of a multivariate probability distribution χ of the outcomes of all the possible measurements, provided that χ does not depend on the measurements a and b.

The distributions $P(r, s|a, b)$ and $\rho(r, s|a, b)$ can be represented as vectors in a $RSAB$-dimensional space. Let us denote them by $\vec{P}$ and $\vec{\rho}$, respectively. Given a positive-definite matrix $\hat{M}$ defining the metrics in the vector space, the computation of the distance from the local polytope is equivalent to the minimization of a functional of the form

$$F[\chi] = \frac{1}{2} \left(\vec{P} - \vec{\rho} \right)^T \hat{M} \left(\vec{P} - \vec{\rho} \right)$$ (10)

with respect to χ, under the constraints that χ is non-negative and normalized. Namely, the distance is the square root of the minimum of $2F$. Hereafter, we choose the metrics so that the functional takes the form

$$F[\chi] \equiv \frac{1}{2} \sum_{r,s,a,b} [P(r, s|a, b) - \rho(r, s|a, b)]^2 W(a, b),$$ (11)

where $W(a, b)$ is some probability distribution. The normalization $\sum_{a,b} W(a, b) = 1$ guarantees that the distance does not diverge in the limit of infinite measurements performed on a given entangled state. In particular, we will consider the case with

$$W(a, b) \equiv \frac{1}{AB}.$$ (12)

Another choice would be to take the distribution $W(a, b)$ maximizing the functional, so that the computation of the distance would be a minimax problem. This case has some interesting advantages, but is more sophisticated and will not be considered here. Since we are interested in a quantity that is equal to zero if and only if $P(r, s|a, b)$ is local, we can simplify the problem of computing the distance by dropping the normalization constraint on χ. Indeed, if the distance is equal to zero, ρ, and thus χ, are necessarily normalized. Conversely, if the distance is different from zero for every normalized local distribution, it is so also for every unnormalized local distribution. Thus, the discrimination between local and non-local correlation is equivalent to the following minimization problem.

Problem 1.

$$\min_{\chi} F[\chi]$$
subject to the constraints
$$\chi(\mathbf{r}, \mathbf{s}) \geq 0.$$

Let us denote the solution of this problem and the corresponding optimal value by χ^{min} and F^{min}, respectively. The associated (unnormalized) local distribution is denoted by $\rho^{min}(r, s|a, b)$. The square root of $2F^{min}$ is the minimal distance of $P(r, s|a, b)$ from the cone defined as the union of all the lines connecting the zero distribution $\rho(r, s|a, b) = 0$ and an arbitrary point of the local polytope. Let us call this set the *local cone*. Hereafter, we will consider the problem of computing the distance from the local cone, but the results can be easily extended to the case of the local polytope, so that we will use "local cone" and "local polytope" as synonyms in the following discussion. Note that there are generally infinite minimizers χ^{min}, since χ lives in a $R^A S^B$-dimensional space, whereas the functional F depends on χ through $\rho(r, s|a, b)$, which lives in a $(d_{NS} + 1)$-dimensional space. In other words, since the local polytope has $R^A S^B$ vertices, but the dimension of the polytope is d_{NS}, a (normalized) distribution ρ has generally infinite representations as convex combination of the vertices, unless ρ is on a face whose dimension plus 1 is equal to the number of vertices defining the face.

At first glance, the computational complexity of this problem seems intrinsically exponential, as the number of real variables defining χ is equal to $R^A S^B$. However, the dimension of the local polytope is d_{NS} and grows polynomially in the number of measurements and outcomes. Thus, by Carathéodory's theorem, a (normalized) local distribution can always be represented as the convex combination of a number of vertices smaller than $d_{NS} + 2$. This implies that there is a minimizer χ^{min} of F whose support contains a number of elements not greater than $d_{NS} + 1$. Therefore, the minimizer can be represented by a number of variables growing polynomially in the input size. The main problem is to find a small set of vertices that are suitable for representing the closest local distribution $\rho^{min}(r, s|a, b)$. In the following, we will show that the computation of the distance from the local cone with an arbitrary fixed accuracy has polynomial complexity, granted the access to the following oracle.

Oracle Max: Given a function $g(r, s; a, b)$, the oracle returns the sequences $\mathbf{r}$ and $\mathbf{s}$ maximizing the function

$$G(\mathbf{r}, \mathbf{s}) \equiv \sum_{a,b} g(r_a, s_b; a, b) W(a, b) \tag{13}$$

and the corresponding maximal value.

Thus, Problem 1 is reduced to determining an efficient simulation of the oracle. Let us consider the case of binary outcomes, with r and s taking values ± 1 ($R = S = 2$). The function $G(\mathbf{r}, \mathbf{s})$ takes the form

$$G(\mathbf{r}, \mathbf{s}) = \sum_{a,b} J_{ab} r_a s_b + \sum_a A_a r_a + \sum_b B_b s_b + G_0, \tag{14}$$

whose minimization falls into the class of spin-glass problems, which are notoriously computationally hard to handle. This suggests that the oracle is generally an intractable problem. Nonetheless, the oracle has a particular structure that can make the problem easier to be solved, in some instances. This will be discussed later, in Sections 6.3 and 7. There, we will show that the oracle can be simulated efficiently in many relevant cases, by using a simple block-maximization strategy. Assuming for the moment that we have access to the oracle, let us introduce the algorithm solving Problem 1.

5. Computing the Distance

The distance from the local polytope can be computed efficiently, once we have a set Ω of vertices that is small enough and suitable for representing the closest distribution $\rho^{min}(r, s|a, b)$. The algorithm introduced in this paper solves Problem 1 by iteratively generating a sequence of sets Ω. At each step, the minimal distance is first computed over the convex hull of the given vertices. Then, the oracle is consulted. If the set does not contain the right vertices, the oracle returns a strictly positive maximal value and a vertex, which is added to the set Ω (after possibly removing vertices with zero weight). The optimization Problem 1 is solved once the oracle returns zero, which guarantees that all the optimality conditions of the problem are satisfied. Before discussing the algorithm, let us derive these conditions.

5.1. Necessary and Sufficient Conditions for Optimality

Problem 1 is a convex optimization problem whose constraints satisfy Slater's condition, requiring the existence of an interior point of the feasible region. This is the case, as a positive χ strictly satisfies all the inequality constraints. Thus, the four Karush–Kuhn–Tucker (KKT) conditions are necessary and sufficient conditions for optimality. Let us briefly summarize these conditions. Given an objective function $F(\vec{x})$ of the variables $\vec{x}$ and equality constraints $G_{k=1,\ldots,n_c}(\vec{x}) = 0$, it is well known that the function F is stationary at $\vec{x}$ if the gradient of the Lagrangian $\mathcal{L}(\vec{x}) \equiv F(\vec{x}) - \sum_{k=1}^{n_c} \eta_k G_k(\vec{x})$ is equal to zero, for some value of the Lagrange multipliers η_k. This is the first KKT condition. The second condition is the feasibility of the constraints; that is, the stationary point $\vec{x}$ must satisfy the constraints $G_k(\vec{x}) = 0$. These two conditions are necessary and sufficient, as there are only equality constraints. If there are also inequalities, two additional conditions on the associated Lagrange multipliers are required. Given inequality constraints $H_k(\vec{x}) \geq 0$, with associated Lagrange multipliers λ_k, the third condition is the non-negativity of the multipliers; that is, $\lambda_k \geq 0$. This condition says that the constraint acts only in one direction, like a floor acts on objects through an upward force, but not with a downward force. The last condition states that the Lagrange multiplier λ_k can differ from zero only if the constraint is active; that is, if $H_k(\vec{x}) = 0$. This is like stating that a floor acts on a body only if they are touching (contact force). This condition can concisely be written as $\lambda_k H_k(\vec{x}) = 0$.

Let us characterize the optimal solution of Problem 1 through the four KKT conditions.

- First KKT condition (*stationarity condition*): The gradient of the Lagrangian is equal to zero. The Lagrangian of Problem 1 is

$$\mathcal{L} = F[\chi] - \sum_{\mathbf{r},\mathbf{s}} \lambda(\mathbf{r},\mathbf{s})\chi(\mathbf{r},\mathbf{s}), \tag{15}$$

  where $\lambda(\mathbf{r},\mathbf{s})$ are the Lagrange multipliers associated with the inequality constraints.
- Second KKT condition (*feasibility of the constraints*): The function χ is non-negative, $\chi(\mathbf{r},\mathbf{s}) \geq 0$.
- Third condition (*dual feasibility*): The Lagrange multipliers λ are non-negative; that is,

$$\lambda(\mathbf{r},\mathbf{s}) \geq 0. \tag{16}$$

- Fourth condition (*complementary slackness*): If $\chi(\mathbf{r},\mathbf{s}) \neq 0$, then the multiplier $\lambda(\mathbf{r},\mathbf{s})$ is equal to zero; that is,

$$\lambda(\mathbf{r},\mathbf{s})\chi(\mathbf{r},\mathbf{s}) = 0. \tag{17}$$

The stationarity condition on the gradient of the Lagrangian gives the equality

$$\sum_{a,b} W(a,b)\left[P(r_a,s_b|a,b) - \rho(r_a,s_b|a,b)\right] + \lambda(\mathbf{r},\mathbf{s}) = 0. \tag{18}$$

Eliminating λ, this equality and the dual feasibility yield the inequality

$$\sum_{a,b} W(a,b)\left[P(r_a, s_b | a, b) - \rho(r_a, s_b | a, b)\right] \leq 0. \tag{19}$$

From Equation (18), we have that the complementary slackness is equivalent to the following condition,

$$\chi(\mathbf{r}, \mathbf{s}) \neq 0 \Rightarrow$$
$$\sum_{a,b} W(a,b)\left[P(r_a, s_b | a, b) - \rho(r_a, s_b | a, b)\right] = 0; \tag{20}$$

that is, the left-hand side of the last inequality is equal to zero if $(\mathbf{r}, \mathbf{s})$ is in the support of χ. The slackness condition (20), the primal constraint and Equation (19) provide necessary and sufficient conditions for optimality. Let us introduce the function

$$g(r, s; a, b) \equiv P(r, s | a, b) - \rho(r, s | a, b), \tag{21}$$

which is the opposite of the gradient of F with respect to ρ, up to the factor $W(a,b)$. Summarizing, the conditions are

$$\sum_{a,b} W(a,b) g(r_a, s_b; a, b) \leq 0, \tag{22}$$
$$\chi(\mathbf{r}, \mathbf{s}) \neq 0 \Rightarrow \sum_{a,b} W(a,b) g(r_a, s_b; a, b) = 0, \tag{23}$$
$$\chi(\mathbf{r}, \mathbf{s}) \geq 0. \tag{24}$$

The second condition can be rewritten in the more concise form

$$\sum_{r,s,a,b} \rho(r, s | a, b) g(r, s | a, b) W(a, b) = 0. \tag{25}$$

Indeed, using Equations (22) and (24), it is easy to show that condition (23) is satisfied if and only if

$$\sum_{\mathbf{r},\mathbf{s}} \chi(\mathbf{r}, \mathbf{s}) \sum_{a,b} W(a,b) g(r_a, s_b; a, b) = 0,$$

which gives equality (25), by definition of ρ (Equation (9)).

Condition (22) can be checked, by consulting the oracle with $g(r, s; a, b)$ as the query. If the oracle returns a non-positive maximal value, then the condition is satisfied. Actually, at the optimal point, the returned value turns out to be equal to zero, as implied by the other optimality conditions.

Similar optimality conditions hold if we force χ to be equal to zero outside some set Ω. Let us introduce the following minimization problem.

Problem 2.

$$\min_\chi F[\chi]$$
subject to the constraints
$$\chi(\mathbf{r}, \mathbf{s}) \geq 0,$$
$$\chi(\mathbf{r}, \mathbf{s}) = 0 \quad \forall (\mathbf{r}, \mathbf{s}) \notin \Omega.$$

The optimal value of this problem gives an upper bound on the optimal value of Problem 1. The two problems are equivalent if the support of a minimizer χ^{min} of Problem 1 is in Ω. The necessary and sufficient conditions for optimality of Problem 2 are the same as of Problem 1, with the only difference that condition (22) has to hold only in the set Ω. That is, the condition is replaced by the weaker condition

$$(\mathbf{r}, \mathbf{s}) \in \Omega \Rightarrow \sum_{a,b} W(a,b) g(r_a, s_b; a, b) \leq 0. \tag{26}$$

Thus, an optimizer of Problem 2 is solution of Problem 1 if the value returned by the oracle with query $g = P - \rho$ is equal to zero.

Hereafter, the minimizer and the minimal value of Problem 2 will be denoted by χ_Ω^{min} and F_Ω^{min}, respectively. The associated optimal local distribution $\rho(r, s | a, b)$, defined by Equation (9), will be denoted by $\rho_\Omega^{min}(r, s | a, b)$.

5.2. Overview of the Algorithm

Problem 1 can be solved iteratively by finding the solution of Problem 2 over a sequence of sets Ω. The sets are built according to the answer of the oracle, which is consulted at each step of the iteration. The procedure stops when a desired accuracy is reached or Ω contains the support of a minimizer χ^{min}, and the solution of Problem 2 is also the solution of Problem 1. Let us outline the algorithm. Suppose that we choose the initial Ω as a set of sequences $(\mathbf{r}, \mathbf{s})$ associated to n_0 linearly independent vertices (n_0 being possibly equal to 1). Let us denote this set by Ω_0. We solve Problem 2 with $\Omega = \Omega_0$ and get the optimal value $F_0^{min} \equiv F_{\Omega_0}^{min}$ with minimizer $\chi_0^{min} \equiv \chi_{\Omega_0}^{min}$. Let us denote the corresponding (unnormalized) local distribution by $\rho_0^{min} \equiv \rho_{\Omega_0}^{min}$. That is,

$$\rho_0^{min}(r, s | a, b) \equiv \sum_{\mathbf{r}, r_a = r} \sum_{\mathbf{s}, s_b = s} \chi_0^{min}(\mathbf{r}, \mathbf{s}). \tag{27}$$

Since the cardinality of Ω_0 is not greater than $d_{NS} + 1$ and the problem is a convex quadratic optimization problem, the corresponding computational complexity is polynomial. Generally, a numerical algorithm provides an optimizer, up to some arbitrarily small but finite error. In Section 5.5, we will provide a bound on the accuracy required for the solution of Problem 2. For now, let us assume that Problem 2 is solved exactly. If the support of χ^{min} is in Ω_0, F_0^{min} is equal to the optimal value of Problem 1, and we have computed the distance from the local polytope. We can verify if this is the case by checking the first optimality condition (22), as the conditions (23) and (24) are trivially satisfied by the optimizer of Problem 2 for every $(\mathbf{r}, \mathbf{s})$. The check is made by consulting the oracle with the function $P(r, s | a, b) - \rho_0^{min}(r, s | a, b)$ as the query. If the oracle returns a maximal value equal to zero, then we have the solution of Problem 1. Note that if the optimal value of Problem 2 is equal to zero, then also the optimal value of the main problem is equal to zero and the conditional distribution $P(r, s | a, b)$ is local. In this case, we have no need of consulting the oracle.

If the optimal value of Problem 2 is different from zero and the oracle returns a maximal value strictly positive, then the minimizer of Problem 2 satisfies all the optimality conditions of Problem 1, except Equation (22) for some $(\mathbf{r}, \mathbf{s}) \notin \Omega$. The next step is to add the pair of sequences $(\mathbf{r}, \mathbf{s})$ returned by the oracle to the set Ω and solve Problem 2 with the new set. Let us denote the new set and the corresponding optimal value by Ω_1 and $F_1^{min} \equiv F_{\Omega_1}^{min}$, respectively. Once we have solved Problem 2 with $\Omega = \Omega_1$, we consult again the oracle to check if we have obtained the solution of Problem 1. If we have not, we add the pair of sequences $(\mathbf{r}, \mathbf{s})$ given by the oracle to the set Ω and we solve Problem 2 with the new set, say Ω_2. We continue until we get the solution of Problem 1 or its optimal value up to some desired accuracy. This procedure generates a sequence of sets $\Omega_{n=1,2,...}$ and values $F_{n=1,2,...}^{min}$. The latter sequence is strictly decreasing, that is, $F_{n+1}^{min} < F_n^{min}$ until Ω_n contains the support of χ^{min} and the oracle returns zero as maximal value. Let us show that. Suppose that χ_n^{min} is the optimizer of Problem 2 with $\Omega = \Omega_n$ and $(\mathbf{r}', \mathbf{s}')$ is the new element in the set Ω_{n+1}. Let us denote by $\rho_n^{min}(r, s | a, b)$ the local distribution associated with χ_n^{min}, that is,

$$\rho_n^{min}(r, s | a, b) \equiv \sum_{\mathbf{r}, r_a = r} \sum_{\mathbf{s}, s_b = s} \chi_n^{min}(\mathbf{r}, \mathbf{s}). \tag{28}$$

The optimal value F_{n+1}^{min} of Problem 2 is bounded from above by the value taken by the function $F[\chi]$ for every feasible χ, in particular, for

$$\chi(\mathbf{r}, \mathbf{s}; \alpha) = \chi_n^{min}(\mathbf{r}, \mathbf{s}) + \alpha \delta_{\mathbf{r}, \mathbf{r}'} \delta_{\mathbf{s}, \mathbf{s}'}, \tag{29}$$

with α positive. Let us set α equal to the value minimizing F; that is,

$$\alpha \equiv \alpha_n = \sum_{ab} W(a, b)[P(r_a', s_b'|a, b) - \rho_n^{min}(r_a', s_b'|a, b)], \tag{30}$$

which is equal to the value returned by the oracle. It is strictly positive, as the oracle returned a positive value—provided that Ω_n does not contain the support of χ^{min}. Hence, $\chi(\mathbf{r}, \mathbf{s}; \alpha_n)$ is a feasible point and, thus, the corresponding value taken by F,

$$F|_{\alpha=\alpha_n} = F_n^{min} - \frac{1}{2}\alpha_n^2, \tag{31}$$

is an upper bound on F_{n+1}^{min}. Hence,

$$F_{n+1}^{min} \leq F_n^{min} - \frac{1}{2}\alpha_n^2, \tag{32}$$

that is, F_{n+1}^{min} is strictly smaller than F_n^{min}.

This procedure generates a sequence F_n^{min} that converges to the optimal value of Problem 1, as shown in Section 6. For any given accuracy, the computational cost of the procedure is polynomial, provided that we have access to the oracle.

To avoid growth of the cardinality of Ω beyond $d_{NS} + 1$ during the iteration and, thus, the introduction of redundant vertices, we have to be sure that the sets $\Omega_0, \Omega_1, \ldots$ contain points $(\mathbf{r}, \mathbf{s})$ associated to linearly independent vertices $\vec{V}(\mathbf{r}, \mathbf{s})$ of the local polytope. This is guaranteed by the following procedure of cleaning up. First, after the computation of χ_n^{min} at step n, we remove the elements in Ω_n where $\chi_n^{min}(\mathbf{r}, \mathbf{s})$ is equal to zero (this can be checked even if the exact χ_n^{min} is not known, as discussed later in Section 5.6). Let us denote the resulting set by Ω_n^{clean}. Then, the set Ω_{n+1} is built by adding the point given by the oracle to the set Ω_n^{clean}. Let us denote by $\mathcal{V}$ the set of vertices associated to the elements in the support of χ_n^{min}. The cleaning up ensures that the optimizer ρ_n^{min} is in the interior of the convex hull of $\mathcal{V}$, up to a normalization constant, and the new vertex returned by the oracle is linearly independent of the ones in $\mathcal{V}$. Indeed, we have seen that the introduction of such a vertex allows us to lower the optimal value of Problem 2. This would not be possible if the added vertex was linearly dependent on the vertices in $\mathcal{V}$, as the (normalized) optimizer ρ_n^{min} of Problem 2 is in the interior of the convex hull of $\mathcal{V}$.

This is formalized in Lemma 1.

Lemma 1. *Let $(\mathbf{r}', \mathbf{s}')$ be a sequence such that*

$$\sum_{a,b} g(r_a', s_b'; a, b) W(a, b) \neq 0. \tag{33}$$

If Ω is a set such that

$$(\mathbf{r}, \mathbf{s}) = \Omega \Rightarrow \sum_{a,b} g(r_a, s_b; a, b) W(a, b) = 0, \tag{34}$$

then the vertex $\vec{V}(\mathbf{r}', \mathbf{s}')$ is linearly independent of the vertices associated to the sequences in Ω.

Proof. The proof is by contradiction. Suppose that the vector $\vec{V}(\mathbf{r}', \mathbf{s}')$ is linearly dependent with the vectors $\vec{V}(\mathbf{r}, \mathbf{s})$ with $(\mathbf{r}, \mathbf{s}) \in \Omega$, then there is a real function $t(\mathbf{r}, \mathbf{s})$ such that

$$\vec{V}(\mathbf{r}', \mathbf{s}') = \sum_{(\mathbf{r}, \mathbf{s}) \in \Omega} t(\mathbf{r}, \mathbf{s}) \vec{V}(\mathbf{r}, \mathbf{s}). \tag{35}$$

By definition of $\vec{V}$, this equation implies that $\sum_{r,s} t(\mathbf{r}, \mathbf{s}) \delta_{r,r_a} \delta_{s,s_b} = \delta_{r,r'_a} \delta_{s,s'_b}$. From this equation and Equation (34), we have

$$\sum_{r,s} \delta_{r,r'_a} \delta_{s,s'_b} \sum_{a,b} g(r, s; a, b) W(a, b) = 0. \tag{36}$$

Summing over r and s, we get a contradiction with Equation (33). $\square$

This lemma and the optimality conditions (22) and (23) imply that the sets $\Omega_0, \Omega_1, \ldots$, built through the previously discussed procedure of cleaning up, always contain points associated to independent vertices and, thus, never contain more than $d_{NS} + 1$ elements. Indeed, the set Ω_n^{clean} contains points $(\mathbf{r}, \mathbf{s})$ where the minimizer χ_n^{min} is different from zero, for which

$$\sum_{a,b} \left[P(r_a, s_b | a, b) - \rho_n^{min}(r_a, s_b | a, b) \right] W(a, b) = 0,$$

as implied by condition (23). Furthermore, given the sequence $(\mathbf{r}', \mathbf{s}')$ returned by the oracle, condition (22) implies that

$$\sum_{a,b} \left[P(r'_a, s'_b | a, b) - \rho_n^{min}(r'_a, s'_b | a, b) \right] W(a, b) > 0$$

until the set Ω_n contains the support of χ^{min} and the iteration generating the sequence of sets Ω is terminated.

The procedure of cleaning up is not strictly necessary for having a polynomial running time, but it can speed up the algorithm. Furthermore, the procedure guarantees that the distribution $\rho(r, s | a, b)$ approaching the minimizer during the iterative computation is always represented as the convex combination of a minimal number of vertices. Thus, we have a minimal representation of the distribution at each stage of the iteration.

5.3. The Algorithm

In short, the algorithm for computing the distance from the local polytope with given accuracy is as follows.

Algorithm 1. *Input: $P(r, s | a, b)$*

1. *Set $(\mathbf{r}', \mathbf{s}')$ equal to the sequences given by the oracle with $P(r, s | a, b)$ as query.*
2. *Set $\Omega = \{(\mathbf{r}', \mathbf{s}')\}$.*
3. *Compute the optimizers $\chi(\mathbf{r}, \mathbf{s})$ and $\rho(r, s | a, b)$ of Problem 2. The associated F provides an upper bound of the optimal value F^{min}.*
4. *Consult the oracle with $g(r, s; a, b) = P(r, s | a, b) - \rho(r, s | a, b)$ as query. Set $(\mathbf{r}', \mathbf{s}')$ and α are equal to the sequences returned by the oracle and the associated maximal value, respectively. That is,*

$$(\mathbf{r}', \mathbf{s}') = argmax \sum_{(\mathbf{r}, \mathbf{s})}^{a,b} g(r_a, s_b; a, b) W(a, b),$$

$$\alpha = \sum_{a,b} g(r'_a, s'_b | a, b) W(a, b),$$

5. *Compute a lower bound on the F^{min} from ρ and α (see following discussion and Section 6.1). The difference between the upper and lower bounds provides an upper bound on the reached accuracy.*

6. If a given accuracy is reached, stop.
7. Remove from Ω the points where χ is zero and add $(\mathbf{r}', \mathbf{s}')$.
8. Go back to Step 3.

The algorithm stops at Step 6 when a desired accuracy is reached. To estimate the accuracy, we need to compute a lower bound on the optimal value F^{min}. To guarantee that the algorithm eventually stops, the lower bound has to converge to the optimal value as the algorithm approaches the solution of Problem 1. We also need a stopping criterion for the numerical routine solving the optimization problem in Step 3. Let us first discuss the stopping criterion for Algorithm 1.

5.4. Stopping Criterion for Algorithm 1

The lower bound on F^{min}, denoted by $F^{(-)}$, is computed by using the dual form of Problem 1. As shown in Section 6.1, any local distribution ρ induces the lower bound

$$F^{(-)} = \frac{1}{2} \sum_{rsab} \left\{ P^2(r,s|a,b) - \left[\rho(r,s|a,b) + \alpha \right]^2 \right\} W(a,b), \tag{37}$$

where α is the maximal value returned by the oracle with $g(r,s;a,b) = P(r,s|a,b) - \rho(r,s|a,b)$ as query. An upper bound on F^{min} is obviously

$$F^{(+)} = F[\chi]. \tag{38}$$

In the limit of ρ equal to the local distribution minimizing F, the lower bound is equal to the optimal value F^{min}. This can be shown by using the optimality conditions. Indeed, conditions (22) and (25) imply the limits

$$\lim_{\chi \to \chi^{min}} \alpha = 0, \tag{39}$$

$$\lim_{\chi \to \chi^{min}} \sum_{r,s,a,b} \rho(r,s|a,b) g(r,s;a,b) W(a,b) = 0, \tag{40}$$

which imply $F^{(-)} \to F^{min}$ as χ approaches the minimizer. This is made even more evident, by computing the difference between the upper bound and the lower bound. Indeed, given the local distribution $\rho(r,s|a,b)$ computed at Step 3 and the corresponding α returned by the oracle at Step 4, the difference is

$$F^{(+)} - F^{(-)} \equiv \Delta F = \frac{RS}{2}\alpha^2 + \sum_{rsab} \rho(r,s|a,b) \left[\alpha - g(r,s;a,b)\right] W(a,b), \tag{41}$$

which evidently goes to zero as χ goes to χ^{min}. Thus, the upper bound ΔF on the accuracy computed in Step 5 goes to zero as $\rho(r,s|a,b)$ approaches the solution. This guarantees that the algorithm stops sooner or later at Step 6, provided that χ converges to the solution. If Problem 2 is solved exactly at Step 3, then the distribution $\rho(r,s|a,b)$ satisfies condition (25), and the upper bound on the reached accuracy takes the form

$$F^{(+)} - F^{(-)} = \frac{RS}{2}\alpha^2 + \alpha \sum_{rsab} \rho(r,s|a,b) W(a,b). \tag{42}$$

Even if Condition (25) is not satisfied, we can suitably normalize $\chi(\mathbf{r},\mathbf{s})$ so that the condition is satisfied.
In the following, we assume that this condition is satisfied.

5.5. Stopping Criterion for Problem 2 (Optimization at Step 3 of Algorithm 1)

In Algorithm 1, Step 3 is completed when the solution of Problem 2 with a given set Ω is found. Optimization algorithms iteratively find a solution $\rho_\Omega^{min}(r,s|a,b)$ up to some accuracy. We can stop when the error is of the order of the machine precision. Here, we will discuss a more effective stopping criterion. This criterion should preserve the two main features previously described:

- The sequence $F_0^{min}, F_1^{min}, \ldots$ of the exact optimal values of Problem 2, with $\Omega = \Omega_0, \Omega_1, \ldots$, is monotonically decreasing.
- The sets $\Omega_0, \Omega_1, \ldots$ contain points associated with linearly independent vertices of the local polytope, implying that the cardinality of Ω_n is never greater than $d_{NS} + 1$.

To guarantee that the first feature is preserved, it is sufficient to compute a lower bound on F_Ω^{min} from a given χ so that the bound approaches F_Ω^{min} as χ approaches the optimizer χ_Ω^{min}. If the lower bound with the set $\Omega = \Omega_n$ is greater than the upper bound $F_n - \alpha_n^2/2$ on F_{n+1}^{min} (see Equation (31)), then $F_{n+1}^{min} < F_n^{min}$. Denoting by $F_\Omega^{(-)}$ the lower bound on the optimal value F_Ω^{min}, the monotonicity of the sequence $F_0^{min}, F_1^{min}, \ldots$ is implied by the inequality

$$F_n - \frac{1}{2}\alpha_n^2 \le F_{\Omega_n}^{(-)}. \tag{43}$$

As shown later, by using dual theory, a lower bound on F_Ω^{min} is

$$F_\Omega^{(-)} = \frac{1}{2}\sum_{rsab}\left\{P^2(r,s|a,b) - \left[\rho(r,s|a,b) + \beta\right]^2\right\}W(a,b), \tag{44}$$

where

$$\beta \equiv \max_{(\mathbf{r},\mathbf{s})\in\Omega}\sum_{ab}W(a,b)[P(r_a,s_b|a,b) - \rho(r_a,s_b|a,b)], \tag{45}$$

and $\rho(r,s|a,b)$ is an unnormalized local distribution, associated to a function $\chi(\mathbf{r},\mathbf{s})$ with support in Ω. This bound becomes equal to F_Ω^{min} in the limit of ρ equal to the minimizer of Problem 2. Equation (43) gives the condition

$$\alpha^2 > RS\beta^2 + \tag{46}$$
$$2\sum_{rsab}[\beta - g(r,s;a,b)]\,\rho(r,s|a,b)W(a,b),$$

where $g(r,s;a,b) = P(r,s|a,b) - \rho(r,s|a,b)$ and $\rho(r,s|a,b)$ is the local distribution computed in Step 3. If this condition is satisfied by the numerical solution found in Step 3, then the series $F_0^{min}, F_1^{min}, \ldots$ is monotonically decreasing. As we will see, to prove that the series converges to the minimizer of Problem 1, we need the stronger condition

$$\gamma\alpha^2 \ge RS\beta^2 + \tag{47}$$
$$2\sum_{rsab}[\beta - g(r,s;a,b)]\,\rho(r,s|a,b)W(a,b),$$

where γ is any fixed real number in the interval $(0,1)$. A possible choice is $\gamma = 1/2$. If this inequality is satisfied in each iteration of Algorithm 1, the sequence $F_0^{min}, F_1^{min}, \ldots$ satisfies the inequality

$$F_{n+1}^{min} \le F_n^{min} - \frac{1-\gamma}{2}\alpha_n^2, \tag{48}$$

which turns out to be equal to Equation (32) in the limit $\gamma \to 0$. The right-hand side of Equation (47) goes to zero as ρ approaches the optimizer, as implied by the optimality conditions of Problem 2. Thus, if the set Ω does not contain all the points where χ^{min} is different from zero, then the inequality is surely satisfied at some point of the iteration solving Problem 2, as α tends to a strictly positive number. When the inequality is satisfied, the minimization at Step 3 of Algorithm 1 is terminated. If Ω is the support of χ^{min}, the inequality will never be satisfied and the minimization at Step 3 will terminate when the desired accuracy on F^{min} is reached.

5.6. Cleaning Up (Step 7)

As previously said, we should also guarantee that the sets Ω_n contain only points associated with linearly independent vertices. This is granted if the procedure in Step 7 of Algorithm 1 successfully removes the points where the exact minimizer χ_n^{min} is equal to zero. How can we find the support of the minimizer from the approximate numerical solution computed in Step 3? Using dual theory, it is possible to prove the following.

Theorem 1. *Let $\chi(\mathbf{r}, \mathbf{s})$ be a non-negative function with support in Ω and $\rho(r, s|a, b)$ be the associated unnormalized local distribution. Then, the inequality*

$$\sum_{a,b} \rho_\Omega^{min}(r_a, s_b|a, b)W(a, b) \geq \sum_{a,b} \rho(r_a, s_b|a, b)W(a, b)$$
$$- \left[2\left(F^{(+)} - F_\Omega^{(-)}\right)\right]^{1/2} \quad (49)$$

holds.

A direct consequence of this theorem and the slackness condition (23) for optimality is the following.

Corollary 1. *Let $\chi(\mathbf{r}, \mathbf{s})$ be a non-negative function with support in Ω and $\rho(r, s|a, b)$ the associated unnormalized local distribution. If the inequality*

$$\sum_{ab} g(r_a, s_b; a, b) \leq \{RS\beta^2 +$$
$$2\sum_{rsab} [\beta - g(r, s; a, b)]\, \rho(r, s|a, b)W(a, b)\}^{1/2} \quad (50)$$

holds, with $g(r, s|a, b) = P(r, s|a, b) - \rho(r, s|a, b)$, then $\chi_\Omega^{min}(\mathbf{r}, \mathbf{s})$ is equal to zero.

Condition (50) is sufficient for having $\chi_\Omega^{min}(\mathbf{r}, \mathbf{s})$ equal to zero, but it is not necessary. A necessary condition can be derived by computing the lowest eigenvalue of the Hessian of the objective function $F[\chi]$. Both the necessary and sufficient conditions allow us to determine the support of the minimizer χ_Ω^{min} once the distribution χ is enough close to χ_Ω^{min}. Thus, the minimization in Step 3 should not stop until each sequence $(\mathbf{r}, \mathbf{s})$ satisfies the sufficient condition or does not satisfy the necessary condition, otherwise the cleaning up could miss some points where the minimizer is equal to zero. However, numerical experiments show that the use of these conditions is not necessary, and the number of elements in the sets Ω_n is generally bounded by $d_{NS} + 1$, provided that Problem 2 is solved by the algorithm described in the following section.

5.7. Solving Problem 2

There are standard methods for solving Problem 2, and numerical libraries are available. The interior point method [19] provides a quadratic convergence to the solution, meaning that the number of digits of accuracy is almost doubled at each iteration step, once χ is sufficiently close to the minimizer. The algorithm uses the Newton method and needs to solve a set of linear equations. Since this can be computationally demanding in terms of memory, we have implemented the solver by using the conjugate gradient method, which does not use the Hessian. Furthermore, if the Hessian turns out to have a small condition number, the conjugate gradient method can be much more efficient than the Newton method, especially if we do not need to solve Problem 2 with high accuracy. This is the case in the initial stage of the computation, when the set Ω is growing and does not contain all the points of the support of χ^{min}.

The conjugate gradient method iteratively performs a one-dimensional minimization, along directions that are conjugate with respect to the Hessian of the objective function [19]. The directions are computed iteratively, by setting the first direction equal to the gradient of the objective function. The conjugate gradient method is generally used with unconstrained problems, whereas Problem 2 has the inequality constraints $\chi(\mathbf{r}, \mathbf{s}) \geq 0$. To adapt the method to our problem, we perform the one-dimensional minimization in the region where χ is non-negative. Whenever an inactive constraint becomes active, or vice versa, we set the search direction equal to the gradient and restart the generation of the directions from that point. Once the procedure terminates, the algorithm provides a list of active constraints with $\chi_n(\mathbf{r}, \mathbf{s}) = 0$. Numerical simulations show that this list is generally complete, and corresponds to the points where the minimizer χ_n^{min} is equal to zero.

In general, the slackness condition (25) is not satisfied by the numerical solution. However, as previously pointed out, we can suitably normalize χ_n so that this condition is satisfied by $\rho_n(r, s|a, b)$. Thus, we will assume that the equality

$$\sum_{r,s,a,b} \rho_n(r, s|a, b) g_n(r, s; a, b) W(a, b) = 0 \tag{51}$$

holds with $g_n = P - \rho_n$. This also implies that

$$\begin{aligned} \alpha_n &= \alpha|_{\rho=\rho_n} \geq 0 \text{ and} \\ \beta_n &= \beta|_{\rho=\rho_n} \geq 0. \end{aligned} \tag{52}$$

6. Convergence Analysis and Computational Cost

Here, we provide a convergence analysis and show that the error on the distance from the local polytope is bounded above by a function decaying at least as fast as $1/n$, where n is the number of iterations. The convergence of this function to zero is sublinear, but its derivation relies on a very rough estimate of a lower bound on the optimal value χ^{min}. Actually, the iteration converges to the solution in a finite number of steps (up to the accuracy of the solver of Problem 2). Indeed, since the number of vertices is finite, also the number of their sets Ω is finite. Thus, the sequence Ω_n converges to the support of the optimizer χ^{min} in a finite number of steps, as the accuracy goes to zero.

We expect that this finite number of steps is of the order of the dimension d_{NS} of the local polytope. Interestingly, the computed bound on the number of required iterations for given error does not depend on the number of measurements. Using this bound, we show that the computational cost for any given error on the distance grows polynomially with the size of the problem input; that is, with A, B, R, and S, provided that the oracle can be simulated in polynomial time.

To prove the convergence, we need to introduce the dual form of Problem 1 (see Ref. [19] for an introduction to dual theory). The dual form of a minimization problem (primal problem) is a maximization problem, whose maximum is always smaller than or equal to the primal minimum, the difference being called the *duality gap*. However, if the constraints of the primal problem satisfy some mild conditions, such as Slater's conditions [19], then the duality gap is equal to zero. As previously said, this is the case of Problem 1.

The dual form is particularly useful for evaluating lower bounds on the optimal value of the primal problem. Indeed, the value taken by the dual objective function in a feasible point of the dual constraints provides such a bound. After introducing the dual form of Problem 1, we derive the lower bound $F^{(-)}$ on F^{min}, given by Equation (37). Then, we use this bound and Equation (48) to prove the convergence.

6.1. Dual Problem

The dual problem of Problem 1 is a maximization problem over the space of values taken by the Lagrange multipliers $\lambda(\mathbf{r}, \mathbf{s})$ subject to the dual constraints $\lambda(\mathbf{r}, \mathbf{s}) \geq 0$. The dual objective function is given by the minimum of the Lagrangian $\mathcal{L}$, defined by Equation (15), with respect to χ. The dual constraint is the non-negativity of the Lagrange multipliers, that is,

$$\lambda(\mathbf{r}, \mathbf{s}) \geq 0. \tag{53}$$

As this minimum cannot be derived analytically, a standard strategy for getting an explicit form of the dual objective function is to enlarge the space of primal variables and, correspondingly, to increase the number of primal constraints. The minimum is then evaluated over the enlarged space. In our case, it is convenient to introduce Equation (9) and $\rho(r, s|a, b)$ as additional constraints and variables, respectively. Thus, F is made independent of χ and expressed as a function of ρ. The new optimization problem, which is equivalent to Problem 1, has Lagrangian

$$\mathcal{L} = F[\rho] - \sum_{\mathbf{r},\mathbf{s}} \lambda(\mathbf{r},\mathbf{s})\chi(\mathbf{r},\mathbf{s}) + \sum_{rsab} W(a,b) \times \eta(r,s,a,b)\left[\rho(r,s|a,b) - \sum_{\mathbf{r},\mathbf{s}} \delta_{r,r_a}\delta_{s,s_b}\chi(\mathbf{r},\mathbf{s})\right], \tag{54}$$

where $\eta(r,s,a,b)$ are the Lagrange multipliers associated with the added constraints. To find the minimum of the Lagrangian, we set its derivative, with respect to the primal variables χ and ρ, equal to zero. We get the equations

$$\sum_{a,b} W(a,b)\eta(r_a,s_b,a,b) = -\lambda(\mathbf{r},\mathbf{s}) \tag{55}$$

$$\rho(r,s|a,b) = P(r,s|a,b) - \eta(r,s,a,b). \tag{56}$$

The first equation does not depend on the primal variables and sets a constraint on the dual variables. If this constraint is not satisfied, the dual objective function is equal to $-\infty$. Thus, its maximum is in the region where Equation (55) is satisfied. Let us add it to the dual constraint (53). The second stationarity condition, Equation (56), gives the optimal ρ. By replacing it in the Lagrangian, we get the dual objective function

$$F_{dual} = \sum_{r,s,a,b} W(a,b)\eta(r,s,a,b) \times \left[P(r,s|a,b) - \frac{\eta(r,s,a,b)}{2}\right]. \tag{57}$$

Eliminating λ, which does not appear in the objective function, the dual constraints (53) and (55) give the inequality

$$\sum_{a,b} W(a,b)\eta(r_a,s_b;a,b) \leq 0. \tag{58}$$

Thus, Problem 1 is equivalent to the following.

Problem 3 (dual problem of Problem 1).

$$\max_\eta F_{dual}[\eta]$$
$$\text{subject to the constraints}$$
$$\sum_{a,b} W(a,b)\eta(r_a,s_b;a,b) \leq 0.$$

The value taken by F_{dual} at a feasible point provides a lower bound on F^{min}. Given any function $\bar{\eta}(r,s;a,b)$, a feasible point is

$$\eta_f(r,s;a,b) \equiv \bar{\eta}(r,s;a,b) - \max_{\mathbf{r},\mathbf{s}} \sum_{\bar{a},\bar{b}} W(\bar{a},\bar{b})\bar{\eta}(r_{\bar{a}},s_{\bar{b}};\bar{a},\bar{b}). \tag{59}$$

Indeed,

$$\sum_{a,b} \eta_f(r_a,s_b;a,b)W(a,b) = \sum_{a,b} \bar{\eta}(r_a,s_b;a,b)$$
$$- \max_{\mathbf{r}',\mathbf{s}'} \sum_{a,b} \bar{\eta}(r'_a,s'_b;a,b)W(a,b) \leq 0. \tag{60}$$

The lower bound turns out to be the optimal value F^{min}, if the distribution $\rho(r,s|a,b)$ given by Equation (56) in terms of $\eta = \eta_f$ is solution of the primal Problem 1. This suggests the transformation

$$\bar{\eta}(r,s;a,b) = P(r,s|a,b) - \rho(r,s|a,b), \tag{61}$$

where $\rho(r,s|a,b)$ is some local distribution up to a normalization constant (in fact, ρ can be any real function). Every local distribution induces a lower bound on the optimal value F^{min}. This lower bound turns out to be an accurate approximation of F^{min} if ρ is close enough to the optimal local distribution. Using the last equation and Equation (59), we get the lower bound (37) from F_{dual}.

The dual problem of Problem 2 is similar to Problem 3, but the constraints have to hold for sequences $(\mathbf{r}, \mathbf{s})$ in Ω.

Problem 4 (dual problem of Problem 2).

$$\max_{\eta} F_{dual}[\eta]$$
$$\text{subject to the constraints}$$
$$(\mathbf{r}, \mathbf{s}) \in \Omega \Rightarrow \sum_{a,b} W(a,b)\eta(r_a, s_b; a, b) \leq 0.$$

This dual problem induces the lower bound F_{Ω}^{min} on the optimal value of Problem 2 (Equation (44)).

6.2. Convergence and Polynomial Cost

Let $\rho_n(r,s|a,b)$ be the local distribution computed in Step 3 of Algorithm 1. From the lower bound (37), we have

$$F^{min} \geq F_n - \frac{RS}{2}\alpha_n^2 + \sum_{r,s,a,b} W(a,b)\rho_n(r,s|a,b)\left[g_n(r,s;a,b) - \alpha_n\right], \tag{62}$$

where α_n is given by Equation (30), and $g_n = P - \rho_n$. The part of the summation linear in g_n is equal to zero, by Equation (51). The remaining part, linear in α_n, is bounded from below by $-\alpha_n[1 + (RS)^{1/2}]$ (α_n is positive). This can be shown by minimizing it under the constraint (51). Thus, we have that

$$F^{min} \geq F_n - \frac{RS}{2}\alpha_n^2 - [1 + (RS)^{1/2}]\alpha_n. \tag{63}$$

As α_n is not greater than 1, the factor α_n^2 in the right-hand side of the inequality can be replaced by α_n, so that we have

$$\alpha_n \geq 2\frac{F_n - F^{min}}{RS + 2 + 2(RS)^{1/2}}, \tag{64}$$

which gives, with Equation (48), the following

$$F_n^{min} - F_{n+1}^{min} \geq 2(1 - \gamma)\left(\frac{F_n - F^{min}}{RS + 2 + 2(RS)^{1/2}}\right)^2. \tag{65}$$

This inequality implies that

$$F_n^{min} - F^{min} \leq \frac{(RS + 2 + 2(RS)^{1/2})^2}{2(1 - \gamma)n}. \tag{66}$$

This can be proved by induction. It is easy to prove that inequality holds for $n = n_0 > 1$, if it holds for $n = n_0 - 1$. Let us prove that it holds for $n = 1$. It is sufficient to prove that $F_1^{min} - F^{min} \leq 1/2$. Using the identity

$$\sum_{r,s,a,b} W(a,b)\rho_1^{min}(r,s|a,b) \times \left[P(r,s|a,b) - \rho_1^{min}(r,s|a,b)\right] = 0, \tag{67}$$

we have

$$\begin{aligned}
F_1^{min} - F^{min} &\leq F_1^{min} = \\
&\sum_{r,s,a,b} W(a,b)\frac{[P(r,s|a,b) - \rho_1^{min}(r,s|a,b)]^2}{2} \\
&= \sum_{r,s,a,b} W(a,b)\frac{P^2(r,s|a,b) - (\rho_1^{min})^2(r,s|a,b)}{2} \\
&\leq \sum_{r,s,a,b} W(a,b)\frac{P^2(r,s|a,b)}{2} \leq \frac{1}{2}.
\end{aligned} \tag{68}$$

Thus, the error decreases at least as fast as $1/n$. Although the convergence of the upper bound is sublinear, we derived this inequality by using Equation (63), which provides a quite loose bound on the optimal value χ^{min}. Nonetheless, the constraint set by Equation (66) on the accuracy is strong enough to imply the polynomial convergence of the algorithm, provided that the oracle can be simulated in polynomial time. Indeed, the inequality implies that the number of steps required to reach a given accuracy does not grow faster than $(RS)^2$. Since the computational cost of completing each step is polynomial, the overall algorithm has polynomial cost. More precisely, each step is completed by solving a quadratic minimization problem. If we do not rely on the specific structure of the quadratic problem, its computational cost does not grow faster than $\max\{n_1^3, n_1^2 n_2, D\}$ [19], where n_1, n_2, and D are the number of variables, the number of constraints, and the cost of evaluating the first and second derivatives of the objective and constraint functions. The numbers n_1 and n_2 are equal, and D is equal to $n_1^2(A + B)$. As the number of vertices in the set Ω_n is not greater than the number of iterations (say, $\bar{n}$), we have that $n_1 \leq \bar{n}$. Furthermore, the number of vertices cannot be greater than d_{NS}. Thus, the number of variables is, in the worst case,

$$n_1 = \min\{\bar{n}, ABRS\}. \tag{69}$$

As implied by Equation (66), about $(RS)^2/\epsilon$ iterations are sufficient for reaching an error not greater than ϵ. Let us set $\bar{n} = (RS)^2/\epsilon$. Denoting the computational cost of Algorithm 1 with accuracy ϵ by C_ϵ, we have that

$$C_\epsilon \leq K\bar{n}\max\{n_1^3, n_1^2(A+B)\} = K\frac{(RS)^2}{\epsilon}n_1^2\max\{n_1, A+B\}, \tag{70}$$

where K is some constant. Let us consider the two limiting cases with $\epsilon(A + B) \geq (RS)^2$ (high number of measurements) and $\epsilon ABRS \leq (RS)^2$ (high accuracy).

In the first case, we have that $A + B \geq \bar{n}$, which also implies that $n_1 = \bar{n}$ (there are at least 2 measurements per party). We have

$$\epsilon(A + B) \geq (RS)^2 \implies C_\epsilon \leq K\frac{(RS)^6}{\epsilon^3}(A + B) \equiv \mathcal{B}_0 \tag{71}$$

Thus, given a fixed error, the computational cost is asymptotically linear in the number of measurements. For $\epsilon = 10^{-2}$ and $R = S = 2$, this bound holds for a number of measurements per party greater than 800. If $A = B = 800$, the computation ends in few hours in the worst case by using available personal computers, provided that the bound $\mathcal{B}_0$ is saturated in the most pessimistic scenario.

In the second case, we have that $ABRS \leq \bar{n}$ and $n_1 = ABRS$. Thus,

$$\epsilon AB \leq RS \implies C_\epsilon \leq K\frac{(RS)^2}{\epsilon}(ABRS)^3 \equiv \mathcal{B}_1. \tag{72}$$

Thus, for a fixed error ϵ and AB smaller than RS/ϵ, the bound on the computational cost scales as the third power of the product AB; that is, the sixth power of the number of measurments, provided that $A = B$. This scaling is in good agreement with the numerical tests, as discussed later. However the tests indicate that the scaling $1/\epsilon$ and, thus, the sublinear convergence is too pessimistic. For example, for $\epsilon = 10^{-3}$, $A = B \leq 40$, and $R = S = 2$, the bound gives a running time of the order of months, whereas the running time in the tests turns out to be less than one hour.

6.3. Simulation of the Oracle

We have shown that the cost of computing the distance from the local polytope grows polynomially, provided that we have access to the oracle. But what is the computational complexity of the oracle? In the case of measurements with two outcomes, we have seen that the solution of the oracle is equivalent to finding the minimal energy of a particular class of Ising spin glasses. These problems are known to be NP-hard. However, the oracle has a particular structure that can make many physically relevant instances numerically tractable. For example, the couplings of the Ising spin model are constrained by the nonsignaling conditions on $P(r, s|a, b)$ and the optimality conditions (22)–(24). Furthermore, the Hamiltonian (14) is characterized by two classes of spins, described by the variables r_k and s_k, respectively, and each element in one class is coupled only to elements in the other class. This particular structure suggests the following block-maximization algorithm for solving the oracle.

Algorithm 2. *Input:* $g(r, s; a, b)$

1. *Generate a random sequence* **r**.
2. *Maximize $\sum_{a,b} g(r_a, s_b; a, b)W(a, b)$ with respect to the sequence **s** (see later discussion).*
3. *Maximize $\sum_{a,b} g(r_a, s_b; a, b)W(a, b)$ with respect to the sequence **r**.*
4. *Repeat from Step 2 until the block-maximizations stop making progress.*

Numerical tests show that this algorithm, when used for computing the distance from the local polytope, stops after a few iterations. Furthermore, only a few trials of the initial random sequence **r** are required for convergence of Algorithm 1. We also note that the probability of a successful simulation of the oracle increases when χ is close to the optimal solution χ^{min}, suggesting that the optimality conditions (22)–(24) play some role in the computational complexity of the oracle. Pragmatically, we have chosen the number of trials equal to d_{NS}, such that the computational cost of simulating the oracle contributes to the overall running time with a constant multiplicative factor and, thus, the sixth-power law of the oracle-assisted algorithm is not affected.

Before discussing the numerical results, let us explain how the maximization on blocks is performed. Let us consider the maximization with respect to **r**, as the optimization with respect to **s** has an identical procedure. We have

$$\max_r \sum_{a,b} W(a, b)g(r_a, s_b; a, b) =$$
$$\sum_a \max_r \sum_b g(r, s_b; a, b)W(a, b) \equiv \qquad (73)$$
$$\sum_a \max_r \tilde{g}(r, \mathbf{s}; a).$$

Thus, the maximum is found by maximizing the function $\tilde{g}(r, \mathbf{s}; a)$, with respect to the discrete variable r for every a. Taking into account the sum over b required for generating $\tilde{g}$, the computational cost of the block-maximization is proportional to RAB. Thus, it does not grow more than linearly with respect to the size of the problem input; that is, $RSAB$.

7. Numerical Tests

In the previous sections, we introduced an algorithm that computes the distance from the local polytope in polynomial time, provided that we have access to oracle Max. Surprisingly, in every simulation performed on entangled qubits, the algorithm implementing the oracle successfully finds the solution in polynomial time. More precisely, the algorithm finds a sequence $(\mathbf{r}, \mathbf{s})$ sufficiently close to the maximum to guarantee convergence of Algorithm 1 to the solution of Problem 1. Interestingly, the probability of a successful simulation of the oracle increases as χ approaches the solution. This suggests that the optimality conditions (22)–(24) play a fundamental role in the computational complexity of the oracle. To check that the algorithm successfully finds the optimizer χ^{min} up to the desired accuracy, we have solved the oracle with a brute-force search at the end of the computation, whenever this was possible in a reasonable time. All of the checks show that the solution is found within the desired accuracy.

In the tests, we considered the case of maximally entangled states, Werner states, and pure non-maximally entangled states. The numerical data are compatible with a running time scaling as the sixth power of the number of measurements. This is in accordance with the theoretical analysis, given in Section 6.2. Furthermore, the simulations show that the sublinear convergence of the upper bound $\mathcal{B}_1$ on the error is very loose, and the convergence turns out to be much faster. Let us discuss the case of entangled qubits in a pure quantum state.

7.1. Maximally Entangled State

In Figure 1, we report the time required for computing the distance from the local polytope as a function of the number of measurements, M, in log-log scale. The distance has been evaluated with accuracy equal to 10^{-3}, 10^{-4}, and 10^{-5} (red, blue, and green points, respectively). We have considered the case of planar measurements on the Bloch sphere. For the sake of comparison, we have also plotted the functions $10^{-6}M^6$ and $10^{-9}M^6$ (dashed lines). The data are compatible with the theoretical power law derived previously. They also show that the sublinear convergence of $\mathcal{B}_1$, derived in Section 6.2, is too pessimistic and the algorithm actually shows better performances. In particular, the bound $\mathcal{B}_1$ says that the running time is not greater than years for $A = B = 40$ and $\epsilon = 10^{-5}$, whereas the observed running time is actually less than one hour. Other simulations have been performed with random measurements. We generated a set of measurements corresponding to random vectors on the Bloch sphere, by considering both the planar and non-planar case. Then, we computed the distance from the local polytope for a different number of measurements. We always observed that the running time scales with the same sixth power law. For a number of measurements below 28, we have solved the oracle with a brute-force search at the end of the computation, and we have always found that Algorithm 1 successfully converged to the solution within the desired accuracy.

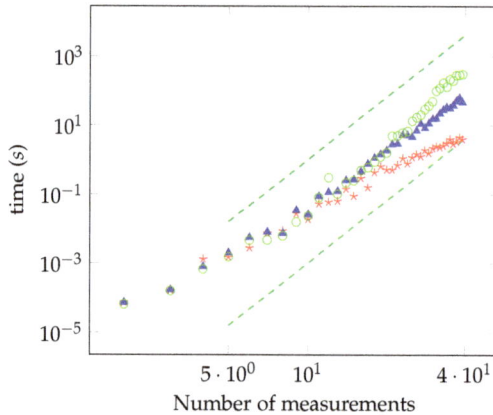

Figure 1. Time required for computing the distance from the local polytope for a maximally entangled state as a function of the number of measurements (log-log scale) with accuracy equal to 10^{-3}, 10^{-4}, and 10^{-5} (red, blue, and green points, respectively).

7.2. Non-Maximally Entangled State

In the case of the non-maximally entangled state

$$|\psi\rangle = \frac{|00\rangle + \gamma|11\rangle}{\sqrt{1 + \gamma^2}}, \tag{74}$$

with $\gamma \in [0, 1]$, we have considered planar measurements orthogonal to the Bloch vector $\vec{v}_z \equiv (0, 0, 1)$ (such that the marginal distributions are unbiased), as well as planar measurements lying in the plane containing $\vec{v}_z$ (biased marginal distributions).

In Figure 2, we report the distance from the local polytope as a function of γ with 10 measurements. The distance changes slightly for higher numbers of measurements. In the unbiased case, the distance goes to zero for γ equal to about 0.4, whereas the correlations become local for $\gamma = 0$ in the biased case.

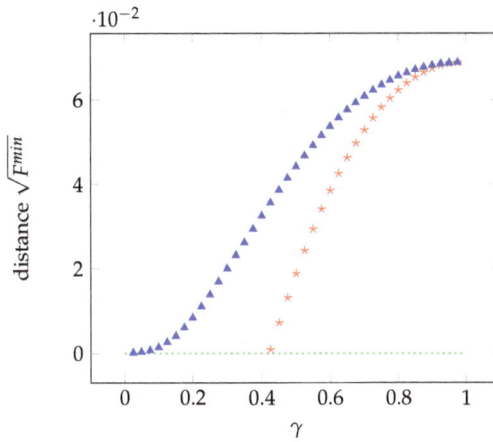

Figure 2. Distance from the local polytope as a function of γ in the unbiased case (red stars) and biased case (blue triangles).

In Figures 3 and 4, the running time as a function of the number of measurements is reported for the unbiased and biased cases, respectively. The power law is, again, in accordance with the theoretical analysis. As done for the maximally entangled case, we have checked the convergence to the solution by solving the oracle with a brute force search for a number of measurements up to 28.

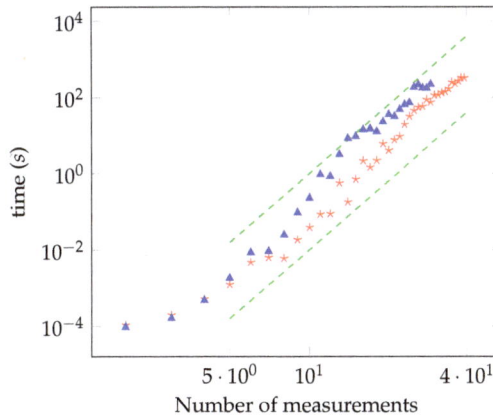

Figure 3. Time required for computing the distance from the local polytope as a function of the number of measurements (log-log scale) in the unbiased case, for $\gamma = 0.8$ (red stars) and $\gamma = 0.6$ (blue triangles). The green lines are the functions $10^{-6} M^6$ and $10^{-8} M^6$. The accuracy is 10^{-5}.

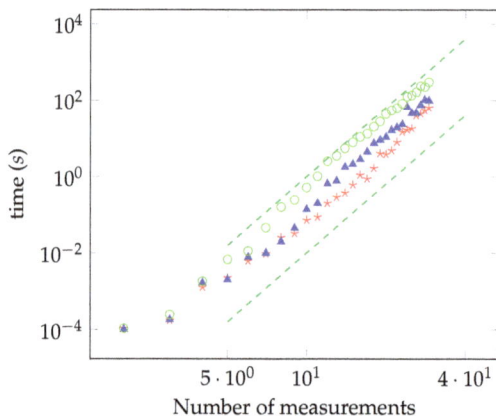

Figure 4. The same as Figure 3 in the biased case, for $\gamma = 0.8$ (red stars), $\gamma = 0.6$ (blue triangles), and $\gamma = 0.4$ (green circles).

8. Conclusions

In conclusion, we have presented an algorithm that computes the distance of a given non-signaling box to the local polytope. The running time, with given arbitrary accuracy, scaled polynomially, granted the access to an oracle determining the optimal locality bound of a Bell inequality. We also proposed an algorithm for simulating the oracle. In all of the numerical tests, the overall algorithm successfully computed the distance with the desired accuracy and a scaling of the running time, in agreement with the bound theoretically derived for the oracle-assisted algorithm. Our algorithm opens the way to tackle many unsolved problems in quantum theory, such as the non-locality of Werner states. Since the non-locality problem is NP-hard, our work and its further refinements could provide alternative algorithms to solve some instances of computationally hard problems.

Author Contributions: Conceptualization, formal analysis, investigation and software, A.M.; draft preparation, supervision, project administration, funding acquisition, A.M. and S.W.

Funding: This research was funded by Hasler Stiftung, grant number 16057, and Swiss National Science Foundation.

Acknowledgments: We wish to thank Arne Hansen for valuable comments and suggestions. This work is supported by the Swiss National Science Foundation, the NCCR QSIT, and the Hasler foundation through the project "Information-Theoretic Analysis of Experimental Qudit Correlations".

Conflicts of Interest: The authors declare no conflict of interest.

References

1. Bell, J. On the einstein podolsky rosen paradox. *Physics* **1964**, *1*, 195. [CrossRef]
2. Barrett, J.; Hardy, L.; Kent, A. No signaling and quantum key distribution. *Phys. Rev. Lett.* **2005**, *95*, 010503. [CrossRef] [PubMed]
3. Acín, A.; Gisin, N.; Masanes, L. From Bell's theorem to secure quantum key distribution. *Phys. Rev. Lett.* **2006**, *97*, 120405. [CrossRef] [PubMed]
4. Scarani, V.; Gisin, N.; Brunner, N.; Masanes, L.; Pino, S.; Acín, A. Secrecy extraction from no-signaling correlations. *Phys. Rev. A* **2006**, *74*, 042339. [CrossRef]
5. Acín, A.; Massar, S.; Pironio, S. Efficient quantum key distribution secure against no-signalling eavesdroppers. *New J. Phys.* **2006**, *8*, 126. [CrossRef]
6. Acín, A.; Brunner, N.; Gisin, N.; Massar, S.; Pironio, S.; Scarani, V. Device-independent security of quantum cryptography against collective attacks. *Phys. Rev. Lett.* **2007**, *98*, 230501. [CrossRef] [PubMed]
7. Masanes, L.; Renner, R.; Christandl, M.; Winter, A.; Barrett, J. Full security of quantum key distribution from no-signaling constraints. *IEEE Trans. Inf. Theory* **2014**, *60*, 4973–4986. [CrossRef]

8. Hänggi, E.; Renner, R.; Wolf, S. The impossibility of non-signaling privacy amplification. *Theor. Comput. Sci.* **2013**, *486*, 27–42. [CrossRef]
9. Colbeck, R.; Renner, R. Free randomness can be amplified. *Nat. Phys.* **2012**, *8*, 450. [CrossRef]
10. Gallego, R.; Masanes, L.; de la Torre, G.; Dhara, C.; Aolita, L.; Acín, A. Full randomness from arbitrarily deterministic events. *Nat. Commun.* **2013**, *4*, 2654. [CrossRef] [PubMed]
11. Buhrman, H.; Cleve, R.; Massar, S.; de Wolf, R. Nonlocality and communication complexity. *Rev. Mod. Phys.* **2010**, *82*, 665. [CrossRef]
12. Pitowski, I. *Quantum Probability—Quantum Logic*; Academic Press: London, UK, 1989.
13. Brierley, S.; Navascues, M.; Vertesi, T. Convex separation from convex optimization for large-scale problems. *arXiv* **2016**, arXiv:1609.05011.
14. Montina, A.; Wolf, S. Can non-local correlations be discriminated in polynomial time? *arXiv* **2016**, arXiv:1609.06269v1.
15. Fine, A. Hidden variables, joint probability, and the Bell inequalities. *Phys. Rev. Lett.* **1982**, *48*, 291. [CrossRef]
16. Collins, D.; Gisin, N. A relevant two qubit Bell inequality inequivalent to the CHSH inequality. *J. Phys. A Math. Theor.* **2004**, *37*, 1775. [CrossRef]
17. Montina, A.; Wolf, S. Information-based measure of nonlocality. *New J. Phys.* **2016**, *18*, 013035. [CrossRef]
18. Bernhard, C.; Bessire, B.; Montina, A.; Pfaffhauser, M.; Stefanov, A.; Wolf, S. Non-locality of experimental qutrit pairs. *J. Phys. A: Math. Theor.* **2014**, *42*, 424013. [CrossRef]
19. Boyd, S.; Vandenberghe, L. *Convex Optimization*; Cambridge University Press: Cambridge, UK, 2004.

entropy

MDPI

Article

Bounding the Plausibility of Physical Theories in a Device-Independent Setting via Hypothesis Testing

Yeong-Cherng Liang [1,*] and **Yanbao Zhang** [2,*]

[1]   Department of Physics and Center for Quantum Frontiers of Research & Technology (QFort), National Cheng Kung University, Tainan 701, Taiwan
[2]   NTT Basic Research Laboratories and NTT Research Center for Theoretical Quantum Physics, NTT Corporation, 3-1 Morinosato-Wakamiya, Atsugi, Kanagawa 243-0198, Japan
*   Correspondence: ycliang@mail.ncku.edu.tw (Y.-C.L.); yanbaoz@gmail.com (Y.Z.)

Received: 15 December 2018; Accepted: 12 February 2019; Published: 15 February 2019

Abstract: The device-independent approach to physics is one where conclusions about physical systems (and hence of Nature) are drawn directly and solely from the observed correlations between measurement outcomes. This operational approach to physics arose as a byproduct of Bell's seminal work to distinguish, via a Bell test, quantum correlations from the set of correlations allowed by local-hidden-variable theories. In practice, since one can only perform a finite number of experimental trials, deciding whether an empirical observation is compatible with some class of physical theories will have to be carried out via the task of hypothesis testing. In this paper, we show that the prediction-based-ratio method—initially developed for performing a hypothesis test of local-hidden-variable theories—can equally well be applied to test many other classes of physical theories, such as those constrained only by the nonsignaling principle, and those that are constrained to produce any of the outer approximation to the quantum set of correlations due to Navascués-Pironio-Acín. We numerically simulate Bell tests using hypothetical nonlocal sources of correlations to illustrate the applicability of the method in both the independent and identically distributed (i.i.d.) scenario and the non-i.i.d. scenario. As a further application, we demonstrate how this method allows us to unveil an apparent violation of the nonsignaling conditions in certain experimental data collected in a Bell test. This, in turn, highlights the importance of the randomization of measurement settings, as well as a consistency check of the nonsignaling conditions in a Bell test.

Keywords: quantum nonlocality; Bell test; device-independent; *p*-value; hypothesis testing; nonsignaling

---

1. Introduction

In physics, the terminology "device-independent" apparently made its first appearance in Ref. [1] where the authors drew a connection between the celebrated discovery by Bell [2] and the vibrant field of quantum cryptography [3]. As of today, device-independent quantum information has become a well-established research area where Bell-inequality-violating correlations find applications not only in the distribution of secret keys [4–6] (see also Ref. [7]), but also in the generation of random bits [8–10], as well as in the assessment of uncharacterized devices (see, e.g., Refs. [11–17]). For a comprehensive review, see Refs. [18,19].

Entropy **2019**, *21*, 185; doi:10.3390/e21020185   www.mdpi.com/journal/entropy

A device-independent approach to physics, however, could be traced back, for example, to the work of Bell [2]. There, he showed that *any* local-hidden-variable (LHV) theory [20] must be incompatible with certain quantum predictions. The proof is "device-independent" in the sense that one needs no further assumption about the nature of the theory (including the detailed functioning of any devices that one may use to test the theory). Rather, the proof relies on a common ingredient of operational physical theories—correlations between measurement outcomes, i.e., the probability of getting particular measurement outcomes conditioned on certain measurement choices being made—to manifest the incompatibility.

By now, this incompatibility has been verified in various loophole-free Bell tests, such as those reported in Refs. [21–25]. Importantly, any real experiments must involve only a finite number of experimental trials. Statistical fluctuations must thus be carefully taken into account in order to draw any conclusion against a hypothetical theory, such as an LHV theory. For example, using the observed relative frequencies as a naïve estimator of the underlying correlations would generically (see, e.g., Refs. [26,27]) lead to a violation of the nonsignaling conditions [28,29]. Since the assumption of nonsignaling is a prerequisite for any Bell tests, it is only natural that a Bell test of LHV theories must also be accompanied by the corresponding test of this assumption [22–25,30] (see also Refs. [31–33]).

The effects of statistical fluctuations in a Bell test were (in fact, still are) often reported in terms of the number of standard deviations the estimated Bell violation exceeds the corresponding local bound (see, e.g., Refs. [34–42]). However, there are several problems with such a statement (see Refs. [19,43] for detailed discussions). Alternatively, as a common practice in hypothesis testing, one could also present the *p*-value according to a certain null hypothesis (e.g., the hypothesis that a LHV theory holds true). The corresponding *p*-value then describes the probability that the statistical model (associated with the null hypothesis) produces some quantity (e.g., the amount of Bell-inequality violation) at least as extreme as that observed.

A pioneering work in this regard is that due to Gill [44] where he presented a *p*-value upper bound according to the hypothesis of a LHV theory based on the violation of the Clauser-Horne-Shimony-Holt (CHSH) [45] Bell inequality. A few years later, a systematic method that works directly on the observed data (without relying on any predetermined Bell inequality)—by the name of the prediction-based-ratio method—was developed by one of the present authors and coworkers [43] (see also Ref. [46]). This method was designed for computing a *p*-value upper bound—based on the data collected in a Bell test—according to LHV theories. As we shall show in this work, essentially the same method can be applied for the hypothesis testing of some other nonlocal physical theories, thus allowing us to bound the plausibility of physical theories beyond LHV theories.

Indeed, since the pioneering work by Popescu and Rohrlich [28], there has been an ongoing effort (see, e.g., Refs. [47–50]) to find well-motivated physical [51,52] or information-theoretic [53–56] principles to recover precisely the set of quantum correlations. Unfortunately, none of these has succeeded. Rather, they each define a set of correlations that outer approximates the quantum set [57]. In other words, they also contain correlations that are more nonlocal than that allowed by quantum theory. For example, the so-called "almost-quantum" [50] set of correlations is one such superset of the quantum set, yet satisfying essentially all the proposed principle known to date. In the rest of this work, it suffices to think of this set as a fairly good outer approximation to the quantum set of correlations.

In this work, we show that the prediction-based-ratio method can be applied to test any physical theory that is constrained to produce correlations that is amenable to a semidefinite programming [58] characterization. In particular, it can be applied to test any physical theory that is constrained to produce nonsignaling [28] correlations, or any theory that respects macroscopic locality [51] or which gives rise to the almost-quantum [50] set of correlations etc.

2. Methods

2.1. Preliminaries

For a complete description of the prediction-based-ratio method and a comparison of its strength against the martingale-based method [44], we refer the reader to Ref. [43]. Here, we merely recall the necessary ingredients of the prediction-based-ratio method and show how it can be used to achieve the purpose of bounding the plausibility of physical theories based on the data collected in a Bell test, with *minimal* assumptions. Making this possibility evident and demonstrating how well it works in practice are the main contributions of the present work.

For simplicity, the following discussions are based on a Bell test that involves two parties (Alice and Bob) who are each allowed to perform one of two measurements randomly selected at each trial, each produces one of two possible outcomes. Generalization to other Bell scenarios will be evident. To this end, let us denote the measurement choice (input) of Alice (Bob) by x (y) and the corresponding measurement outcome (output) by a (b), where $a, b, x, y \in \{0, 1\}$. The extent to which the distant measurement outcomes are correlated is then succinctly summarized by the collection of joint conditional probability distributions $\vec{P} = \{P(a, b|x, y)\}_{a,b,x,y}$.

In an LHV theory, the outcome probability distributions can be produced with the help of some LHV λ (distributed according to q_λ) via the local response functions satisfying $0 \le P_\lambda^A(a|x), P_\lambda^B(b|y) \le 1$ and $\sum_a P_\lambda^A(a|x) = \sum_b P_\lambda^B(b|y) = 1$ such that [2]:

$$P(a, b|x, y) = \sum_\lambda q_\lambda P_\lambda^A(a|x) P_\lambda^B(b|y). \tag{1}$$

Hereafter, we refer to any $\vec{P}$ that can be decomposed in the above manner as a (Bell-) local correlation and denote the set of such correlations as $\mathcal{L}$.

In contrast, if Alice and Bob conduct the experiment by performing local measurements on some shared quantum state ρ, quantum theory predicts setting-dependent outcome distributions for all a, b, x, y of the form:

$$P(a, b|x, y) = \text{tr}(\rho \, M_{a|x}^A \otimes M_{b|y}^B), \tag{2}$$

where $M_{a|x}^A$ and $M_{b|y}^B$ denote, respectively, the local positive-operator-value-measure element associated with the a-th outcome of Alice's x-th measurement and the b-th outcome of Bob's y-th measurement. Accordingly, we refer to any $\vec{P}$ that can be written in the form of Equation (2) as a quantum correlation and the set of such correlations as $\mathcal{Q}$.

Importantly, both local and quantum correlations satisfy the nonsignaling conditions [29]:

$$\begin{aligned} P_A(a|x, y) &= P_A(a|x, y') := P_A(a|x) \quad \forall \, a, x, y, y', \\ P_B(b|x, y) &= P_B(b|x', y) := P_B(b|y) \quad \forall \, b, x, x', y, \end{aligned} \tag{3}$$

where $P_A(a|x, y) := \sum_b P(a, b|x, y)$ and $P_B(b|x, y) := \sum_a P(a, b|x, y)$ are marginal probability distributions of $P(a, b|x, y)$. Should (any of) these conditions be violated in a way that is independent of spatial separation, Alice and Bob would be able to communicate faster-than-light [28] via the choice of measurement x, y. We shall denote the set of $\vec{P}$ satisfying Equation (3) as $\mathcal{NS}$. It is known that $\mathcal{L}, \mathcal{Q}$, and $\mathcal{NS}$ are convex sets and that they satisfy the strict inclusion relations $\mathcal{L} \subset \mathcal{Q} \subset \mathcal{NS}$ (see, e.g., Ref. [19] and references therein).

A few other convex sets of correlations are worth mentioning for the purpose of subsequent discussions. To this end, note that the problem of deciding if a given $\vec{P}$ is in $\mathcal{Q}$ is generally a difficult problem. However, the characterization of $\mathcal{Q}$ can, in principle, be achieved by solving a converging hierarchy of semidefinite programs [58] due to Nacascués, Pironio, and Acín (NPA) [59,60] (see also Ref. [61,62]). The lowest level outer approximation of $\mathcal{Q}$ in this hierarchy, often denoted by $\mathcal{Q}_1 \supset \mathcal{Q}$, happens to be exactly the set of correlations that is characterized by the physical principle of macroscopic locality [51]. A finer outer approximation of $\mathcal{Q}$ corresponding to the lowest-level hierarchy of Ref. [62], which we denote by $\tilde{\mathcal{Q}}$, is known in the literature as the almost-quantum set [50], as it appears to satisfy all the physical principles that have been proposed to characterize $\mathcal{Q}$. In Section 3, we use $\tilde{\mathcal{Q}}$ and $\mathcal{NS}$ as examples to illustrate how the prediction-based-ratio method can be adapted to test physical theories that are constrained to produce correlations from these sets.

2.2. Finite Statistics and the Prediction-Based-Ratio Method

Coming back to an actual Bell test, let N_{total} be the total number of experimental trials carried out during the course of the experiment. During each experimental trial, x and y are to be chosen randomly according to some fixed probability distribution P_{xy} (This distribution may be varied from one trial to another but for simplicity of discussion, we consider in this work only the case where this is fixed once and for all before the experiment begins). From the data collected in a Bell test, a naïve (but very commonly-adopted) way to estimate the correlation $\vec{P}$ between measurement outcomes is to compute the relative frequencies $\vec{f}$ that each combination of outcomes (a, b) occurs given the choice of measurement (x, y), i.e.,

$$f(a, b|x, y) = \frac{N_{a,b,x,y}}{N_{x,y}},\tag{4}$$

where $N_{a,b,x,y}$ is the total number of trials the events corresponding to (a, b, x, y) are registered and $N_{x,y} = \sum_{a,b} N_{a,b,x,y}$ is the number of times the particular combination of measurement settings (x, y) is chosen. By definition, $N_{\text{total}} = \sum_{x,y} N_{x,y}$.

If the experimental trials are independent and identically distributed (i.i.d.) corresponding to a *fixed* state ρ with *fixed* measurement strategies $\{M^A_{a|x}\}_{a,x}, \{M^B_{b|y}\}_{b,y}$, then in the asymptotic limit, $\lim_{N_{\text{total}} \to \infty} f(a, b|x, y) = P(a, b|x, y)$ where $\vec{P}$ here would satisfy Equation (2). In this limit, the amount of statistical evidence in the data against a particular hypothesis $\mathfrak{H}$ can be quantified by the Kullback-Leibler (KL) divergence [63] (also known as the relative entropy) from $\vec{P}$ to $\mathcal{L}$, see Refs. [64,65] for a detailed explanation with quantum experiments. We remark that the KL divergence is directly related with the Fisher information metric and so it measures the distinguishability of a distribution from its neighborhood. This provides a motivation for using the KL divergence as a measure of statistical evidence.

In the (original) prediction-based-ratio method of Ref. [43] (see also Ref. [66]), the hypothesis of interest is that the experimental data can be produced using an LHV theory, in other words, that the underlying correlation $\vec{P} \in \mathcal{L}$. For convenience, we shall refer to this hypothesis as $\mathfrak{L}$. In this case, given $\vec{f}$ and P_{xy}, the relevant KL divergence from $\vec{f}$ to $\mathcal{L}$ reads as

$$D_{\text{KL}}\left(\vec{f}||\mathcal{L}\right) := \min_{\vec{P} \in \mathcal{L}} \sum_{a,b,x,y} P_{xy} f(a, b|x, y) \log \left[\frac{f(a, b|x, y)}{P(a, b|x, y)}\right]\tag{5}$$

As the objective function in Equation (5) is strictly convex in $\vec{P}$ and the feasible set $\mathcal{L}$ is convex, the minimizer of the above optimization problem—which we shall denote by $\vec{P}_{\mathrm{KL}}^{\mathcal{L},*}$—is *unique* (see, e.g., Ref. [27]). It follows from the results presented in Ref. [43] that this unique minimizer $\vec{P}_{\mathrm{KL}}^{\mathcal{L},*}$ can be used to construct a Bell inequality:

$$\sum_{a,b,x,y} R(a,b,x,y) P_{xy} P(a,b|x,y) \overset{\mathcal{L}}{\leq} 1, \tag{6a}$$

where the non-negative coefficients of the Bell inequality are defined via the ratios

$$R(a,b,x,y) := \frac{f(a,b|x,y)}{\vec{P}_{\mathrm{KL}}^{\mathcal{L},*}(a,b|x,y)}. \tag{6b}$$

This Bell inequality is the key ingredient of the prediction-based-ratio method and is ideally suited for performing a hypothesis test of $\mathcal{L}$.

To understand the method, we introduce the random variables X and Y to denote the random inputs and the variables A and B to denote the random outputs of Alice and Bob at a trial. The ability to select measurement settings randomly, in particular, is an indispensable prerequisite of the prediction-based-ratio method, or more generally, a proper Bell test (see, e.g., Ref. [20]). We further denote the possible values of inputs and outputs by the respective lower-case letters. Then we can think of the ratio R in Equation (6) as a non-negative function of the inputs X, Y and outputs A, B at each experimental trial such that its expectation according to an arbitrary $\vec{P} \in \mathcal{L}$ with the fixed input distribution P_{xy} satisfies

$$\langle R(A,B,X,Y) \rangle \overset{\mathcal{L}}{\leq} 1. \tag{7}$$

Equation (7) is an alternative way of expressing the Bell inequality of Equation (6). A real experiment necessarily involves only a finite number $N_{\mathrm{total}} = (N_{\mathrm{est}} + N_{\mathrm{test}})$ of experimental trials in time order. Here, we have split the experimental data into two sets: the data from the first N_{est} trials as the *training data* and the data from the remaining N_{test} trials as the *hypothesis-testing data*. In practice, we first construct the function R using the training data and then perform a hypothesis test with the test data. Since the ratio R is determined before the hypothesis test based on the prediction according to the training data, R is called a *prediction-based ratio*.

Given a prediction-based ratio and a finite number N_{test} of test data, we can quantify the evidence against the hypothesis $\mathcal{L}$ by a *p*-value. For concreteness, suppose that the actual measurements chosen at the i-th test trial are x_i, y_i and the corresponding measurement outcomes observed are a_i, b_i. Then the value of the prediction-based ratio at the i-th test trial is $R(a_i, b_i, x_i, y_i)$, abbreviated as r_i. We introduce a test static T as the product of the possible values of the prediction-based ratio at all test trials, so the observed value of the test statistic is $t = \prod_{i=1}^{N_{\mathrm{test}}} r_i$. If we denote by $N'_{a,b,x,y}$ the total number of counts registered for the input-output combination (a,b,x,y) in the test data, then t can be expressed also as

$$t = \prod_{a,b,x,y} R(a,b,x,y)^{N'_{a,b,x,y}}. \tag{8}$$

According to Ref. [43], the *p*-value, which is defined as the maximum probability according to the hypothesis $\mathcal{L}$ of obtaining a value of T at least as high as t actually observed in the experiment, is bounded by

$$p \leq \min\{1/t, 1\}. \tag{9}$$

The smaller the *p*-value, the stronger the evidence against the hypothesis $\mathfrak{L}$ is, in other words, the less plausible LHV theories are. It is worth noting that the *p*-value bound computed in this manner remains valid even if the experimental trials are not i.i.d., while when the experimental trials are i.i.d., the *p*-value bound is asymptotically optimal (or tight) [43].

2.3. Generalization for Hypothesis Testing Beyond LHV Theories

The following two simple observations, which allow one to apply the prediction-based-ratio method to test physical theories beyond those described by LHV, are where our novel contribution enters. Firstly, we make the observation that in the above arguments leading to the *p*-value bound of Equation (9), the actual hypothesis $\mathfrak{L}$ only enters at Equation (6) via the set of correlations $\mathcal{L}$ compatible with the hypothesis $\mathfrak{L}$. In particular, if we are to consider the hypothesis $\mathfrak{H}$ that the data observed is produced by a physical theory H (e.g., a nonsignaling theory), then we merely have to replace $\mathcal{L}$ by the (convex) set of correlations $\mathcal{H}$ (e.g., $\mathcal{NS}$) associated with H in the optimization problem of Equation (5). The method then allows us to bound the plausibility of the hypothesis $\mathfrak{H}$ via the *p*-value bound in Equation (9) with the possible values of the prediction-based ratio given by

$$R(a, b, x, y) := \frac{f(a, b|x, y)}{\vec{P}_{\mathrm{KL}}^{\mathcal{H},*}(a, b|x, y)}, \tag{10}$$

where $\vec{P}_{\mathrm{KL}}^{\mathcal{H},*}$ is the unique minimizer of the optimization problem:

$$D_{\mathrm{KL}}\left(\vec{f}||\mathcal{H}\right) := \min_{\vec{P} \in \mathcal{H}} \sum_{a,b,x,y} P_{xy} f(a, b|x, y) \log\left[\frac{f(a, b|x, y)}{P(a, b|x, y)}\right]. \tag{11}$$

Although Equation (8), Equation (9) and Equation (10) together provide us, in principle, a recipe to test the plausibility of a general physical theory H, its implementation depends on the nature of the set of correlations associated with the hypothesis. Indeed, a crucial part of the procedure is to solve the optimization problem of Equation (11) for the convex set of correlations $\mathcal{H}$ compatible with H, which is generally far from trivial. If $\mathcal{H}$ is a convex polytope, such as $\mathcal{L}$ and $\mathcal{NS}$, or the set of correlations associated with the models considered in Refs. [67,68]), it is known [43] that Equation (11) can indeed be solved numerically.

Our second observation is that for the convex sets of correlations that are amenable to a semidefinite programming characterization, such as those considered in Refs. [59,62,69,70], Equation (11) is an instance of a conic program [58] that can be efficiently solved using a freely available solver, such as PENLAB [71]. To see this, one first notes that, apart from the constant factor P_{xy}, the optimization of Equation (11) is essentially the same as that considered in Ref. [27]. A straightforward adaptation of the argument presented in Appendix D 2 of Ref. [27] would then allow us to complete the aforementioned observation. The data observed in a Bell test can thus be used to test not only $\mathfrak{L}$, but also $\mathfrak{N}$ and even the hypothesis $\mathfrak{Q}$ that the observation is compatible with Born's rule, cf. Equation (2), via outer approximations of $\mathcal{Q}$ (such as $\mathcal{Q}_1$ and $\tilde{\mathcal{Q}}$).

A remark is now in order. In order to avoid so-called *p*-value hacking, it is essential that the test data used in the computation of the test statistic T is not used to determine $\vec{f}$, and hence the values of the prediction-based ratio R in Equation (10). In this work, for simplicity we use the first N_{est} trials of an experiment as the training data for estimating $\vec{f}$ and further constructing a prediction-based ratio R that is applied for all test trials. In principle, we can use different training data for different test trials. For example, we can define the training data for a test trial as the data from all trials performed before this test trial, and then we can adapt the construction of the prediction-based ratio for each individual test trial. We refer to Ref. [43] for more details on the adaptability of the prediction-based ratio.

3. Results

To illustrate how well the prediction-based-ratio method works in identifying data that are *not* even explicable by some nonlocal physical theories, such as quantum theory, we now consider a few examples of applications of the method. As above, we restrict our attention to a bipartite Bell test, where each party performs two binary-outcome measurements randomly selected at each trial. Throughout this section, we assume that the input distribution is uniform, specifically $P_{xy} = \frac{1}{4}$ for all combinations of $x, y \in \{0, 1\}$. In Sections 3.2 and 3.3 we study the behaviour of numerically simulated Bell tests based on hypothetical sources of correlations described in Section 3.1, while in Section 3.4, we analyze the real experimental data reported in Ref. [72].

3.1. Modeling a Bell Test

For our numerical simulations, we consider a $\vec{P}$ that resembles a nonlocal source targeted at in various actual Bell experiments [35–37,72,73]:

$$\vec{P}(v) := v\vec{P}_{PR} + (1 - v)\vec{P}_{\mathbb{1}}, \tag{12}$$

where $v \in [0, 1]$, $\vec{P}_{PR}$ is the Popescu-Rohrlich (PR) correlation [28] $P_{PR}(a, b|x, y) = \frac{1}{2}\delta_{a \oplus b, xy}$ with $a, b, x, y \in \{0, 1\}$, and $P_{\mathbb{1}}(a, b|x, y) = \frac{1}{4}$ for all a, b, x, y is the white-noise distribution. In Equation (12), the real parameter v can be seen as the weight associated with $\vec{P}_{PR}$ in the convex mixture. Importantly, the nonlocal source represented by such a mixture can (in principle) be produced by performing appropriate local measurements on a maximally entangled two-qubit state if and only if $v \le v_c := \frac{1}{\sqrt{2}} \approx 0.71$ (see, e.g., Refs. [27,57]). In particular, when $v = v_c$—corresponding to an ideal nonlocal source—the mixture gives rise to the maximal quantum violation of the CHSH [45] Bell inequality.

To mimic an experimental scenario with noise (something unavoidable in practice), we shall introduce a slight perturbation to the ideal source $\vec{P}(v)$ of Equation (12). Specifically, we require the measurement outcomes observed at each trial in the simulated Bell test to be governed by the nonlocal source $(1 - \epsilon)\vec{P}(v) + \epsilon\vec{P}_{noise}$, where $\epsilon \ll 1$ is the weight associated with the noise term $\vec{P}_{noise}$. Moreover, for the purpose of illustrating the effectiveness of the method in identifying non-quantum-compatible data, we set $v > v_c$. In our simulations, we set $\epsilon = 0.01$ and $v = 0.72 > v_c$. However, as long as the given mixture lies outside $\tilde{Q}$ (and hence also outside Q), the actual choices of $\epsilon \ll 1$ and $v \in (v_c, 1]$ are irrelevant. The only impact that these choices may have is the number of trials N_{total} needed to falsify the hypothesis

"The observed data is compatible with a physical theory that is constrained to produce only the almost-quantum set of correlations."

with the same level of confidence. Inspired by the experiments of Ref. [72] where $N_{total} = 10^5 \sim 10^6$, we set in our simulations $N_{total} = 10^6$. Note also that instead of $\tilde{Q}$, we can equally well choose another set of correlations that admits a semidefinite programming characterization, such as those described in Refs. [59,62].

Since we are interested to model a nonlocal source that obeys the nonsignaling conditions of Equation (3), there is no loss in generality by considering $\vec{P}_{noise} \in \mathcal{NS}$. To this end, let $\vec{P}_j^{Ext}$ be the j-th extreme point of the nonsignaling polytope [29], then we may write $\vec{P}_{noise} = \sum_j p_j \vec{P}_j^{Ext}$ where p_j is the weight associated with $\vec{P}_j^{Ext}$ in the convex decomposition of $\vec{P}_{noise}$. We may thus write the nonlocal source of interest as:

$$\vec{P}(v, \epsilon, \{p_j\}) := (1 - \epsilon)\vec{P}(v) + \epsilon \sum_j p_j \vec{P}_j^{Ext}. \tag{13}$$

Finally, to simulate the raw data $\{(a_i, b_i, x_i, y_i)\}_{i=1}^N$ obtained in an N-trial Bell test for any given input distribution P_{xy} and correlation $\vec{P}$, we make use of the MATLAB toolbox Lightspeed developed by Minka [74].

3.2. Simulations of Bell Tests with an i.i.d. Nonlocal Source

Let us begin with the case of i.i.d. trials, corresponding to a source of correlation that remains unchanged throughout the experiment, and where the inputs at each trial are independent of the inputs of the previous trials. To this end, we first sample the weights $\{p_j\}_j$ uniformly from the interval $[0, 1]$ and renormalize them such that $\sum_j p_j = 1$. With our choice of $v = 0.72$ and $\epsilon = 0.01$, it is easy to find such a randomly generated correlation $\vec{P}(v, \epsilon, \{p_j\})$ that lies outside $\tilde{\mathcal{Q}}$. (Verifying that any given $\vec{P}$ is (not) in $\tilde{\mathcal{Q}}$ can be carried out by solving a semidefinite program. Specifically, for any given correlation $\vec{P}$, if the maximal white-noise visibility v such that $v\vec{P} + (1 - v)\vec{P}_{\mathbb{1}} \in \tilde{\mathcal{Q}}$ is smaller than 1, then $\vec{P} \notin \tilde{\mathcal{Q}} \supset \mathcal{Q}$, and hence outside $\mathcal{Q}$, otherwise $\vec{P} \in \tilde{\mathcal{Q}}$.) For convenience, we denote by $\mathcal{P}$ the specific set of $\{p_j\}_j$ employed in our simulation of 500 Bell tests, each with $N_{\text{total}} = 10^6$ trials. In Figure 1, we summarize the steps involved in our analysis of the numerically simulated data using the prediction-based-ratio method. The resulting p-value upper bounds are summarized in Table 1.

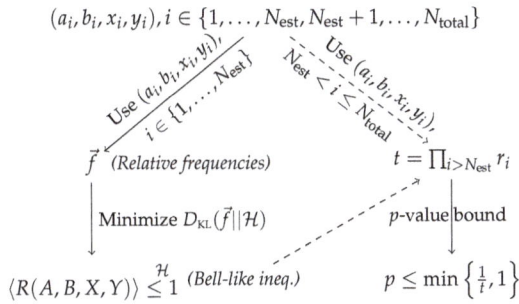

Figure 1. Flowchart summarizing the steps involved in our application of the prediction-based-ratio method on the simulated data $\{(a_i, b_i, x_i, y_i)\}_{i=1}^{N_{\text{total}}}$ of a *single* Bell test. In the first step, we separate the data into two sets, with the data collected from the first N_{est} trials serving as the training data while the rest is used for the actual hypothesis testing. Specifically, the training data is used to compute the relative frequencies $\vec{f}$ and to minimize the KL divergence $D_{\text{KL}}(\vec{f}||\mathcal{H})$ with respect to the set of correlations $\mathcal{H} \in \{\mathcal{NS}, \tilde{\mathcal{Q}}\}$ associated, respectively, with the hypothesis of $\mathfrak{N}$ and $\tilde{\mathfrak{Q}}$. The correlation $\vec{P}_{\text{KL}}^{\mathcal{H},*} \in \mathcal{H}$ that minimizes $D_{\text{KL}}(\vec{f}||\mathcal{H})$ gives rise to a Bell-like inequality with coefficients $\{R(A = a, B = b, X = x, Y = y)\}_{x,y,a,b}$. The remaining data is then used to compute $t = \prod_{i>N_{\text{est}}} r_i$ where $r_i := R(a_i, b_i, x_i, y_i)$. Finally, a p-value bound according to the hypothesis is obtained by computing $\min\{\frac{1}{t}, 1\}$.

As expected, despite statistical fluctuations, the data does not suggest any obvious evidence against the nonsignaling hypothesis. In fact, among the 500 p-value bounds obtained, 97% of them are trivial (i.e., equal to unity), while the smallest non-trivial p-value bound obtained is approximately 0.14. On the contrary, for the hypothesis test of the almost-quantum set of correlations, more than half of the simulated Bell tests give a p-value upper bound that is less than 10^{-10}. Although there are also 5.8% of these simulated Bell tests that give a trivial p-value bound according to the almost-quantum hypothesis, we see that the method generally works very well in falsifying this hypothesis. In fact, a separate calculation (not shown in the table) shows that when we increase N_{total} to 10^7, all the 500 p-value upper bounds obtained according to the almost-quantum hypothesis are less than or equal to 10^{-10}.

Table 1. Summary of frequency distributions of the p-value upper bounds obtained from 500 numerically simulated Bell tests, each consists of $N_{\text{est}} = 10^6$ trials and assumes the same i.i.d. nonlocal source $\vec{P}(v, \epsilon, \{p_j\})$ of Equation (13) that lies *outside* $\tilde{\mathcal{Q}}$. The second and third row give, respectively, the frequency distributions according to the hypothesis associated with $\mathcal{NS}$ (nonsignaling) and $\tilde{\mathcal{Q}}$ (almost-quantum). For these hypotheses, the smallest p-value upper bound found among these 500 Bell tests are, respectively, 0.14 and 5.7×10^{-20}. The second to the fifth column give, respectively, the fraction of simulated Bell tests having a p-value upper bound (for each hypothesis) that satisfies the given (increasing) threshold (e.g., 10^{-10} for the second column). Similarly, in the last column, we give the fraction of instances where the p-value upper bound obtained is trivial, i.e., exactly equals to 1. The smaller the p-value upper bound, the less likely it is that a physical theory associated with the hypothesis produces the observed data. Thus, the larger the value in the second (to the fourth) column, the less likely it is that the assumed physical theory holds true. In contrast, the larger the value in the rightmost column, the weaker the empirical evidence against the assumed theory is.

p-Value Bound	$\leq 10^{-10}$	$\leq 10^{-4}$	$\leq 10^{-2}$	$\leq 10^{-1}$	Trivial
$\mathcal{NS}$	0	0	0	0	97%
$\tilde{\mathcal{Q}}$	58%	85%	90%	93%	5.8%

3.3. Simulations of Bell tests with a non-i.i.d. Nonlocal Source

In a real experiment, the assumption that the experimental trials are i.i.d is often far from justifiable, as that would require, for example, that the experimental setup remain as it is over the entire course of the experiment. As a result, we also consider here the case where the source that generates the data actually varies from one trial to another. To this end, for the i-th trial of the Bell test, we simulate according to the conditional outcome distributions:

$$\vec{P}_i(v, \epsilon, n_i) = (1 - \epsilon)\vec{P}(v) + \epsilon \vec{P}_{n_i}^{\text{Ext}}, \tag{14}$$

where $n_i = 1, 2, \ldots, 24$ labels the *single* nonsignaling extreme point used to mix with $\vec{P}(v)$ at this trial, cf. Equation (13) with $p_j = 1$ if $j = n_i$ but vanishes otherwise. Moreover, to facilitate a comparison with the i.i.d. case, before the i-th trial, we randomly pick n_i according to the probability $P(n_i = j) = p_j$ where $p_j \in \mathcal{P}$ is exactly the probability employed in the simulation of Section 3.2. With this choice, the outcome distributions governed by the nonlocal source of Equation (14) (for the i-th trial) averages to that of Equation (13) when the number of trials $N_{\text{total}} \to \infty$. Again, we follow the steps summarized in Figure 1 to compute the relevant p-value upper bounds using the prediction-based-ratio method. The resulting p-value upper bounds are summarized in Table 2.

Table 2. Summary of frequency distributions of the *p*-value upper bounds obtained from 500 numerically simulated Bell tests. Each of these Bell tests involves $N_{est} = 10^6$ trials and each trial assumes a varying source $\vec{P}_i(v, \epsilon, n_i)$ of Equation (14). For the hypothesis of $\mathfrak{N}$ and $\tilde{\mathfrak{Q}}$, associated with $\mathcal{NS}$ (second row) and $\tilde{\mathfrak{Q}}$ (third row), respectively, the smallest *p*-value upper bound found among these 500 instances are 0.21 and 1.3×10^{-15}. The significance of each column follows that described in the caption of Table 1.

p-Value Bound	$\leq 10^{-10}$	$\leq 10^{-4}$	$\leq 10^{-2}$	$\leq 10^{-1}$	Trivial
$\mathcal{NS}$	0	0	0	0	97%
$\tilde{\mathfrak{Q}}$	17	59%	69%	72	24%

As with the i.i.d. case, for these 500 simulated Bell tests, our application of the prediction-based-ratio method does not lead to any obvious evidence against the nonsignaling hypothesis $\mathfrak{N}$. However, for the hypothesis associated with the almost-quantum set $\tilde{\mathfrak{Q}}$, our results (last row of Table 2) give more than half of the *p*-value upper bounds that are less than 10^{-4} (accordingly, 17% if we set the cutoff at 10^{-10}). Although there are 24% of these instances where the returned *p*-value upper bound for the same hypothesis is trivial, we see that, as with the i.i.d. case, the method remains very effective in showing that the observed data cannot be entirely accounted for using a theory that is constrained to produce only almost-quantum correlations. In addition, as with the i.i.d. case, our separate calculation shows that the effectiveness of this method can be substantially improved when we increase N_{total} to 10^7: all the 500 *p*-value upper bounds obtained according to the almost-quantum hypothesis become less than or equal to 10^{-10}.

3.4. Application to Some Real Experimental Data

Armed with the experience gained in the above analyses, let us now analyze the experimental results presented in Figure 3 of Ref. [72] using the prediction-based-ratio method. One of the goals of Ref. [72] was to experimentally approach the boundary of the quantum set of correlations in the two-dimensional subspace spanned by the two Bell parameters:

$$\begin{aligned}
\mathcal{S}_{\text{CHSH}} &= E_{00} + E_{01} + E_{10} - E_{11}, \\
\mathcal{S}'_{\text{CHSH}} &= -E_{00} + E_{01} + E_{10} + E_1,
\end{aligned} \tag{15}$$

where $E_{xy} := \sum_{a,b=0}^{1}(-1)^{a+b}P(a,b|x,y)$ is the correlator. To this end, the Bell parameter $\mathcal{S}_{\text{CHSH}} \cos\theta + \mathcal{S}'_{\text{CHSH}} \sin\theta$ for 180 uniformly-spaced values of $\theta \in \{\theta_1, \theta_2, \ldots, \theta_{180}\} \subset [0, 2\pi)$ were estimated by performing the measurements presented in Appendix A of Ref. [72] on a two-qubit maximally entangled state.

Unfortunately, only the total counts for each combination of input-output $N_{a,b,x,y}$ (rather than the time sequences of raw data) given the value of θ are available [75]. Therefore, in analogy with the analyses presented above, we use the relative frequencies obtained for θ_k as the training data to derive a prediction-based ratio (which corresponds to a Bell-like inequality) for the hypothesis test using the data associated with θ_{k+1} (for the case of $k = 180$, the hypothesis test uses the data associated with θ_1). The analysis therefore essentially follows the steps outlined in Figure 1, but with the computation of t carried out using Equation (8) instead, since we do not have the time sequences of raw data. Moreover, to apply the prediction-based-ratio method, we *assume*, as with the numerical experiments reported earlier that the input distributions are uniform, i.e., $P_{xy} = \frac{1}{4}$ for all combinations of $x, y \in \{0, 1\}$. A summary of the *p*-value upper bounds obtained from these 180 Bell tests is given in Table 3.

For both hypotheses, approximately half of the *p*-value upper bounds obtained are trivial. At the same time, about the same fraction of the *p*-value bounds obtained are less than 10^{-2} (with the majority of them being less than 10^{-4}). In fact, the smallest of the *p*-value upper bounds are remarkably small: 3.2×10^{-55} for the hypothesis of nonsignaling $\mathfrak{N}$ and 2.7×10^{-55} for the hypothesis of almost-quantum $\tilde{\mathfrak{Q}}$. These results strongly suggest that under the *assumption* that the measurement settings were *randomly* chosen according to a uniform input distribution, it is extremely unlikely that a physical theory associated with each of these hypotheses can produce the observed relative frequencies.

These conclusions that the observed data are incompatible with the fundamental principle of nonsignaling or with quantum theory (via the almost-quantum hypothesis), however, turn out to be *flawed*, as it was brought to our attention [75] that during the course of the experiment, the measurement bases were not at all randomized—the measurements were carried out in blocks using the same combination of (x, y) before moving to another. Why should this pose a problem? In the extreme scenario, if the measurement settings were fully correlated to some *local* hidden variable, it is known that the the resulting correlation between measurement outcomes can violate the nonsignaling conditions of Equation (3), see, e.g., Ref. [76]. Consequently, it is not surprising that in the prediction-based-ratio method (as well as any other methods employed for the statistical analysis of a Bell test), the measurement inputs (x_i, y_i) during the *i*-th trial, as discussed in Section 2, ought to be randomly chosen.

Table 3. Summary of frequency distributions of the *p*-value upper bounds obtained from the 180 Bell tests of Ref. [72] according to the hypothesis of $\mathfrak{N}$ and $\tilde{\mathfrak{Q}}$ (associated, respectively, with $\mathcal{NS}$, the second row, and $\tilde{\mathcal{Q}}$, the third row) under the assumption that the measurement settings were randomly chosen according to a uniform distribution. The significance of each column follows that described in the caption of Table 1.

p-Value Bound	$\leq 10^{-10}$	$\leq 10^{-4}$	$\leq 10^{-2}$	$\leq 10^{-1}$	Trivial
$\mathcal{NS}$	38%	45%	48%	51%	48%
$\tilde{\mathcal{Q}}$	35%	44%	47%	49%	49%

4. Discussion

As discussed in the last section, the conclusion that "the experimental data of Ref. [72] show a violation of the nonsignaling principle" based on an erroneous application of the prediction-based-ratio method is unfounded. The results are nonetheless thought-provoking. For example, suppose for now that we had access to the raw data for all trials. Since the analysis was flawed because of the nonrandomnization of measurement settings, one can imagine that—under the assumption that the trials are exchangeable—we first artificially randomize the hypothesis-testing trials to simulate the randomization of measurement settings in the experiment. Should we then expect to obtain *p*-value bounds with fundamentally different features? The answer is negative. The reason is that in our crude application of the method, only the number of counts $N'_{a,b,x,y}$ for each input-output combination matters, see Equation (8). In particular, the actual trials in which a particular combination of (a, b, x, y) appears are irrelevant in such an analysis.

So, if one holds the view that the nonsignaling principle cannot be flawed, then one must come to the conclusion that "should the measurement choices be randomized, it would be impossible to register the same number of counts $N'_{a,b,x,y}$ for each input-output combination". A plausible cause for this is that the experimental setup suffered from some systematic drift during the course of the experiment, which is exactly a manifestation that the experimental trials are not i.i.d. It might then appear that a hypothesis test of the nonsignaling principle is hopeless in such a scenario. However, as mentioned above, the prediction-based-ratio method is applicable even for *non*-i.i.d. experimental trials. Indeed, as we illustrate in Section 3.3 (see, specifically Table 2), such fluctuations have not lead to any false positive in the sense of giving very small *p*-value upper bound according to the nonsignaling hypothesis.

More generally, as the above example of Section 3.4 illustrates, an unexpectedly small *p*-value upper bound according to the nonsignaling hypothesis may be a consequence that certain premises needed to perform a sensible Bell test are violated. In other words, an *apparent violation* as such does not necessarily pose a problem to any physical principle, such as the nonsignaling principle that is rooted in the theory of relativity. However, as nonlocal correlations also find applications in device-independent quantum information processing [18,19], it is important to carry out such consistency checks alongside the violation of a Bell inequality before one applies the estimated nonlocal correlation in any such protocols.

Of course, an unexpectedly small *p*-value upper bound according to the nonsignaling hypothesis could also be a consequence of mere statistical fluctuation. Indeed, our results in Sections 3.2 and 3.3 show that when a null hypothesis indeed holds *true*, it can still happen that one obtains a relatively small *p*-value upper bound (of the order of 10^{-1}) even after a large number of trials ($N_{\text{total}} = 10^6$). However, as explained in Appendix 1 of Ref. [43], if a null hypothesis is correct, the probability of obtaining a *p*-value upper bound smaller than *q* with the prediction-based-ratio method is no larger than *q*. Indeed, in each of these instances, *p*-value upper bounds that are less than 10^{-1} occur way less than 50 times among the 500 simulated experiments. In any case, this means that even though the prediction-based-ratio method already gets rids of the often unjustifiable i.i.d. assumption involved in such an analysis, the interpretation of the significance of a small *p*-value upper bound must still be carried out with care, as advised, for example, in Refs. [77–79].

5. Conclusion

In this work, we revisited the prediction-based-ratio method developed [43]—in the context of a Bell test—for performing hypothesis tests of LHV theories. We showed that with the two observations presented in Section 2.3, the method can equally well be applied to perform hypothesis tests of *other* physical theories, specifically those that are constrained to produce correlations amenable to a semidefinite programming characterization. Prime examples of such theories include those that obey the principle of nonsignaling [28], those that satisfy the principle of macroscopic locality [51], the so-called *v*-causal models [67], as well as physical theories that are constrained to produce the almost-quantum set [50] or any other outer approximations [59,62,69] of the quantum set of correlations.

To illustrate the effectiveness of the method, we first numerically simulated 500 Bell tests using a hypothetical source of correlations that lies somewhat outside the almost-quantum set of correlations. We then applied the method to obtain a *p*-value upper bound according to both the almost-quantum hypothesis and the nonsignaling hypothesis for the simulated data obtained in each of these Bell tests. In the majority (> 90%) of these 500 instances, the *p*-value upper bound according to the almost-quantum hypothesis is less than 10^{-2}. Since a *p*-value upper bound quantifies the evidence against the assumed (almost-quantum) theory given the observed data, these results show that in most of these simulated Bell tests, the data is unlikely to be explicable by the assumed theory. In a similar manner, we numerically simulated another 500 Bell tests using a hypothetical source that *varies* from one trial to another. Again, the method remained very effective (giving a *p*-value upper bound that is less than 10^{-2} for 69% of the instances) in identifying the incompatibility between the observed data and the assumed (almost-quantum) theory in such a non-i.i.d. scenario.

Finally, we applied the prediction-based-ratio method to the experimental data of Ref. [72]. To this end, we assumed that the measurement settings were randomly chosen with uniform distributions. An application of the method under this assumption again led to very small p-value upper bounds (10^{-4}) for more than 40% of the 180 Bell tests analyzed—not only for the almost-quantum hypothesis, but also for the nonsignaling hypothesis. Such a violation of the nonsignaling conditions, however, is apparent, as we learned after the analysis that the measurement settings were *not* randomized during the course of the experiments, thereby invalidating one of the basic assumptions needed in the application of the prediction-based-ratio method. Nonetheless, as we remarked in the Discussion section, the analysis nevertheless unveils that the possibility of using the prediction-based-ratio method to identify a situation where a certain premise is needed to perform a proper Bell test, such as the randomization of settings, is invalidated.

Note added: While preparing this manuscript, we became aware of the work of Smania et al. [80], which also discussed, among others, the implication of not randomizing the settings in a Bell test, and its relevance in quantitative applications.

Author Contributions: Both authors contributed toward the computation of the numerical results and the preparation of the manuscript.

Funding: This work is supported by the Ministry of Science and Technology, Taiwan (Grants No. 104-2112-M-006-021-MY3, 107-2112-M-006-005-MY2, 107-2627-E-006-001) and the National Center for Theoretical Science, Taiwan (R.O.C.).

Acknowledgments: Y.C.L. is grateful to Adán Cabello, Bradley Christensen, Ehtibar Dzhafarov, Nicolas Gisin, Scott Glancy, Paul Kwiat, Jan-Åke Larsson, Denis Rosset, and Lev Vaidman for useful discussions.

Conflicts of Interest: The authors declare no conflict of interest.

References

1. Acín, A.; Gisin, N.; Masanes, L. From Bell's Theorem to Secure Quantum Key Distribution. *Phys. Rev. Lett.* **2006**, *97*, 120405. [CrossRef]
2. Bell, J.S. On the Einstein-Podolsky-Rosen paradox. *Physics* **1964**, *1*, 195. [CrossRef]
3. Gisin, N.; Ribordy, G.; Tittel, W.; Zbinden, H. Quantum cryptography. *Rev. Mod. Phys.* **2002**, *74*, 145–195. [CrossRef]
4. Barrett, J.; Hardy, L.; Kent, A. No Signaling and Quantum Key Distribution. *Phys. Rev. Lett.* **2005**, *95*, 010503. [CrossRef]
5. Acín, A.; Brunner, N.; Gisin, N.; Massar, S.; Pironio, S.; Scarani, V. Device-Independent Security of Quantum Cryptography against Collective Attacks. *Phys. Rev. Lett.* **2007**, *98*, 230501. [CrossRef] [PubMed]
6. Vazirani, U.; Vidick, T. Fully Device-Independent Quantum Key Distribution. *Phys. Rev. Lett.* **2014**, *113*, 140501. [CrossRef] [PubMed]
7. Ekert, A.K. Quantum cryptography based on Bell's theorem. *Phys. Rev. Lett.* **1991**, *67*, 661–663. [CrossRef]
8. Colbeck, R. Quantum and Relativistic Protocols for Secure Multi-Party Computation. *arXiv* **2009**, arXiv:0911.3814.
9. Pironio, S.; Acín, A.; Massar, S.; de La Giroday, A.B.; Matsukevich, D.N.; Maunz, P.; Olmschenk, S.; Hayes, D.; Luo, L.; Manning, T.A.; Monroe, C. Random numbers certified by Bell's theorems theorem. *Nature (London)* **2010**, *464*, 1021. [CrossRef]
10. Colbeck, R.; Kent, A. Private randomness expansion with untrusted devices. *J. Phys. A Math. Theor.* **2011**, *44*, 095305. [CrossRef]
11. Mayers, D.; Yao, A. Self Testing Quantum Apparatus. *Quantum Inf. Comput.* **2004**, *4*, 273.
12. Brunner, N.; Pironio, S.; Acín, A.; Gisin, N.; Méthot, A.A.; Scarani, V. Testing the Dimension of Hilbert Spaces. *Phys. Rev. Lett.* **2008**, *100*, 210503. [CrossRef] [PubMed]
13. Reichardt, B.W.; Unger, F.; Vazirani, U. Classical command of quantum systems. *Nature (London)* **2013**, *496*, 456. [CrossRef] [PubMed]

14. Yang, T.H.; Vértesi, T.; Bancal, J.D.; Scarani, V.; Navascués, M. Robust and Versatile Black-Box Certification of Quantum Devices. *Phys. Rev. Lett.* **2014**, *113*, 040401. [CrossRef] [PubMed]
15. Liang, Y.C.; Rosset, D.; Bancal, J.D.; Pütz, G.; Barnea, T.J.; Gisin, N. Family of Bell-like Inequalities as Device-Independent Witnesses for Entanglement Depth. *Phys. Rev. Lett.* **2015**, *114*, 190401. [CrossRef] [PubMed]
16. Coladangelo, A.; Goh, K.T.; Scarani, V. All pure bipartite entangled states can be self-tested. *Nat. Comm.* **2017**, *8*, 15485. [CrossRef] [PubMed]
17. Sekatski, P.; Bancal, J.D.; Wagner, S.; Sangouard, N. Certifying the Building Blocks of Quantum Computers from Bell's Theorem. *Phys. Rev. Lett.* **2018**, *121*, 180505. [CrossRef]
18. Scarani, V. The device-independent outlook on quantum physics. *Acta Phys. Slovaca* **2012**, *62*, 347.
19. Brunner, N.; Cavalcanti, D.; Pironio, S.; Scarani, V.; Wehner, S. Bell nonlocality. *Rev. Mod. Phys.* **2014**, *86*, 419–478. [CrossRef]
20. Bell, J.S. *Speakable and Unspeakable in Quantum Mechanics: Collected Papers on Quantum Philosophy*, 2nd ed.; Cambridge University Press: Cambridge, UK, 2004.
21. Hensen, B.; Bernien, H.; Dreau, A.E.; Reiserer, A.; Kalb, N.; Blok, M.S.; Ruitenberg, J.; Vermeulen, R.F.L.; Schouten, R.N.; Abellan, C.; et al. Loophole-free Bell inequality violation using electron spins separated by 1.3 kilometres. *Nature* **2015**, *526*, 682–686. [CrossRef]
22. Shalm, L.K.; Meyer-Scott, E.; Christensen, B.G.; Bierhorst, P.; Wayne, M.A.; Stevens, M.J.; Gerrits, T.; Glancy, S.; Hamel, D.R.; Allman, M.S.; et al. Strong Loophole-Free Test of Local Realism. *Phys. Rev. Lett.* **2015**, *115*, 250402. [CrossRef] [PubMed]
23. Giustina, M.; Versteegh, M.A.M.; Wengerowsky, S.; Handsteiner, J.; Hochrainer, A.; Phelan, K.; Steinlechner, F.; Kofler, J.; Larsson, J.A.; Abellán, C.; et al. Significant-Loophole-Free Test of Bell's Theorem with Entangled Photons. *Phys. Rev. Lett.* **2015**, *115*, 250401. [CrossRef] [PubMed]
24. Rosenfeld, W.; Burchardt, D.; Garthoff, R.; Redeker, K.; Ortegel, N.; Rau, M.; Weinfurter, H. Event-Ready Bell Test Using Entangled Atoms Simultaneously Closing Detection and Locality Loopholes. *Phys. Rev. Lett.* **2017**, *119*, 010402. [CrossRef] [PubMed]
25. Li, M.H.; Wu, C.; Zhang, Y.; Liu, W.Z.; Bai, B.; Liu, Y.; Zhang, W.; Zhao, Q.; Li, H.; Wang, Z.; et al. Test of Local Realism into the Past without Detection and Locality Loopholes. *Phys. Rev. Lett.* **2018**, *121*, 080404. [CrossRef] [PubMed]
26. Schwarz, S.; Bessire, B.; Stefanov, A.; Liang, Y.C. Bipartite Bell inequalities with three ternary-outcome measurements - from theory to experiments. *New J. Phys.* **2016**, *18*, 035001. [CrossRef]
27. Lin, P.S.; Rosset, D.; Zhang, Y.; Bancal, J.D.; Liang, Y.C. Device-independent point estimation from finite data and its application to device-independent property estimation. *Phys. Rev. A* **2018**, *97*, 032309. [CrossRef]
28. Popescu, S.; Rohrlich, D. Quantum nonlocality as an axiom. *Found. Phys.* **1994**, *24*, 379–385. [CrossRef]
29. Barrett, J.; Linden, N.; Massar, S.; Pironio, S.; Popescu, S.; Roberts, D. Nonlocal correlations as an information-theoretic resource. *Phys. Rev. A* **2005**, *71*, 022101. [CrossRef]
30. Liu, Y.; Zhao, Q.; Li, M.H.; Guan, J.Y.; Zhang, Y.; Bai, B.; Zhang, W.; Liu, W.Z.; Wu, C.; Yuan, X.; et al. Device-independent quantum random-number generation. *Nature* **2018**, *562*, 548–551. [CrossRef]
31. Adenier, G.; Khrennikov, A.Y. Test of the no-signaling principle in the Hensen loophole-free CHSH experiment. *Fortschr. Phys.* **2017**, *65*, 1600096. [CrossRef]
32. Bednorz, A. Analysis of assumptions of recent tests of local realism. *Phys. Rev. A* **2017**, *95*, 042118. [CrossRef]
33. Kupczynski, M. Is Einsteinian no-signalling violated in Bell tests? *Open Phys.* **2017**, *15*, 739. [CrossRef]
34. Aspect, A.; Dalibard, J.; Roger, G. Experimental Test of Bell's Inequalities Using Time-Varying Analyzers. *Phys. Rev. Lett.* **1982**, *49*, 1804–1807. [CrossRef]
35. Tittel, W.; Brendel, J.; Zbinden, H.; Gisin, N. Violation of Bell Inequalities by Photons More Than 10 km Apart. *Phys. Rev. Lett.* **1998**, *81*, 3563–3566. [CrossRef]
36. Weihs, G.; Jennewein, T.; Simon, C.; Weinfurter, H.; Zeilinger, A. Violation of Bell's Inequality under Strict Einstein Locality Conditions. *Phys. Rev. Lett.* **1998**, *81*, 5039–5043. [CrossRef]

37. Rowe, M.A.; Kielpinski, D.; Meyer, V.; Sackett, C.A.; Itano, W.M.; Monroe, C.; Wineland, D.J. Experimental violation of a Bell's inequality with efficient detection. *Nature* **2001**, *409*, 791–794. [CrossRef] [PubMed]

38. Giustina, M.; Mech, A.; Ramelow, S.; Wittmann, B.; Kofler, J.; Beyer, J.; Lita, A.; Calkins, B.; Gerrits, T.; Nam, S.W.; et al. Bell violation using entangled photons without the fair-sampling assumption. *Nature* **2013**, *497*, 227. [CrossRef]

39. Christensen, B.G.; McCusker, K.T.; Altepeter, J.B.; Calkins, B.; Gerrits, T.; Lita, A.E.; Miller, A.; Shalm, L.K.; Zhang, Y.; Nam, S.W.; et al. Detection-Loophole-Free Test of Quantum Nonlocality, and Applications. *Phys. Rev. Lett.* **2013**, *111*, 130406. [CrossRef]

40. Erven, C.; Meyer-Scott, E.; Fisher, K.; Lavoie, J.; Higgins, B.L.; Yan, Z.; Pugh, C.J.; Bourgoin, J.P.; Prevedel, R.; Shalm, L.K.; et al. Experimental three-photon quantum nonlocality under strict locality conditions. *Nature Photonics* **2014**, *8*, 292. [CrossRef]

41. Lanyon, B.P.; Zwerger, M.; Jurcevic, P.; Hempel, C.; Dür, W.; Briegel, H.J.; Blatt, R.; Roos, C.F. Experimental Violation of Multipartite Bell Inequalities with Trapped Ions. *Phys. Rev. Lett.* **2014**, *112*, 100403. [CrossRef]

42. Shen, L.; Lee, J.; Thinh, L.P.; Bancal, J.D.; Cerè, A.; Lamas-Linares, A.; Lita, A.; Gerrits, T.; Nam, S.W.; Scarani, V.; et al. Randomness Extraction from Bell Violation with Continuous Parametric Down-Conversion. *Phys. Rev. Lett.* **2018**, *121*, 150402. [CrossRef] [PubMed]

43. Zhang, Y.; Glancy, S.; Knill, E. Asymptotically optimal data analysis for rejecting local realism. *Phys. Rev. A* **2011**, *84*, 062118. [CrossRef]

44. Gill, R.D. Time, Finite Statistics, and Bell's Fifth Position. *arXiv* **2003**, arXiv:quant-ph/0301059.

45. Clauser, J.F.; Horne, M.A.; Shimony, A.; Holt, R.A. Proposed Experiment to Test Local Hidden-Variable Theories. *Phys. Rev. Lett.* **1969**, *23*, 880–884. [CrossRef]

46. Zhang, Y.; Glancy, S.; Knill, E. Efficient quantification of experimental evidence against local realism. *Phys. Rev. A* **2013**, *88*, 052119. [CrossRef]

47. Cavalcanti, D.; Salles, A.; Scarani, V. Macroscopically local correlations can violate information causality. *Nat. Commun.* **2010**, *1*, 136. [CrossRef] [PubMed]

48. Fritz, T.; Sainz, A.B.; Augusiak, R.; Brask, J.B.; Chaves, R.; Leverrier, A.; Acín, A. Local orthogonality as a multipartite principle for quantum correlations. *Nat. Commun.* **2013**, *4*, 2263. [CrossRef] [PubMed]

49. Amaral, B.; Cunha, M.T.; Cabello, A. Exclusivity principle forbids sets of correlations larger than the quantum set. *Phys. Rev. A* **2014**, *89*, 030101. [CrossRef]

50. Navascués, M.; Guryanova, Y.; Hoban, M.J.; Acín, A. Almost quantum correlations. *Nat. Commun.* **2015**, *6*, 6288. [CrossRef]

51. Navascués, M.; Wunderlich, H. A glance beyond the quantum model. *Proc. R. Soc. A* **2010**, *466*, 881. [CrossRef]

52. Rohrlich, D. PR-Box Correlations Have No Classical Limit. In *Quantum Theory: A Two-Time Success Story*; Struppa, D.C., Tollaksen, J.M., Eds.; Springer Milan: Milano, Italy, 2014; pp. 205–211.

53. Van Dam, W. Implausible consequences of superstrong nonlocality. *Nat. Comput.* **2013**, *12*, 9–12. [CrossRef]

54. Brassard, G.; Buhrman, H.; Linden, N.; Méthot, A.A.; Tapp, A.; Unger, F. Limit on Nonlocality in Any World in Which Communication Complexity Is Not Trivial. *Phys. Rev. Lett.* **2006**, *96*, 250401. [CrossRef]

55. Linden, N.; Popescu, S.; Short, A.J.; Winter, A. Quantum Nonlocality and Beyond: Limits from Nonlocal Computation. *Phys. Rev. Lett.* **2007**, *99*, 180502. [CrossRef] [PubMed]

56. Pawłowski, M.; Paterek, T.; Kaszlikowski, D.; Scarani, V.; Winter, A.; Żukowski, M. Information causality as a physical principle. *Nature* **2009**, *461*, 1101. [CrossRef] [PubMed]

57. Goh, K.T.; Kaniewski, J.; Wolfe, E.; Vértesi, T.; Wu, X.; Cai, Y.; Liang, Y.C.; Scarani, V. Geometry of the set of quantum correlations. *Phys. Rev. A* **2018**, *97*, 022104. [CrossRef]

58. Boyd, S.; Vandenberghe, L. *Convex Optimization*, 1st ed.; Cambridge University Press: Cambridge, UK, 2004.

59. Navascués, M.; Pironio, S.; Acín, A. Bounding the Set of Quantum Correlations. *Phys. Rev. Lett.* **2007**, *98*, 010401. [CrossRef] [PubMed]

60. Navascués, M.; Pironio, S.; Acín, A. A convergent hierarchy of semidefinite programs characterizing the set of quantum correlations. *New J. Phys.* **2008**, *10*, 073013. [CrossRef]

61. Doherty, A.C.; Liang, Y.C.; Toner, B.; Wehner, S. The Quantum Moment Problem and Bounds on Entangled Multi-prover Games. In Proceedings of the 2008 23rd Annual IEEE Conference on Computational Complexity, College Park, MD, USA, 23–26 June 2008; pp. 199–210.

62. Moroder, T.; Bancal, J.D.; Liang, Y.C.; Hofmann, M.; Gühne, O. Device-Independent Entanglement Quantification and Related Applications. *Phys. Rev. Lett.* **2013**, *111*, 030501. [CrossRef] [PubMed]

63. Kullback, S.; Leibler, R.A. On Information and Sufficiency. *Ann. Math. Statist.* **1951**, *22*, 79–86. [CrossRef]

64. Van Dam, W.; Gill, R.D.; Grunwald, P.D. The statistical strength of nonlocality proofs. *IEEE Trans. Inf. Theor.* **2005**, *51*, 2812–2835. [CrossRef]

65. Acín, A.; Gill, R.; Gisin, N. Optimal Bell Tests Do Not Require Maximally Entangled States. *Phys. Rev. Lett.* **2005**, *95*, 210402. [CrossRef]

66. Zhang, Y.; Knill, E.; Glancy, S. Statistical strength of experiments to reject local realism with photon pairs and inefficient detectors. *Phys. Rev. A* **2010**, *81*, 032117. [CrossRef]

67. Bancal, J.D.; Pironio, S.; Acin, A.; Liang, Y.C.; Scarani, V.; Gisin, N. Quantum non-locality based on finite-speed causal influences leads to superluminal signalling. *Nat. Phys.* **2012**, *8*, 867–870. [CrossRef]

68. Barnea, T.J.; Bancal, J.D.; Liang, Y.C.; Gisin, N. Tripartite quantum state violating the hidden-influence constraints. *Phys. Rev. A* **2013**, *88*, 022123. [CrossRef]

69. Chen, S.L.; Budroni, C.; Liang, Y.C.; Chen, Y.N. Natural Framework for Device-Independent Quantification of Quantum Steerability, Measurement Incompatibility, and Self-Testing. *Phys. Rev. Lett.* **2016**, *116*, 240401. [CrossRef] [PubMed]

70. Chen, S.L.; Budroni, C.; Liang, Y.C.; Chen, Y.N. Exploring the framework of assemblage moment matrices and its applications in device-independent characterizations. *Phys. Rev. A* **2018**, *98*, 042127. [CrossRef]

71. Fiala, J.; Kočvara, M.; Stingl, M. PENLAB: A MATLAB solver for nonlinear semidefinite optimization. *arXiv* **2013**, arXiv:1311.5240.

72. Christensen, B.G.; Liang, Y.C.; Brunner, N.; Gisin, N.; Kwiat, P.G. Exploring the Limits of Quantum Nonlocality with Entangled Photons. *Phys. Rev. X* **2015**, *5*, 041052. [CrossRef]

73. Poh, H.S.; Joshi, S.K.; Cerè, A.; Cabello, A.; Kurtsiefer, C. Approaching Tsirelson's Bound in a Photon Pair Experiment. *Phys. Rev. Lett.* **2015**, *115*, 180408. [CrossRef]

74. Minka, T. The Lightspeed Matlab Toolbox. Available online: https://github.com/tminka/lightspeed (accessed on 18 June 2017).

75. Christensen, B.G. (University of Wisconsin-Madison, Madison, WI, USA). Personal communication, 2017.

76. Pütz, G.; Rosset, D.; Barnea, T.J.; Liang, Y.-C.; Gisin, N. Arbitrarily Small Amount of Measurement Independence Is Sufficient to Manifest Quantum Nonlocality. *Phys. Rev. Lett.* **2014**, *113*, 190402. [CrossRef]

77. Nuzzo, R. Statistical errors: P values, the 'gold standard' of statistical validity, are not as reliable as many scientists assume. *Nature* **2014**, *506*, 150. [CrossRef]

78. Leek, J.T.; Peng, R.D. P values are just the tip of the iceberg. *Nature* **2015**, *520*, 612. [CrossRef] [PubMed]

79. Wasserstein, R.L.; Lazar, N.A. The ASA's Statement on p-Values: Context, Process, and Purpose. *Am. Stat.* **2016**, *70*, 129–133. [CrossRef]

80. Smania, M.; Kleinmann, M.; Cabello, A.; Bourennane, M. Avoiding apparent signaling in Bell tests for quantitative applications. *arXiv* **2018**, arXiv:1801.05739.

![entropy logo] *entropy*

MDPI

Article

Efficient Quantum Teleportation of Unknown Qubit Based on DV-CV Interaction Mechanism

Sergey A. Podoshvedov

Department of Computer Modeling and Nanotechnology, Institute of Natural and Exact Sciences, South Ural State University, Lenin Av. 76, 454080 Chelyabinsk, Russia; sapodo68@gmail.com

Received: 13 December 2018; Accepted: 1 February 2019; Published: 5 February 2019

Abstract: We propose and develop the theory of quantum teleportation of an unknown qubit based on the interaction mechanism between discrete-variable (DV) and continuous-variable (CV) states on highly transmissive beam splitter (HTBS). This DV-CV interaction mechanism is based on the simultaneous displacement of the DV state on equal in absolute value, but opposite in sign displacement amplitudes by coherent components of the hybrid in such a way that all the information about the displacement amplitudes is lost with subsequent registration of photons in the auxiliary modes. The relative phase of the displaced unknown qubit in the measurement number state basis can vary on opposite, depending on the parity of the basis states in the case of the negative amplitude of displacement that is akin to action of nonlinear effect on the teleported qubit. All measurement outcomes of the quantum teleportation are distinguishable, but the teleported state at Bob's disposal may acquire a predetermined amplitude-distorting factor. Two methods of getting rid of the factors are considered. The quantum teleportation is considered in various interpretations. A method for increasing the efficiency of quantum teleportation of an unknown qubit is proposed.

Keywords: discrete-variable states; continuous-variable states; quantum teleportation of unknown qubit; hybrid entanglement; collapse of the quantum state

1. Introduction

Quantum nonlocality is a property of the universe that is independent of our description of nature. Quantum mechanical predictions on entangled quantum states cannot be simulated by any local hidden variable theory [1] that is confirmed in the experiments [2,3]. Bell's theorem [1] rules out local hidden variables to explain observed results. Although, in the general case, quantum nonlocality is not equivalent to notion of entanglement, and the pure bipartite quantum state can most obviously manifest its nonlocal correlations. An example of the manifestation of the nonlocal nature of quantum objects is quantum teleportation [4]. The quantum entangled state, connecting the sender and receiver of quantum information, is used. In the protocol, an unknown quantum state of a physical system and a part of an entangled state are measured in base of some states and, subsequently, reconstructed at a remote location (the physical components of the original system remain at the sending location) due to nonlocal nature of quantum channel. Quantum nonlocality does not allow for faster-than-communication [5], and hence is compatible with special relativity. Quantum teleportation can be reviewed as a protocol that most clearly demonstrates the nonlocal trait of quantum entanglement. Quantum teleportation protocol is of interest as a concept, as well as a basis, for many other quantum protocols. Quantum teleportation protocol is used in schemes with quantum repeaters [6], serving as the main ingredient for quantum communication over large distances. Quantum teleportation protocol underlies quantum gate teleportation [7] and measurement-based computing [8]. The quantum teleportation protocol is demonstrated in experiments using different physical systems and technologies. The quantum teleportation with polarization qubits is shown in

Reference [9]. The teleportation of unknown qubits of various nature through two-mode squeezed vacuum was demonstrated in References [10–12]. Also, quantum teleportation was achieved in laboratories including nuclear magnetic resonance [13], atomic ensembles [14], trapped atoms [15], and solid-state systems [16].

Traditionally, when we talk about quantum teleportation, we mean quantum teleportation for a two-level system called the qubit [4]. Alice performs a joint quantum measurement, called Bell detection, which projects her unknown qubit and half-quantum channel into one of the states $(\sigma_i \otimes I)|\Psi\rangle$, where σ_i is Pauli operator, I is identical operator, $|\Psi\rangle$ is one of the four Bell states, $i = 0, \ldots, 3$ and the symbol $\otimes$ means tensor product. Alice's state of an unknown qubit disappears at her disposal, but in return, Bob simultaneously receives a state $\sum_{i=0}^{3} \sigma_i^{+} \varrho \sigma_i$, where ϱ is teleported qubit and σ_i^{+} means Hermitian adjoint Pauli operator. Alice must communicate her measurement outcome k to Bob, who then applies σ_i and recovers the original unknown qubit ϱ. Despite its mathematical simplicity, the implementation of the complete Bell-states measurement faces a fundamental limitation [17]. Only two Bell states can be distinguished by linear optics methods, which limits the probability of the success of quantum teleportation and the implementation of a controlled $-X$ gate by 0.5 and 0.25 [7], respectively. Attempts to circumvent this limitation are hardly possible due to the increasing difficulties in implementation [18–20]. Therefore, the multiparticle quantum entangled channel, which can hardly be generated in practice, with the subsequent registration of measurement outcomes exceeding 2 bits of classical information, is required for teleportation of an unknown qubit with the success probability approaching unity in case of a significant increase of the number of the particles [18].

Quantum teleportation can also be extended to transmit information about quantum systems living in infinite-dimensional Hilbert space, known as continuous-variable (CV) systems. Vaidman proposed the teleportation of state of one-dimensional particle and CV quantum system using (Einstein-Podolsky-Rosen) EPR-Bohm pair [21]. Later, this idea was developed in representation of position- and momentum-like quadrature operators [22], now known as CV teleportation. CV teleportation can be made in a deterministic manner, but with limited fidelity, in contrast to discrete-variable (DV) teleportation, with the fidelity of the output state equal to one in ideal conditions. CV teleportation is applicable to transmitting both CV [10,11] and DV [12] states. Details of the CV states, including CV quantum teleportation, can be found in Reference [23].

It was shown in Reference [24] that one cannot perform complete Bell-states measurement without a "quantum-quantum" interaction, which implies consideration of a hybrid physical system consisting of different ingredients, for example, atom and electromagnetic field in cavities. In general, a hybrid system may consist of components that may differ in nature, in size, or in description. So, in the case of using light, we can consider hybrid systems that are formed by DV and CV states [25]. Recently, the possibility of generating [26] and manipulating [27] hybrid entangled states was shown. The hybrid entangled states that are formed from number states and their displaced analogues or the same displaced number states [28–30] are of interest. The implementation of the displaced states of light was discussed in References [31,32]. Here, we offer a new type of quantum teleportation of an unknown qubit, which is based on nonlinear effect of interaction of DV and CV states on a highly transmissive beam splitter (HTBS). Such an approach aims to make use of advantages of DV and CV states to teleport an unknown qubit with larger success probability and high fidelity. The proposed approach differs from DV and CV teleportation, but can be recognized as being closer to CV one. Hybrid entanglement, formed by coherent components with different in sign amplitudes and dual-rail single photon, is used for transmission of quantum information from sender to receiver. The nonlinear effect on the target state in Bob's hands is realized due to interaction of CV and DV states on HTBS [26,33,34] (DV-CV interaction mechanism). Various interpretations of the DV-CV quantum teleportation of an unknown qubit are reviewed and found, to date, the best strategies for increasing its efficiency in terms of success probability.

2. DV-CV Quantum Teleportation of Unknown Qubit via Hybrid Non-Maximally Entangled State

Consider the following hybrid entangled state as quantum channel for the quantum teleportation of unknown qubit

$$|\Psi\rangle_{156} = (|-\beta\rangle_1|01\rangle_{56} + |\beta\rangle_1|10\rangle_{56})/\sqrt{2}, \tag{1}$$

where the subscript denotes the number of the mode as indicated in Figure 1. The hybrid entangled state consists of the coherent components with opposite in sign amplitudes (here and in the following the amplitude is assumed to be positive $\beta > 0$) and the single photon taking simultaneously two modes (dual-rail single photon). The state (1) is non-maximally entangled state due to the non-orthogonality of the coherent states. Negativity, which is easy to compute in four-dimensional Hilbert space, can be taken as a measure of the quantum entanglement [35]. The quantity is derived from (Positive Partial Transposition) PPT criterion for separability [36] and possesses all proper properties for the entanglement measure. The negativity of the composed system can be defined in terms of the density matrix ϱ as $\tau = (\|\varrho^{T_A}\| - 1)/2$, where ϱ^{T_A} is the partial transpose of ϱ with respect to subsystem A of two-partite system AB and $\|\varrho^{T_A}\| = tr|\varrho^{T_A}| = tr\sqrt{(\varrho^{T_A})^+ \varrho^{T_A}}$ is the trace norm of the sum of the singular values of the operator ϱ^{T_A}, where $(\varrho^{T_A})^+$ means Hermitian conjugate operator of original ϱ^{T_A}. The negativity takes the maximum value $\tau_{max} = 1$ for maximally entangled states. Doing the calculations for the state (1), one obtains

$$\tau = \sqrt{1 - exp(-4|\beta|^2)}. \tag{2}$$

The negativity of the hybrid state (1) attains maximal value $\tau \to \tau_{max}$ in the case of an infinitely large value of the amplitude of the coherent states $\beta \to \infty$. Otherwise, the hybrid state (1) is non-maximally entangled state. Although for sufficiently large values of the amplitude β of the coherent states, the hybrid state (1) can be considered as almost maximal one $\tau \approx \tau_{max}$ since the exponential factor decreases rapidly enough.

Now, we are going to use non-maximally entangled state (1) to teleport unknown qubit, in general case, represented by the following superposition (Figure 1)

$$|\varphi^{(lk)}\rangle_{34} = a_0|lk\rangle_{34} + a_1|kl\rangle_{34}, \tag{3}$$

satisfying the normalization condition $|a_0|^2 + |a_1|^2 = 1$ with qubit's amplitudes a_0 and a_1 unknown to anyone, where $|l\rangle$ and $|k\rangle$ are the arbitrary number (Fock) states. In particular, we have unknown dual-rail single photon

$$|\varphi^{(01)}\rangle_{34} = a_0|01\rangle_{34} + a_1|10\rangle_{34}, \tag{4}$$

in the case of $l = 0$ and $k = 1$. Consider the optical scheme in Figure 1 adjusted for teleportation of the unknown qubit. Alice and Bob are the standard participants of the protocol who can be at considerable distance from each other. The hybrid entangled state $|\Psi\rangle_{156}$ (Equation (1)) in modes 1, 5 and 6 is used as quantum channel for the quantum teleportation, where the coherent part in mode 1 belongs to Alice, while the single photon taking simultaneously both fifth and sixth modes is in Bob's location. An unknown qubit $|\varphi^{(lk)}\rangle_{34}$ is at the disposal of Alice. In addition to the states, Alice uses ancillary coherent state with real amplitude $\beta_1 > 0 |-\beta_1\rangle_2$ taking the second mode to mix it with one of modes of the unknown qubit of beam splitter, where, in the general case, is $\beta_1 \neq \beta$. The optical scheme in Figure 1 operates in linear optics domain with optical elements and photodetectors. Key moment of the quantum teleportation implementation is to provide discrete-continuous interaction between coherent components and unknown qubit. The discrete-continuous interaction is realized on highly transmissive beam splitter, which is described by the following unitary matrix

$$BS_{13} = \begin{bmatrix} t & -r \\ r & t \end{bmatrix},$$

(5a)

where the real parameters $t > 0, r > 0$ are the transmittance $t \rightarrow 1$ and reflectance $r \rightarrow 0$, respectively, satisfying the normalization condition $t^2 + r^2 = 1$. Here, subscripts 13 imply the first mode of the state (1) and third mode of the unknown qubit (3) are mixed on the HTBS. Another HTBS

$$BS_{24} = \begin{bmatrix} t_1 & -r_1 \\ r_1 & t_1 \end{bmatrix},$$

(5b)

is used to mix ancillary coherent state, with fourth mode of the teleported qubit (the subscript 24 is used in (5b) to discriminate the beam splitter from one (5a)). Here, the real beam splitter parameters obey the similar condition $t_1^2 + r_1^2 = 1$ and, in the general case, $t_1 \neq t$ and $r_1 \neq r$. Interaction of discrete- and continuous-variable states ends in measurements performed in the modes 1, 3, and 4 leaving the state in mode 2 untouched to collapse Bob's state into a new in dependence on Alice's measurement outcomes. All information about the teleported qubit disappears in measurement process. Alice can communicate with Bob with negligible number of the classical information to help him to recover the original state.

Strong coherent pumping $|\beta\rangle$ displaces an arbitrary state ρ by some amount, provided that the beam splitter transmits a significant part of the input light $t \rightarrow 1$ [37]

$$BS(\rho \otimes |\beta\rangle\langle\beta|)BS^+ \approx D(\alpha)\rho D^+(\alpha) \otimes |\beta\rangle\langle\beta|,$$

(6)

where the displacement operator $D(\alpha)$ [37] with displacement amplitude α is used, symbol $\otimes$ means tensor product of two operators and $D^+(\alpha)$ is Hermitian conjugate of the operator $D(\alpha)$. The amplitude of the displacement is given by

$$\alpha = \beta r/t \approx \beta r,$$

(7)

in the case of $t \approx 1$. The same reasoning is applicable to interaction of arbitrary state ρ with the coherent state $|-\beta\rangle$ with output approximate state

$$BS(\rho \otimes |-\beta\rangle\langle-\beta|)BS^+ \approx D(-\alpha)\rho D^+(-\alpha) \otimes |-\beta\rangle\langle-\beta|$$

(8)

Note the condition (7) means that amplitude of the coherent state must tend to infinity $\beta \rightarrow \infty$ if $r \rightarrow 0$ to keep exact condition $\alpha = \beta r = const$. However, in a real experiment with the non-zero reflectance $r \neq 0$, the amplitude of the coherent states takes large, but nevertheless, finite values sufficient to satisfy the condition (7). For this reason, approximate equality is used in Equations (6) and (8) which goes into the exact equality in the limit case of $t \rightarrow 1$. The better we fulfill the condition $r \rightarrow 0$ and $\beta \rightarrow \infty$ with higher fidelity, the closer the output states are to the ideal ones on the right-hand side of the Equations (6) and (8).

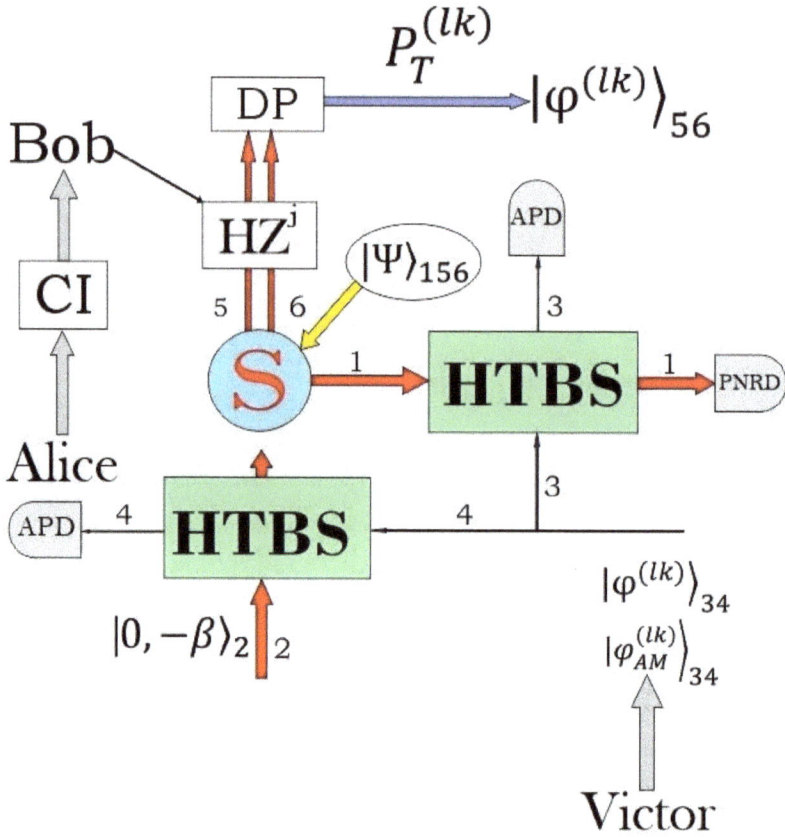

Figure 1. A schematic representation of implementation of DV-CV quantum teleportation with help of the hybrid non-maximally entangled state (1). Coherent components interact with unknown qubit in an indistinguishable manner on the HTBS. Measurements made at a microscopic level allow for Bob to obtain (after the corresponding unitary transformations initiated by the classical information (CI) from Alice) set of the states depending on Alice's measurement outcomes due to quantum nonlocality. Part of states is the original unknown states, while the others acquire additional amplitude known factors. DV-CV quantum teleportation can be performed in various interpretations in order to influence which part of the teleported qubits is original unknown state and which are (amplitude-modulated) AM states. Different implementation schemes also determine the amplitude-distorting factors of the output states. So, if the scheme involves additional HTBS for interaction of coherent state $|0,-\beta\rangle_2$ with the original state (4), then we deal with amplitude-distorting factors (16). If the scheme without additional HTBS is used, then recipient obtains the states (37) with corresponding known amplitude-distorting factors. Another interpretation includes the third party that initially generates AM unknown qubit. The considered schemes should also include a demodulation procedure (DP) in order to get rid of amplitude-distorting factors. Commercially achievable avalanche photodiode (APD), being highly sensitive semiconductor electronic devices, are used for registration of the measurement outcomes. Photon number resolving detector is used in first (coherent) mode to determine the parity of the (superposition of coherent states) SCS. S means a source of the hybrid entangled state (6).

Now, we are going to make use of mathematical apparatus developed in References [28,33,34] with displaced number states defined with help of the displacement operator $|n,\alpha\rangle = D(\alpha)|n\rangle$ [28]. The states are orthogonal $\langle n,\alpha|m,\alpha\rangle = \delta_{nm}$ with δ_{nm} being Kronecker delta [28]. The displaced number states are defined by two numbers: Quantum discrete number n and classical continuous parameter

α which can be recognized as their size. The partial case is the infinite set of the number states $\{|n\rangle, n = 0, 1, 2, \ldots, \infty\}$ with $\alpha = 0$. Here, we are going to make use of the completeness of the Fock's states $\sum_{n=0}^{\infty} |n\rangle\langle n| = I$ to decompose arbitrary displaced number states $|l, \alpha\rangle$ over them [34]

$$|l, \alpha\rangle = F \sum_{n=0}^{\infty} c_{ln}(\alpha)|n\rangle, \tag{9}$$

where the overall multiplier $F(\alpha) = exp(-|\alpha|^2/2)$ is introduced. The matrix elements $c_{ln}(\alpha)$ satisfy the normalization condition $F^2 \sum_{n=0}^{\infty} |c_{ln}(\alpha)|^2 = 1$ [34]. In particular, the matrix elements $c_{0n}(\alpha) = \alpha^n/\sqrt{n!}$ are the amplitudes of the coherent state $|\alpha\rangle \equiv |0, \alpha\rangle$ [28]. All other matrix elements with $l \neq 0$ are presented in Reference [34].

The realization of the nonlinear effect in DC interaction is ensured by the property of matrix elements to change their sign when changing the displacement amplitude on opposite in sing $\alpha \rightarrow -\alpha$. The matrix elements change as

$$c_{ln}(-\alpha) = (-1)^{n-l} c_{ln}(\alpha). \tag{10}$$

under change of the displacement amplitude on opposite $\alpha \rightarrow -\alpha$ [34]. In particular, we have the following relation for the matrix elements of even $l = 2m$ displaced number states

$$c_{2mn}(-\alpha) = (-1)^n c_{2mn}(\alpha), \tag{11}$$

and for the matrix elements of odd $l = 2m + 1$ displaced number states and

$$c_{2m+1n}(-\alpha) = (-1)^{n-1} c_{2m+1n}(\alpha), \tag{12}$$

for the decomposition of odd $l = 2m + 1$ displaced number states. In particular, we have $c_{0n}(-\alpha) = (-1)^n c_{0n}(\alpha)$ for the amplitudes of the coherent state. This difference in the behavior of the matrix elements when changing parity of the displaced number states (Equations (11) and (12)) is similar to a nonlinear action of two-qubit gate controlled $-Z$ gate. Coherent components of the hybrid entangled state (1) simultaneously displace the unknown teleported qubit (3) in an indistinguishable manner on HTBS, as shown in Equations (6) and (8), respectively, by the values that differ from each other only by sign. All information about value of the displacement of the teleported qubit (either by α or $-\alpha$) disappears. Measurement of the unknown teleported state and coherent part of the state (1) collapses the original state $BS_{13}BS_{24}\left(|\Psi\rangle_{156}|0, -\beta_1\rangle_2|\varphi^{(lk)}\rangle_{34}\right)$ onto state at Bob's disposal subject controlled $-Z$ operation in the case of corresponding parity of the number states $|l\rangle$ and $|k\rangle$ in (3) and the teleported state can be recovered through classical communication.

Let us present mathematical details of interaction of hybrid non-maximally entangled state (1) and ancillary coherent state with unknown qubit on two HTBS (5a) and (5b) as shown in Figure 1. Due to linearity of the beam splitter operators, we have

$$BS_{13}BS_{24}\left(|\Psi\rangle_{156}|0, -\beta_1\rangle_2|\varphi^{(lk)}\rangle_{34}\right) = \left(1/\sqrt{2}\right)$$
$$\left(BS_{13}BS_{24}\left(|0, -\beta\rangle_1|0, -\beta\rangle_1|\varphi^{(lk)}\rangle_{34}\right) + BS_{13}BS_{24}\left(|0, -\beta\rangle_1|0, -\beta\rangle_1|\varphi^{(lk)}\rangle_{34}\right)\right), \tag{13}$$

where the hybrid non-maximally entangled state (1) is considered to take modes 1, 5, and 6, the teleported unknown qubit is located in modes 3 and 4, while ancillary coherent state is used in second mode. Consider action of the beam splitters on the states separately. Then, we have [38]

$$
\begin{aligned}
&BS_{13}BS_{24}\Big(|0,-\beta\rangle_1|0,-\beta\rangle_2|\varphi^{(lk)}\rangle_{34}\Big)|01\rangle_{56} = \\
&BS_{13}BS_{24}(|0,-\beta\rangle_1|0,-\beta\rangle_2(a_0|01\rangle_{34}+a_1|10\rangle_{34}))|01\rangle_{56} \to \\
&F^2|0,-\beta/t\rangle_1|0,-\beta/t_1\rangle_2 \sum_{n=0}^{\infty}\sum_{m=0}^{\infty} t^{n+m}c_{ln}(\alpha)c_{km}(\alpha_1) \\
&\Big(a_0+a_1 A_{nm}^{(lk)}\Big)|nm\rangle_{34}|01\rangle_{56}+ \\
&rF^2\left(\begin{array}{c}\sum_{n=0}^{\infty}\sum_{m=0}^{\infty} t^{n+m-1}c_{ln}(\alpha)c_{km}(\alpha_1)\Big(a_0+a_1 A_{nm}^{(lk)}\Big) \\ \left(\begin{array}{c}\sqrt{n}|1,-\beta/t\rangle_1|0,-\beta/t_1\rangle_2|n-1m\rangle_{34}+ \\ \sqrt{m}|0,-\beta/t\rangle_1|1,-\beta/t_1\rangle_2|nm-1\rangle_{34}\end{array}\right)\end{array}\right)|01\rangle_{56},
\end{aligned}
\tag{14}
$$

for the first term in Equation (13) and

$$
\begin{aligned}
&BS_{13}BS_{24}\Big(|0,\beta\rangle_1|0,-\beta\rangle_2|\varphi^{(lk)}\rangle_{34}\Big)|10\rangle_{56} = \\
&BS_{13}BS_{24}(|0,\beta\rangle_1|0,-\beta\rangle_2(a_0|01\rangle_{34}+a_1|10\rangle_{34}))|10\rangle_{56} \to \\
&F^2|0,\beta/t\rangle_1|0,-\beta/t_1\rangle_2\sum_{n=0}^{\infty}\sum_{m=0}^{\infty}(-1)^{n-l}t^{n+m}c_{ln}(\alpha)c_{km}(\alpha_1)\Big(a_0+(-1)^{l-k}a_1 A_{nm}^{(lk)}\Big)|nm\rangle_{34}|10\rangle_{56}, \\
&rF^2\left(\begin{array}{c}\sum_{n=0}^{\infty}\sum_{m=0}^{\infty}(-1)^{n-l}t^{n+m-1}c_{ln}(\alpha)c_{km}(\alpha_1)\Big(a_0+(-1)^{l-k}a_1 A_{nm}^{(lk)}\Big) \\ \left(\begin{array}{c}\sqrt{n}|1,\beta/t\rangle_1|0,-\beta/t_1\rangle_2|n-1m\rangle_{34}+ \\ \sqrt{m}|0,\beta/t\rangle_1|1,-\beta/t_1\rangle_2|nm-1\rangle_{34}\end{array}\right)\end{array}\right)|01\rangle_{56},
\end{aligned}
\tag{15}
$$

for the second term in Equation (13), where amplitude-distorting coefficients $A_{nm}^{(lk)}$ are given by

$$
A_{nm}^{(lk)}(\alpha,\alpha_1) = \frac{c_{kn}(\alpha)c_{lm}(\alpha_1)}{c_{ln}(\alpha)c_{km}(\alpha_1)}.
\tag{16}
$$

Note the displacement amplitude α_1 is determined by $\alpha_1 = \beta_1 r/t \approx \beta_1 r$ (Equation (7)). Here, we limited ourselves by the first two terms in order of smallness $r \ll 1$ neglecting members of higher order of smallness in the reflectance proportional to $\sim r^n$ with $n > 1$. First terms of zeroth order in r give maximal contribution, while influence of the second terms proportional to $\sim r$ goes to zero in the case of $r \to 0$.

Consider output state in ideal case of $t = 1$ and $r = 0$ in terms of even/odd superposition of coherent states (SCS) defined by

$$
|even\rangle = N_+(|-\beta\rangle + |\beta\rangle),
\tag{17}
$$

$$
|odd\rangle = N_-(|-\beta\rangle - |\beta\rangle),
\tag{18}
$$

where the factors $N_{\pm} = \Big(2(1\pm exp(-2|\beta|^2))\Big)^{-1/2}$ are the normalization parameters. Then, we can approximate the state $BS_{13}BS_{24}\Big(|\Psi\rangle_{156}|0,-\beta_1\rangle_2|\varphi^{(lk)}\rangle_{34}\Big)$ in zeroth order on parameter $r \ll 1$

$$
\begin{aligned}
&BS_{13}BS_{24}\Big(|\Psi\rangle_{156}|0,-\beta_1\rangle_2|\varphi^{(lk)}\rangle_{34}\Big) \approx (F^2/2)|0,-\beta_1\rangle_2 \\
&\sum_{n=0}^{\infty}\sum_{m=0}^{\infty} c_{ln}(\alpha)c_{km}(\alpha_1)N_{nm}^{(lk)-1}\left(\frac{|even\rangle_1}{N_+}|\Psi_{nm}^{(lk)}\rangle_{56}+\frac{|odd\rangle_1}{N_-}|\Psi_{n+1m}^{(lk)}\rangle_{56}\right)|nm\rangle_{34},
\end{aligned}
\tag{19}
$$

where the state at Bob's location (Bob's states) becomes

$$
|\Psi_{nm}^{(lk)}\rangle_{56} = \frac{N_{nm}^{(lk)}}{\sqrt{2}}\Big(\Big(a_0+a_1 A_{nm}^{(lk)}\Big)|01\rangle_{56}+(-1)^{n-l}\Big(a_0+(-1)^{l-k}a_1 A_{nm}^{(lk)}\Big)|10\rangle_{56}\Big),
\tag{20}
$$

$$|\Psi_{n+1m}^{(lk)}\rangle_{56} = \frac{N_{nm}^{(lk)}}{\sqrt{2}}\left(\left(a_0 + a_1 A_{nm}^{(lk)}\right)|01\rangle_{56} + (-1)^{n+1-l}\left(a_0 + (-1)^{l-k}a_1 A_{nm}^{(lk)}\right)|10\rangle_{56}\right), \qquad (21)$$

where the normalization factor $N_{nm}^{(lk)}$ is given by

$$N_{nm}^{(lk)} = \left(|a_0|^2 + |a_1|^2 |A_{nm}^{(lk)}|^2\right)^{-1/2} = \left(1 + \left(\left|A_{nm}^{(lk)}\right|^2 - 1\right)|a_1|^2\right)^{-1/2}. \qquad (22)$$

To provide the performance of nonlinear action of controlled $-Z$ gate

$$(-1)^{l-k} = -1, \qquad (23)$$

we need to impose additional requirement on the teleported qubit (3), namely, difference $l - k$ must be an odd number for used displacement amplitude $\beta > 0$ of the hybrid non-maximally entangled state (1). For example, if we take $l = 0$ and $k = 1$ (dual-rail unknown single photon), we provide performance of the condition (23).

Now, Alice must do the parity measurement at first mode to recognize even/odd SCS and registers the measurement outcome $|nm\rangle_{34}$ in measured third and fourth modes. Then, Bob obtains one of the two states either $|\Psi_{nm}^{(lk)}\rangle_{56}$ (Equation (20)) or $|\Psi_{n+1m}^{(lk)}\rangle_{56}$ (Equation (21)) in dependence on parity of the measured photons at mode 1. Assume that Alice registers only definite measurement outcome (nm) and informs Bob about it. Then, Bob can apply sequence of operators of Hadamard gate and $Z-$ gate in some power to get

$$HZ^{n-l}|\Psi_{nm}^{(lk)}\rangle = N_{nm}^{(lk)}\begin{bmatrix} a_0 \\ a_1 A_{nm}^{(lk)} \end{bmatrix}, \qquad (24)$$

$$HZ^{n-l+1}|\Psi_{n+1m}^{(lk)}\rangle = N_{nm}^{(lk)}\begin{bmatrix} a_0 \\ a_1 A_{nm}^{(lk)} \end{bmatrix}. \qquad (25)$$

$Z-$ gate is applied in dependence on the parity of the numbers $n - l$ and $n - l + 1$ as $Z^2 = I$, where I is an identical operator. Hadamard operation is applied regardless of whether Bob should initially use $Z-$ gate or not. These operations (Hadamard gate and $Z-$ gate) are easily implemented by linear optics devices on single photon [39]. Obtained state contains amplitude-distorting factor $A_{nm}^{(lk)}$ defined by Equation (16). We are going to consider such states to be amplitude-modulated (AM) states. The presence of this additional factor $A_{nm}^{(lk)}$ is a distinctive feature inherent to DV-CV interaction. One can even say that the CV state leaves its imprint in the teleporting DV state. The success probability for Alice to register the measurement outcome $|nm\rangle_{34}$ not depending on parity of the states in first mode is given by

$$P_{nm}^{(lk)} = \frac{F^4 |c_{ln}(\alpha)|^2 |c_{km}(\alpha_1)|^2}{N_{nm}^{(lk)2}}, \qquad (26)$$

where the probabilities are normalized

$$\sum_{n=0}^{\infty}\sum_{m=0}^{\infty} P_{nm}^{(lk)} = 1, \qquad (27)$$

not depending on the numbers l and k that can be directly checked using normalization of the matrix elements $c_{nm}(\alpha)$. It is worth noting the success probabilities of the measurement outcomes $P_{nm}^{(lk)}$ depend on the displacement amplitudes α and α_1 and can change in wide diapason. In other words, Alice has additional parameters which she manipulates to vary the success probabilities of her measurement outcomes.

Consider the case of $\alpha = \alpha_1$ that can be produced by application of coherent states with equal displacement amplitudes $\beta = \beta_1$ that displace the teleported qubit on equivalent HTBS (Equations (5a) and (5b)). Then, by definition (16), we have

$$A_{nn}^{(lk)} = 1. \tag{28}$$

This means that the probabilistic protocol of the DV-CV quantum teleportation of an unknown qubit can be realized if Alice registers only the same measurement outcomes $n = m$ together with parity measurement at first mode by discarding all other $n \neq m$. Moreover, Alice must transmit one bit of classical information over the classical communication channel to indicate to Bob whether he should apply $Z-$ transformation in the probabilistic teleportation. The success probability of the event is equal to

$$P_T^{(lk)} = \sum_{n=0}^{\infty} P_{nn}^{(lk)} = F^4 \sum_{n=0}^{\infty} |c_{ln}(\alpha)|^2 |c_{kn}(\alpha)|^2. \tag{29}$$

In all remaining cases $n \neq m$, the Bob's qubit receives an additional amplitude-distorting factor $A_{nm}^{(lk)}$ not equal to one being a price for implementation of controlled $-Z$ operation in DV-CV interaction. But the factor is known to both participants of the protocol provided that they know the displacement amplitude α and measurement outcomes n and m. The probability for Bob to receive AM qubit (after receiving relevant auxiliary classical information from Alice) is equal to

$$P_{AM}^{(lk)} = \sum_{n} \sum_{m,n \neq m}^{\infty} P_{nm}^{(lk)}. \tag{30}$$

Thus, the total probability can be divided into two categories: the success probability to perfectly teleport unknown qubit (29) with only one bit of assisting classical information and probability to transmit to Bob AM qubit with some amount of auxiliary classical information

$$P_T^{(lk)} + P_{AM}^{(lk)} = 1. \tag{31}$$

It is worth noting that both $P_{nm}^{(lk)}$ with $n \neq m$ (26) also depend (in addition to dependence on the displacement amplitude α) on the parameters of the teleported qubit (3), namely on the amplitude $|a_1|$ due to the amplitude-distorting factor $A_{mn}^{(lk)}$ in the normalization multiplier $N_{nm}^{(lk)}$. When receiving AM qubits, Bob can take certain measures to get rid of the amplitude-distorting factors.

Note only the amplitude factor obey the condition

$$A_{mn}^{(lk)} = \left(A_{nm}^{(lk)} \right)^{-1}, \tag{32}$$

in the case of $\alpha = \alpha_1$. Using the relation, it is possible to show that sum of two probabilities $P_{nm}^{(lk)}$ and $P_{mn}^{(lk)}$ does not depend on the amplitude $|a_1|$ of the teleported unknown qubit

$$P_{nm}^{(lk)S} = P_{nm}^{(lk)} + P_{mn}^{(lk)} = F^4 (|c_{ln}|^2 |c_{km}|^2 + |c_{lm}|^2 |c_{kn}|^2) = F^4 |c_{ln}|^2 |c_{km}|^2 \left(1 + \left| A_{nn}^{(lk)} \right|^2 \right), \tag{33}$$

where superscript S concerns the sum of two probabilities. It proves the fact that the total probability $P_{AM}^{(lk)}$ (Equation (30)) also does not depend on the parameter $|a_1|$ of the teleported qubit in spite of the fact that each member $P_{nm}^{(lk)}$ of this sum still depends on the parameters $|a_1|$ of the teleported qubit. Finally, the probability for Bob to obtain AM originally unknown qubit can be rewritten as

$$P_{AM}^{(lk)} = \sum_{n} \sum_{m,n \neq m}^{\infty} P_{nm}^{(lk)S} = F^4 \sum_{n} \sum_{m,n \neq m}^{\infty} (|c_{ln}|^2 |c_{km}|^2 + |c_{lm}|^2 |c_{kn}|^2). \tag{34}$$

The proposed method of implementing DV-CV quantum teleportation can also be used for the unknown single-rail unknown qubit composed of $|l\rangle$ and $|k\rangle$ photons

$$|\phi^{(lk)}\rangle_2 = a_0|l\rangle_2 + a_1|k\rangle_2. \tag{35}$$

In particular, the unknown single-rail qubit $|\phi^{(01)}\rangle$ is the superposition of vacuum and single photon. The same state in Equation (1) is used as quantum channel for quantum teleportation of unknown qubit (35). In this case, we can also use the scheme in Figure 1, but only without interacting with the additional coherent state $|-\beta_1\rangle_2$. Then, following the same technique, we obtain

$$BS_{12}\left(|\Psi\rangle_{134}|\phi^{(lk)}\rangle_2\right) \to (F/2)\sum_{n=0}^{\infty} c_{ln}(\alpha)N_n^{(lk)-1}\left(\frac{|even\rangle_1}{N_+}|\Psi_n^{(lk)}\rangle_{34} + \frac{|odd\rangle_1}{N_-}|\Psi_{n+1}^{(lk)}\rangle_{56}\right)|n\rangle_2, \tag{36}$$

in the case of $t \to 1$ and $r \to 0$. Another difference from the formula (19) is that the real amplitude-distorting factors $A_n^{(lk)}$ in the states $|\Psi_n^{(lk)}\rangle$ and $|\Psi_{n+1}^{(lk)}\rangle$ are determined by

$$A_n^{(lk)} = \frac{c_{kn}(\alpha)}{c_{ln}(\alpha)}, \tag{37}$$

where the states in Bob's location are the same as in Equations (20) and (21) with the normalization factors $N_n^{(lk)} = \left(1 + \left(\left|A_n^{(lk)}\right|^2 - 1\right)|a_1|^2\right)^{-1/2}$. If Alice performs the parity measurement in the first mode and determines the number of photons in the second measurement mode, then she collapses the initial state into one of the possible states either (20) or (21). Then, she can send Bob additional classic information so that he can make corresponding unitary transformations with his qubit to get the AM state with known factor $A_n^{(lk)}$

$$N_n^{(lk)}\begin{bmatrix} a_0 \\ a_1 A_n^{(lk)} \end{bmatrix}, \tag{38}$$

with success probability

$$P_n^{(lk)} = \frac{F^2|c_{ln}(\alpha)|^2}{N_n^{(lk)2}}. \tag{39}$$

From the comparison of amplitude-distorting coefficients $A_{nm}^{(lk)}$ and $A_n^{(lk)}$, we can see a difference in the two types of DV-CV quantum teleportation of unknown qubit. Registration of identical outcomes $n = m$ in two auxiliary measurement modes leads to the fact that Bob's state gets rid of these additional parameters $A_{nn}^{(lk)}$ (Equation (28)). Teleportation of the single-rail initial state (35) without amplitude-distorting parameter $A_n^{(lk)} = \pm 1$ is possible if $c_{kn}(\alpha) = \pm c_{ln}(\alpha)$.

3. Methods to Increase the Success Probabilities of the DV-CV Quantum Teleportation

In the previous section, we showed that the DV-CV quantum teleportation protocol allows us to transfer to Bob either the original unknown qubit or its amplitude-distorted version. All measurement outcomes give different states and all amplitude-distorting coefficients are known in advance. The implementation of the DV-CV protocol takes place in a deterministic manner, but the fidelity of the output state, in the general case, is not ideal equal to one. Therefore, our efforts are now focused on the opportunity for Bob to restore the initial state from AM qubit with help of communication with Alice. To consider methods to increase the success probabilities of the quantum teleportation, let us present matrix elements for the first six displaced number states. So, we have for the coherent state $|0, \alpha\rangle$

$$c_{0n}(\alpha) = \alpha^n / \sqrt{n!}, \tag{40}$$

for the displaced singe photon $|1, \alpha\rangle$

$$c_{1n}(\alpha) = \alpha^{n-1}(n - |\alpha|^2)/\sqrt{n!}, \tag{41}$$

for the displaced two-photon state $|2, \alpha\rangle$

$$c_{2n}(\alpha) = \alpha^{n-2}(n(n-1) - 2n|\alpha|^2 + |\alpha|^4)/(\sqrt{2}\sqrt{n!}), \tag{42}$$

for the displaced three-photon state $|3, \alpha\rangle$

$$c_{3n}(\alpha) = \alpha^{n-3}(n(n-1)(n-2) - 3n(n-1)|\alpha|^2 + 3n|\alpha|^4 - |\alpha|^6)/(\sqrt{3!}\sqrt{n!}), \tag{43}$$

for the displaced four-photon state $|4, \alpha\rangle$

$$c_{4n}(\alpha) = \alpha^{n-4}\left(\begin{array}{c} n(n-1)(n-2)(n-3) - 4n(n-1)(n-2)|\alpha|^2 + \\ 6n(n-1)|\alpha|^4 - 4n|\alpha|^6 + |\alpha|^8 \end{array} \right)/\left(\sqrt{4!}\sqrt{n!}\right), \tag{44}$$

for the displaced state with five photons $|5, \alpha\rangle$

$$c_{5n}(\alpha) = \alpha^{n-5}\left(\begin{array}{c} n(n-1)(n-2)(n-3)(n-4) - \\ 5n(n-1)(n-2)(n-3)|\alpha|^2 + \\ 10n(n-1)(n-2)|\alpha|^4 - 10n(n-1)|\alpha|^6 + \\ 5n|\alpha|^8 - |\alpha|^{10} \end{array} \right)/\left(\sqrt{5!}\sqrt{n!}\right). \tag{45}$$

Using the expressions and formulas $A_{nm}^{(lk)}$ (Equation (16)) and $A_n^{(lk)}$ (Equation (37)), we can calculate any amplitude-distorting factor for any teleported unknown qubit.

Suppose that Bob can demodulate his AM unknown qubit either $N_{nm}^{(lk)}(a_0|01\rangle + a_0 A_{nm}^{(lk)}|01\rangle)$ or $N_n^{(lk)}(a_0|01\rangle + a_0 A_n^{(lk)}|01\rangle)$ with the probability $q_{nm}^{(lk)}$. Then, we get the next addition to the overall success probability of DV-CV quantum teleportation

$$\delta P_T^{(lk)} = F^4 \sum_n \sum_{m,n \neq m}^{\infty} \left(q_{nm}^{(lk)}|c_{ln}|^2|c_{km}|^2 + q_{mn}^{(lk)}|c_{lm}|^2|c_{kn}|^2 \right), \tag{46}$$

where the overall success probability $P_T^{(lk)O}$ becomes

$$P_T^{(lk)O} = P_T^{(lk)} + \delta P_T^{(lk)}. \tag{47}$$

Here, the normalization factor $N_{nm}^{(lk)}$ in expression for the success probability disappears as we get rid of the amplitude-distorting factor $A_{nm}^{(lk)}$. Similar addition to the success probability can be obtained in the case of amplitude demodulation of an unknown qubit $N_n^{(lk)}(a_0|01\rangle + a_0 A_n^{(lk)}|01\rangle)$.

Amplitude demodulation of an unknown qubit (or the same deliverance from amplitude-distorting factor) may not be an easy task. It seems that this operation could be performed at the next conversion: $|01\rangle \rightarrow |01\rangle$ and $|10\rangle \rightarrow \exp(\pm\Gamma)|10\rangle$, where either $\exp(\pm\Gamma) = A_{nm}^{(lk)-1}$ or $\exp(\pm\Gamma) = A_n^{(lk)-1}$ in dependency on $A_{nm}^{(lk)} < 1$, $A_n^{(lk)} < 1$ or $A_{nm}^{(lk)} > 1$, $A_n^{(lk)} > 1$ with Γ being some either amplifying or weakening parameter. The conversion is not unitary. Consider more realistic scheme for amplitude demodulation of unknown qubit $N_{nm}^{(lk)}(a_0|01\rangle_{12} + a_1 A_{nm}^{(lk)}|10\rangle_{12})$. Reconstruction of the original state [40] is probabilistic provided that some measurement outcome is fixed in auxiliary mode. The mode 2 in the state is auxiliary. The displacement operator $D_2(\gamma)$ with amplitude γ acts on second mode of the state producing

$$D_2(\gamma)N_{nm}^{(lk)}\left(a_0|01\rangle_{12} + a_1 A_{nm}^{(lk)}|10\rangle_{12}\right) = $$
$$F(\gamma)\sum_{n=0}^{\infty}c_{1n}(\gamma)\left(a_0|0\rangle_1 + a_1 A_{nm}^{(lk)}(c_{0n}(\gamma)/c_{1n}(\gamma))|1\rangle_1\right)|n\rangle_2. \tag{48}$$

Measurement of the $|n\rangle$ photons in second mode generates the following state (leaving out normalization factor) $a_0|0\rangle_1 + a_1 A_{nm}^{(lk)}(c_{0n}(\gamma)/c_{1n}(\gamma))|1\rangle_1$ [40] which is converted into original one provided the following condition

$$A_{nm}^{(lk)}(\alpha)(c_{0n}(\gamma)/c_{1n}(\gamma)) = \pm 1, \tag{49}$$

is satisfied. Then, the success probability of the amplitude demodulation through the displacement operator is given by

$$q_{nm}^{(lk)D} = F^2(\gamma)|c_{1n}(\gamma)|^2, \tag{50}$$

where value of the parameter γ follows from (49) and superscript D means the original state is obtained with help of mixing it with coherent state.

Consider another way to get rid of amplitude factor $A_{nm}^{(lk)}$ in the unknown qubit. To do this, we are going to make use of quantum swapping method [41] when AM unknown qubit $N_{nm}^{(lk)}\left(a_0|01\rangle_{12} + a_1 A_{nm}^{(lk)}|10\rangle_{12}\right)$ interacts with the prearranged state

$$|\Psi_{nm}^{(lk)'}\rangle_{34} = N_n^{(lk)'}\left(A_{nm}^{(lk)}|01\rangle_{34} + |10\rangle_{34}\right), \tag{51}$$

where $N_n^{(lk)'} = \left(1 + \left|A_n^{(lk)}\right|^2\right)^{-1/2}$ is a normalization factor. Here, modes 2 and 3 are mixed on balanced beam splitter (5a) with $t = r = 1/\sqrt{2}$ with subsequent registration of outcomes either $|01\rangle_{23}$ or $|10\rangle_{23}$ that leads to production of original unknown qubit with success probability

$$q_{nm}^{(lk)S} = \frac{\left|A_{nm}^{(lk)}\right|^2}{1 + \left|A_{nm}^{(lk)}\right|^2}, \tag{52}$$

where subscript S concerns the fact that an unknown qubit was restored by the quantum swapping method. We note only the fact that amplitude demodulation by using amplitude displacement allows us to continue this procedure with the remaining states not satisfying the condition (49), while quantum swapping procedure can only be performed once.

The same demodulation methods are applicable to the states (38). Then, we have the success probability for Bob to restore original unknown qubit from AM one

$$\delta P_T^{(lk)D} = F^2(\alpha)\sum_n^{\infty}\sum_p^{\infty}F^2(\gamma)|c_{ln}(\alpha)|^2|c_{1p}(\gamma)|^2, \tag{53}$$

where parameter γ follows from relation

$$A_n^{(lk)}(\alpha)(c_{0n}(\gamma)/c_{1n}(\gamma)) = \pm 1. \tag{54}$$

Another way to demodulate AM unknown qubits (38) allows for us to perform it with success probability

$$\delta P_T^{(lk)S} = F^2(\alpha)|c_{ln}(\alpha)|^2 \frac{\left|A_n^{(lk)}\right|^2}{1 + \left|A_n^{(lk)}\right|^2}. \tag{55}$$

We consider the case of $l < k$ and $n < m$. Let us start with the case of $l = 0$ and $k = 1$. Corresponding curves of $P_T^{(01)}$ and $P_{nm}^{(01)S}$ for different n and m in dependency on α are shown in the left part of the Figure 2. Success probability to teleport unknown qubit without amplitude demodulation procedures takes maximal value $\left(P_T^{(01)}\right)_{max} = 0.2637$ under $\alpha = 0.628482$. The condition $A_{01}^{(01)} = A_{01}^{(01)} = -1$ is turned out to be satisfied in the case of $\alpha = 1/\sqrt{2}$. This allows us to increase the success probability $P_T^{(01)}\left(\alpha = 1/\sqrt{2}\right) = 0.2578$ by 0.18394. Thus, the success probability for Alice to directly teleport to Bob unknown qubit becomes $P_S = P_T^{(01)}\left(\alpha = 1/\sqrt{2}\right) + P_{01}^{(01)S}\left(\alpha = 1/\sqrt{2}\right) = 0.441789$ as shown on the right side of the Figure 2. At the same time, the probability $P_S = P_T^{(01)}(\alpha = 0.628482) + P_{01}^{(01)S}(\alpha = 0.628482) = 0.500673$ takes on greater value for the displacement amplitude corresponding to maximal value of $P_T^{(01)}$. But this probability consists of two events: the direct teleportation of an unknown qubit (without amplitude-distorting factor) and the teleportation with output AM qubit which needs an amplitude demodulation procedure. Consider the case of $l = 1$ and $k = 2$, whose functions $P_T^{(12)}$ and $P_{nm}^{(12)S}$ for different n and m in dependency on α are shown in the left part of the Figure 3. The success probability $P_T^{(12)}$ has its maximum under $= 0.4072$ $\left(P_T^{(01)}\right)_{max} = 0.24371$. If we consider the contribution from the realization of the AM states with $A_{12}^{(12)}(\alpha = 0.4072) = A_{12}^{(12)-1}(\alpha = 0.4072)$, then this adds a value $P_{12}^{(12)S}(\alpha = 0.4072) = 0.2883$ to $\left(P_T^{(01)}\right)_{max}$, finally, resulting in $\left(P_T^{(01)}\right)_{max} + P_{12}^{(12)S}(\alpha = 0.4072) = 0.5317$ as shown on the right side of the Figure 3. We have $A_{12}^{(12)}(\alpha = 0.5053) = A_{12}^{(12)-1}(\alpha = 0.5053) = -1$. Thus, choosing the value of $\alpha = 0.5053$, we get the probability of success of quantum teleportation of an unknown state (without amplitude-distorting factor) equal to $P_S = P_T^{(12)}(\alpha = 0.5053) + P_{12}^{(12)S}(\alpha = 0.5053) = 0.4014$.

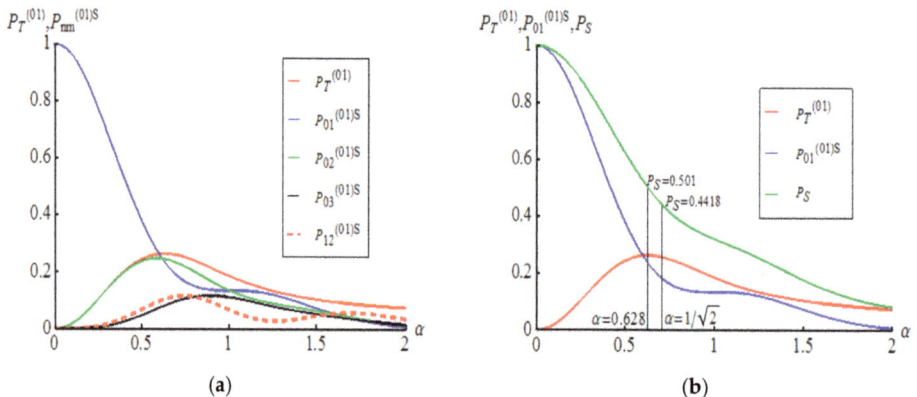

Figure 2. (a) Plots of the success probabilities $P_T^{(01)}$ and $P_{nm}^{(01)}$ for different n and m in dependency on the displacement amplitude α. (b) Only three graphs of probabilities $P_T^{(01)}$, $P_{01}^{(01)S}$, giving the maximum contribution, and $P_S = P_T^{(01)} + P_{01}^{(01)S}$ are shown. The value $P_S = 0.4418$ corresponds to quantum teleportation of unknown qubit without amplitude-distorting factor.

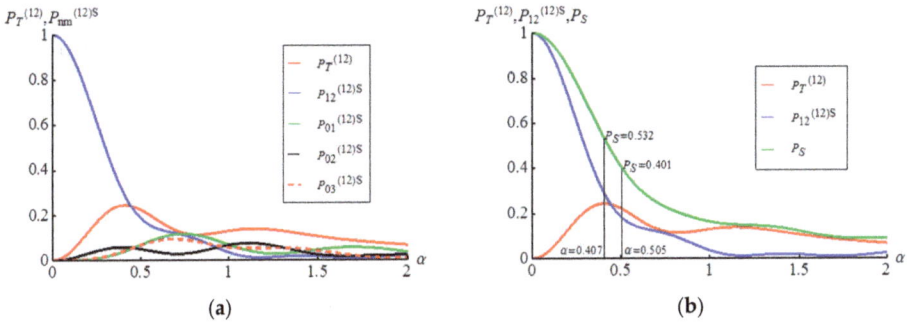

Figure 3. (a) Plots of the success probabilities $P_T^{(12)}$ and $P_{nm}^{(12)}$ for different n and m in dependency on the displacement amplitude α. (b) Only three graphs of probabilities $P_T^{(12)}$, $P_{12}^{(12)S}$, giving the maximum contribution, and $P_S = P_T^{(12)} + P_{12}^{(12)S}$ are shown. The value $P_S = 0.401$ corresponds to quantum teleportation of unknown qubit without amplitude-distorting factor.

Let us analyze the amplitude-distorting factors $A_{nm}^{(lk)}$. Two examples of the values of this parameter are given in Table 1 for $l = 0$ and $k = 1$ and Table 2 for $l = 1$ and $k = 2$, respectively.

Table 1. Values of amplitude-distorting factors $A_{nm}^{(01)}\left(\alpha = 1/\sqrt{2}\right)$ for different values of n and m.

n	0	0	1	0	1	0	1	0
m	2	3	2	4	3	5	4	5
$A_{nm}^{(01)}$	$-1/3$	-0.2	$1/3$	$-1/7$	0.2	$-1/9$	$1/7$	$3/5$
$A_{mn}^{(01)}$	-3	-5	3	-7	5	-9	7	$5/3$

Table 2. Values of amplitude-distorting factors $A_{nm}^{(12)}(\alpha = 0.5053)$ for different values of n and m.

n	0	0	0	0	1
m	1	2	3	4	3
$A_{nm}^{(12)}$	0.427	-0.427	-0.155	-0.0954	-0.362
$A_{mn}^{(12)}$	2.343	-2.343	-6.468	-10.481	-2.76

Amplitude-distorting factors can be divided into two types: $A_{nm}^{(lk)} < 1$ and $A_{mn}^{(lk)} > 1$, provided that $n < m$. It follows from Equation (52) the probability $q_{nm}^{(lk)S} \approx 1$ in the case of $A_{mn}^{(lk)} > 1$ that means quantum swapping procedure can be used to restore the original unknown qubit from AM one with high probability. In the opposite case of AM state with amplitude-distorting factor $A_{nm}^{(lk)}$, the probability $q_{nm}^{(lk)S}$ takes small values. It turns out that the probability $F^4|c_{ln}|^2|c_{km}|^2$ can be much larger than $F^4|c_{lm}|^2|c_{kn}|^2$ i.e., $F^4|c_{ln}|^2|c_{km}|^2 > F^4|c_{lm}|^2|c_{kn}|^2$. Then, the main task is to search for demodulation procedure of the AM state with $A_{nm}^{(lk)} < 1$ which, for the time being, is quite a difficult problem. So, we have observed that overall success probability to teleport unknown qubit only using those two proposed demodulation methods becomes $P_T^{(01)O} = 0.522765$ and $P_T^{(12)O} = 0.4968$.

Similar difficulties occur in the demodulation of AM states (38) with amplitude-distorting factors $A_n^{(lk)}$ (Equation (37)). Again, states with $A_n^{(lk)} > 1$ can be restored by quantum swapping procedure with probability (52) close to 1. The corresponding success probabilities $\delta P_T^{(lk)S}$ (Equation (55)) and $\delta P_T^{(lk)D}$ (Equation (53)) for teleporting and restoring AM unknown qubit depending on the displacement amplitude α are shown in Figure 4. It is worth noting Bob can continue the demodulation procedure in the case of use of method with displacement operator.

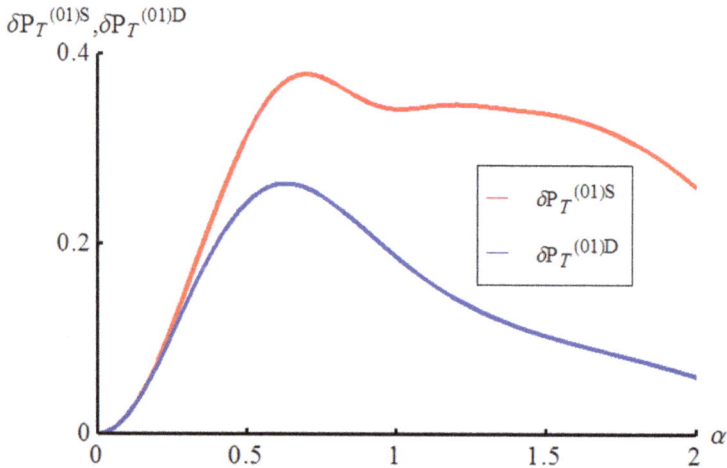

Figure 4. Plots of the success probabilities $\delta P_T^{(01)S}$ and $\delta P_T^{(01)D}$ to teleport and restore (get rid of amplitude-distorting factor by one two proposed methods) an unknown qubit in dependency on the displacement amplitude α.

4. DV-CV Quantum Teleportation of Unknown initially Amplitude-Distorting Qubit

In the previous part, we showed the possibility for Bob to restore the original unknown qubit from the AM states with previously known amplitude-distorting factors $A_{nm}^{(lk)}$ and $A_n^{(lk)}$. These methods are probabilistic and allow us to demodulate the unknown qubit in the case of $A_{nm}^{(lk)} > 1$ and $A_n^{(lk)} > 1$ with high fidelity (52). In order to significantly increase the probability of success of the DV-CV quantum teleportation, we must increase the probability of demodulation of AM states with amplitude-distorting factors $A_{nm}^{(lk)} < 1$ and $A_n^{(lk)} < 1$.

Consider quantum teleportation of unknown qubit which was originally subjected to amplitude modulation by a third person, for example, Victor. The third-party scheme is the most common. Victor prepares an unknown qubit and then checks the quality of the teleported qubit. Suppose, he prepares the following qubit

$$|\varphi_{AM}^{(01)}\rangle_{12} = N_{AM}^{(01)} \left(a_0 |01\rangle_{12} + a_1 A_{01}^{(01)-1} |10\rangle_{12} \right),\tag{56}$$

with known amplitude-distorting factor $A_{01}^{(01)}$ and a_0, a_1 being the unknown amplitudes, where $N_{AM}^{(01)} = \left(1 + \left(\left| A_{01}^{(01)} \right|^{-2} - 1 \right) |a_1|^2 \right)^{-0.5}$ is a normalization factor. After preparing the AM qubit, Victor hands over it to Alice. The same entangled state (1) is used to implement DV-CV quantum teleportation of initially AM unknown qubit. Using the same mathematical apparatus, we can get similar expressions (19) but with different states $|\Psi_{nm}^{(lk)}\rangle$ (Equations (20) and (21)). After Alice makes the parity measurement in the first mode and fixes n and m photons in the third and fourth modes, she can send information about them to Bob so that he can carry out unitary transformations (23) and (24) over his photon. Finally, Bob obtains the state

$$N_{nm}^{(01)'} \left[\begin{array}{c} a_0 \\ a_1 A_{01}^{(01)-1} A_{nm}^{(01)} \end{array} \right],\tag{57}$$

where $N_{nm}^{(01)'} = \left(1 + \left(\left|A_{01}^{(01)}\right|^{-2}\left|A_{nm}^{(01)}\right|^2 - 1\right)|a_1|^2\right)^{-0.5}$ is a normalization factor with probability

$$P_{nm}^{(01)} = \frac{F^4|c_{ln}(\alpha)|^2|c_{km}(\alpha)|^2 N_{AM}^{(01)2}}{N_{nm}^{(01)'2}}. \tag{58}$$

In this case, the probability of success depends on the parameter $|a_1|$ of the unknown qubit due to the presence of members $N_{AM}^{(01)}$ and $N_{nm}^{(01)'}$ in formula (58).

The advantage of the initial modulation of an unknown qubit is that when fixing certain measurement outcomes, Bob gets the original unknown qubit with higher success probability. So, if Alice registers the measurement outcomes $|01\rangle_{34}$, then Bob (after applying unitary transformations) receives the original unknown qubit as $A_{01}^{(01)-1}A_{01}^{(01)} = 1$ with the success probability

$$P_{01}^{(01)} = F^4|c_{ln}(\alpha)|^2|c_{km}(\alpha)|^2 N_{AM}^{(01)2}. \tag{59}$$

All other states resulting from the registration of other measurement outcomes contain an amplitude-distorting factor $A_{01}^{(01)-1}A_{nm}^{(01)}$. Bob can proceed to the demodulation procedure using the methods discussed above. So, if he uses the quantum swapping method (Equations (51) and (52)) to get rid of amplitude-distorting factor, then, in the general case, the probability of success for Bob to get the original unknown quantum qubit becomes

$$P_T^{(01)} = F^4 N_{AM}^{(01)2} \begin{pmatrix} |c_{00}|^2|c_{11}|^2 + |c_{01}|^2|c_{10}|^2\dfrac{\left|A_{10}^{(01)}\right|^4}{1+\left|A_{10}^{(01)}\right|^4} + \\[2ex] \dfrac{\left|A_{10}^{(01)}\right|^2}{1+\left|A_{10}^{(01)}\right|^2}\sum_{n=0}^{\infty}|c_{0n}|^2|c_{1n}|^2 + \\[2ex] \sum_{n}^{\infty}\sum_{m,n\neq m,n+m>1}^{\infty}|c_{0n}|^2|c_{1m}|^2\dfrac{\left|A_{10}^{(01)}\right|^{-2}\left|A_{nm}^{(01)}\right|^2}{1+\left|A_{10}^{(01)}\right|^{-2}\left|A_{nm}^{(01)}\right|^2} \end{pmatrix}. \tag{60}$$

The contribution of only a few events in is significant. The contribution of the overwhelming number of events in (60) is very small and can be neglected. The corresponding plots of the success probability $P_T^{(01)}$ depending on the parameter $|a_1|$ of unknown qubit are shown in Figure 5 (left side of the figure) for different values of the displacement amplitude α. As can be seen from these plots, there are values of $|a_1| \ll 1$, for which the probability of success can take values close to one. Thus, the method of initial amplitude modulation of an unknown qubit can lead to an increase in the efficiency of the DV-CV quantum teleportation.

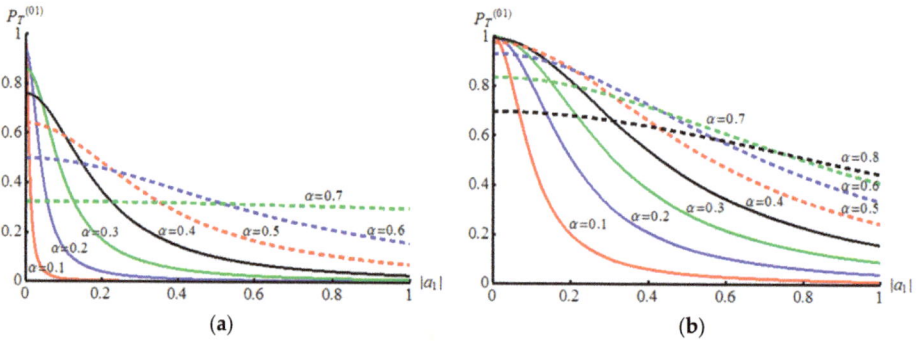

Figure 5. Plots of the success probabilities $P_T^{(01)}$ to teleport and restore (get rid of amplitude-distorting factor) unknown initially AM qubit in dependency on $|a_1|$ for the different values of the displacement amplitude α. (**a**) The plot shows the success probability when using the initially modulated unknown qubit (56) with amplitude-distorting factor $A_{01}^{(01)-1}$. The quantum swapping method (Equations (51) and (52)) is used to get rid of amplitude-distorting factors (Equation (60)). (**b**) The plot shows the success probability of teleporting AM unknown qubit (61), where original state is restored with help of mixing AM unknown qubit with coherent state (Equations (48) and (49)).

Consider another possibility to implement DV-CV quantum teleportation of initially AM unknown qubit. Suppose Victor prepares the next unknown qubit

$$|\varphi_{AM}^{(01)}\rangle_{12} = N_{AM}^{(01)}\left(a_0|0\rangle_1 + a_1 A_0^{(01)-1}|1\rangle_1\right), \tag{61}$$

where $N_{AM}^{(01)} = \left(1 + \left(\left|A_0^{(01)}\right|^{-2} - 1\right)|a_1|^2\right)^{-0.5}$ is a normalization factor, and transmit it to Alice. Amplitudes a_0 and a_1 of the state (61) are unknown to anyone, while amplitude-distorting factor $A_0^{(01)-1}$ follows from (37). Alice's unknown AM qubit interacts with an entangled hybrid state (1) on HTBS, as shown in Figure 1. After Alice performs the measurement in the auxiliary modes (36) and sends the measurement results to Bob on the classical channel, he can implement the corresponding unitary transformations on his dual-rail single photon. The result of this procedure is the following state

$$N_n^{(01)'}\left[\begin{array}{c} a_0 \\ a_1 A_0^{(01)-1} A_n^{(01)} \end{array}\right], \tag{62}$$

where $N_n^{(01)'} = \left(1 + \left(\left|A_0^{(01)}\right|^{-2}\left|A_n^{(01)}\right|^2 - 1\right)|a_1|^2\right)^{-0.5}$ is the normalization factor of obtained state. Success probability to get the state is

$$P_n^{(01)} = \frac{F^2|c_{0n}(\alpha)|^2 N_{AM}^{(01)2}}{N_n^{(01)'2}}. \tag{63}$$

The probability of success includes the normalization parameters $N_{AM}^{(01)}$ and $N_n^{(01)'}$, so it depends on the parameter $|a_1|$ of the unknown qubit. If Alice registered the vacuum in the auxiliary second mode, then Bob has the initial unknown qubit (after the implementation of the corresponding unitary transformations), since $A_0^{(01)-1}A_0^{(01)} = 1$. Success probability for Alice to register such outcome becomes

$$P_0^{(01)} = F^2|c_{0n}(\alpha)|^2 N_{AM}^{(01)2}, \tag{64}$$

as $N_0^{(01)'} = 1$. If Alice registers a non-vacuum outcome $|n$ with $n \neq 0$, then Bob's state contains amplitude-distorting factor $A_0^{(01)-1} A_n^{(01)}$. In the case, Bob can use one of the two considered methods for demodulating the AM states with corresponding success probabilities. Consider the method of demodulation of the AM states using a coherent state of large amplitude (Equation (48)). To use this method, one needs to find the value of the parameter γ (Equation (54)) which greatly complicates the analytical view of the probability of success to teleport unknown qubit and get rid of the amplitude-distorting factor. The corresponding dependences of the success probability $P_T^{(01)}$ of the initially AM unknown qubit depending on the parameter $|a_1|$ of the unknown qubit for different values of the displacement amplitudes α are shown in the right part of Figure 5. As can be seen from these graphs, the probability of success can be significantly increased compared to the case discussed above.

5. Results

We considered the ability to teleport an unknown qubit using DV-CV interaction mechanism. This mechanism is implemented in the interaction of CV and DV states on HTBS. A non-maximally entangled hybrid state, composed of coherent components with opposite in sign amplitudes and DV state, is used to perform DV-CV quantum teleportation of an unknown qubit. The coherent components of the state (1) displace the unknown state to equal modulo, but opposite in sign amplitudes in an indistinguishable manner so that all information about the value of the displacement disappears. The unknown state can be displaced by both positive and negative values. If an unknown qubit is displaced by a positive value, then the relative phase of the decomposition coefficients of the displaced states in the measurement basis does not change regardless of the parity of the basic states. On the contrary, the relative phase of the displaced unknown qubit in the measurement basis can vary on the opposite depending on the parity of the basis states in the case of the negative amplitude of displacement. This nonlinear effect akin to the action of controlled-Z gate is a base of DV-CV quantum teleportation of unknown qubit. Bob, having received a limited amount of classical information about Alice's results of the measurements, can perform the appropriate set of unitary transformations over his single photon. Since the amplitudes of the decomposition of the displaced states of light in the measuring basis are not equal to each other, the teleported states acquire additional known amplitude-distorting coefficients. The presence of an amplitude-distorting factor in the teleported qubit can be recognized as an inherent trait of the DV-CV quantum teleportation. It may recall the CV deterministic quantum teleportation of an unknown qubit whose output fidelity suffers due to the absence of the maximally entangled quantum channel. On the contrary, DV teleportation allows us to get output state with a high degree of fidelity (in the ideal case with unit fidelity), but the implementation of the full Bell-states measurement using linear optics is impossible, which reduces the probability of success up to 0.5. All measurement outcomes in DV-CV teleportation are distinguishable. But the fidelity of the output state in Bob's hands is also not ideal as in CV teleportation. Although it is worth noting that the protocol participants know the values of amplitude-distorting factors. Notice the difference of quantum channels used in CV and DV-CV quantum teleportation. A two-mode squeezed vacuum is used in CV teleportation. The description and predictions of the protocol are based on quantum nonlocality of entangled quantum state (1) and cannot be simulated by any local hidden variable theory.

The key issue for increasing the protocol's efficiency is resolving the demodulation problem or getting rid of an unknown quantum state from a previously known amplitude-distorting factor, preferably with a success rate close to unity. We considered only two such probabilistic possibilities, which are based on the displacement of the state in the auxiliary mode with the subsequent registration of some events in the measurement number state basis and the quantum swapping procedure. Moreover, it is worth noting that this mechanism can be implemented in various interpretations, some of which we considered. Each of the considered schemes allows us to calculate the amplitude-distorting factors (Equations (16) and (37)). So, the optical scheme shown in Figure 1 with the additional interaction of an unknown qubit with coherent state allows us to teleport it without demodulation procedures if detectors register the same number of photons in the auxiliary modes. Other interpretations that could increase the efficiency of the DV-CV quantum teleportation are possible. We did not consider amplitude-distorting factors (16) with different displacement amplitudes $\alpha \neq \alpha_1$. We used the hybrid state (1) with dual-rail single-photon at Bob's location. In fact, the same universal mechanism works if we use a quantum state (1) with different state in Bob's hands, including states from other physical systems, which could increase the success probability of the demodulation procedure. Consideration of these issues requires separate investigation. Within the considered interpretations of the DV-CV quantum teleportation of unknown qubit $|\varphi^{(lk)}\rangle$ (Equation (3)) with small values of l and k, it is necessary to recognize that the scheme with the initial amplitude modulation of the unknown qubit is the most effective (Figure 5).

Funding: The work was supported by Act 211 Government of the Russian Federation, contract No 02.A03.21.0011.

Conflicts of Interest: The author declares no conflict of interest.

References

1. Bell, J. *Speakable and Unspeakable in Quantum Mechanics*; Cambridge University Press: Cambridge, UK, 1987.
2. Aspect, A.; Grangier, P.; Roger, G. Experimental realization of Einstein-Podoslsky-Rosen-Bohm gedankenexperiment: A new violation of Bell's inequalities. *Phys. Rev. Lett.* **1982**, *49*, 91–94. [CrossRef]
3. Rowe, M.A.; Kielpinski, D.; Meyer, V.; Sackett, C.A.; Itano, W.M.; Monroe, C.; Wineland, D.J. Experimental violation of a Bell's inequality with efficient detection. *Nature* **2001**, *409*, 791–794. [CrossRef] [PubMed]
4. Bennett, C.H.; Brassard, G.; Crepeau, C.; Jozsa, R.; Peres, A.; Wootters, W.K. Teleporting an unknown quantum state via dual classical and Einstein-Podolsky-Rosen channels. *Phys. Rev. Lett.* **1993**, *70*, 1895–1899. [CrossRef] [PubMed]
5. Ghirardi, G.C.; Rimini, A.; Weber, T. A general argument against superluminal transmission through the quantum mechanical measurement process. *Lettere Al Nuovo Climento* **1980**, *27*, 293–298. [CrossRef]
6. Briegel, H.-J.; Dür, W.; Cirac, J.I.; Zoller, P. Quantum repeaters: The role of imperfect local operations in quantum communication. *Phys. Rev. Lett.* **1998**, *81*, 5932–5935. [CrossRef]
7. Gottesman, D.; Chuang, I.L. Demonstrating the viability of universal quantum computation using teleportation and single-qubit operations. *Nature* **1999**, *402*, 390–393. [CrossRef]
8. Rausendorf, R.; Briegel, H.J. A one-way quantum computer. *Phys. Rev. Lett.* **2001**, *86*, 5188–5191. [CrossRef] [PubMed]
9. Bouwmeester, D.; Pan, J.W.; Mattle, K.; Eible, M.; Weinfurter, H.; Zeilinger, A. Experimental quantum teleportation. *Nature* **1997**, *390*, 575–579. [CrossRef]
10. Furusawa, A.; Sørensen, J.L.; Braunstein, S.L.; Fuchs, C.A.; Kimble, H.J.; Polzik, E.S. Unconditional quantum teleportation. *Science* **1998**, *252*, 706–707. [CrossRef]
11. Lee, N.; Benichi, H.; Takeno, Y.; Takeda, S.; Webb, J.; Huntington, E.; Furusawa, A. Teleportation of nonclassical wave packets of light. *Science* **2011**, *352*, 330–333. [CrossRef]
12. Takeda, S.; Mizuta, T.; Fuwa, M.; vby van Loock, P.; Furusawa, A. Deterministic quantum teleportation of photonic quantum bits by a hybrid technique. *Nature* **2013**, *500*, 315–318. [CrossRef] [PubMed]
13. Nielsen, M.A.; Knill, E.; Laflamme, R. Complete quantum teleportation using nuclear magnetic resonance. *Nature* **1998**, *396*, 52–55. [CrossRef]
14. Krauter, H.; Salart, D.; Muschik, C.A.; Petersen, J.M.; Shen, H.; Fernholz, T.; Polzik, E.S. Deterministic quantum teleportation between distant atomic objects. *Nat. Phys.* **2013**, *9*, 400–404. [CrossRef]
15. Barrett, M.D.; Chiaverini, J.; Schaetz, T.; Britton, J.; Itano, W.M.; Jost, J.D.; Knill, E.; Lager, C.; Leibfried, D.; Ozeri, R.; Wineland, D.J. Deterministic quantum teleportation of atomic qubits. *Nature* **2004**, *429*, 737–739. [CrossRef] [PubMed]
16. Gao, W.B.; Fallahi, P.; Togan, E.; Delteil, A.; Chin, Y.S.; Miguel-Sanchez, J.; Imamoğlu, A. Quantum teleportation from a photon to a solid-state spin qubit. *Nat. Commun.* **2013**, *4*, 1–8. [CrossRef] [PubMed]
17. Lutkenhaus, N.; Calsamiglia, J.; Suominen, K.A. Bell measurements for teleportation. *Phys. Rev. A* **2000**, *59*, 3295–3300. [CrossRef]

18. Knill, E.; Laflamme, L.; Milburn, G.J. A scheme for efficient quantum computation with linear optics. *Nature* **2001**, *409*, 46–52. [CrossRef] [PubMed]

19. Grice, W.P. Arbitrarily complete Bell-state measurement using only linear optical elements. *Phys. Rev. A* **2011**, *84*, 042331. [CrossRef]

20. Zaidi, H.A.; van Loock, P. Beating the one-half limit of ancilla-free linear optics Bell measurements. *Phys. Rev. Lett.* **2013**, *110*, 260501. [CrossRef]

21. Vaidman, L. Teleportation of quantum states. *Phys. Rev. A* **1994**, *49*, 1473–1476. [CrossRef]

22. Braunstein, S.L.; Kimble, H.J. Teleportation of continuous quantum variables. *Phys. Rev. Lett.* **1998**, *80*, 869–872. [CrossRef]

23. Braunstein, S.; van Loock, P. Quantum information with continuous variable states. *Rev. Mod. Phys.* **2005**, *77*, 513–577. [CrossRef]

24. Vaidman, L.; Yoran, N. Methods for reliable teleportation. *Phys. Rev. A* **1999**, *59*, 116–125. [CrossRef]

25. van Loock, P. Optical hybrid approaches to quantum information. *Laser Photonics. Rev.* **2011**, *5*, 167–200. [CrossRef]

26. Morin, O.; Haung, K.; Liu, J.; Jeannic, H.L.; Fabre, C.; Laurat, J. Remote creation of hybrid entanglement between particle-like and wave-like optical qubits. *Nature Photonics* **2014**, *8*, 570–574. [CrossRef]

27. Le Jeannic, H.; Cavailles, A.; Raskop, J.; Huang, K.; Laurat, J. Remote preparation of continuous-variable qubits using loss-tolerant hybrid entanglement of light. *Optica* **2018**, *5*, 1012–1015. [CrossRef]

28. Podoshvedov, S.A. Generation of displaced squeezed superpositions of coherent states. *J. Exp. Theor. Phys.* **2012**, *114*, 451–464. [CrossRef]

29. Podoshvedov, S.A.; Kim, J.; Kim, K. Elementary quantum gates with Gaussian states. *Quantum. Inf. Proc.* **2014**, *13*, 1723–1749. [CrossRef]

30. Lvovsky, A.I.; Ghobadi, R.; Chandra, A.; Prasad, A.S.; Simon, C. Observation of micro-macro entanglement of light. *Nature Phys.* **2013**, *9*, 541–544. [CrossRef]

31. Sekatski, P.; Sangouard, N.; Stobinska, M.; Bussieres, F.; Afzelius, M.; Gisin, N. Proposal for exploring macroscopic entanglement with a single photon and coherent states. *Phys. Rev. A* **2012**, *86*, 060301. [CrossRef]

32. Bruno, S.; Martin, A.; Sekatski, P.; Sangouard, N.; Thew, R.; Gisin, N. Displacement of entanglement back and forth between the micro and macro domains. *Nat. Phys.* **2013**, *9*, 545–550. [CrossRef]

33. Podoshvedov, S.A. Building of one-way Hadamard gate for squeezed coherent states. *Phys. Rev. A* **2013**, *87*, 012307. [CrossRef]

34. Podoshvedov, S.A. Elementary quantum gates in different bases. *Quantum. Inf. Proc.* **2016**, *15*, 3967–3993. [CrossRef]

35. Vidal, G.; Werner, R.F. Computable measure of entanglement. *Phys. Rev. A* **2002**, *65*, 032314. [CrossRef]

36. Peres, A. Separability criterion for density matrices. *Phys. Rev. Lett.* **1996**, *77*, 1413–1415. [CrossRef] [PubMed]

37. Paris, M.G.A. Displacement operator by beam splitter. *Phys. Lett. A* **1996**, *217*, 78–80. [CrossRef]

38. Podoshvedov, S.A. Extraction of displaced number state. *JOSA B* **2014**, *31*, 2491–2503. [CrossRef]

39. Reck, M.; Zeilinger, A.; Bernstein, H.J.; Bertani, P. Experimental realization of any discrete unitary operator. *Phys. Rev. Lett.* **1994**, *73*, 58–61. [CrossRef]

40. Podoshvedov, S.A.; Kim, J.; Lee, J. Generation of a displaced qubit and entangled displaced photon state via conditional measurement and their properties. *Opt. Commun.* **2008**, *281*, 3748–3754. [CrossRef]

41. Pan, J.-W.; Bouwmeester, D.; Weinfurter, H.; Zeilinger, A. Experimental entanglement swapping: Entangling photons that never interacted. *Phys. Rev. Lett.* **1998**, *80*, 3891–3894. [CrossRef]

Article

Bell Inequalities with One Bit of Communication

Emmanuel Zambrini Cruzeiro * and Nicolas Gisin

Department of Applied Physics, University of Geneva, 1211 Geneva, Switzerland; nicolas.gisin@unige.ch
* Correspondence: emmanuel.zambrinicruzeiro@unige.ch

Received: 13 December 2018; Accepted: 6 February 2019; Published: 13 February 2019

Abstract: We study Bell scenarios with binary outcomes supplemented by one bit of classical communication. We developed a method to find facet inequalities for such scenarios even when direct facet enumeration is not possible, or at least difficult. Using this method, we partially solved the scenario where Alice and Bob choose between three inputs, finding a total of 668 inequivalent facet inequalities (with respect to relabelings of inputs and outputs). We also show that some of these inequalities are constructed from facet inequalities found in scenarios without communication, that is, the well-known Bell inequalities.

Keywords: quantum nonlocality; communication complexity

1. Introduction

Bell nonlocality [1,2] is one of the most intriguing phenomena encountered in modern physics. Nonlocality was discovered more than 50 years ago, and there are still simple well-posed fundamental questions about nonlocality that remain unanswered. In this article, we focus on one of these questions, which is impressively simple to state but has proven very hard to answer. In the interest of quantifying and understanding nonlocality, one can create variations of Bell's original local hidden variable (LHV) model by adding a nonlocal resource. A nonlocal resource is any resource that establishes correlations at a distance. A PR box [3–5] is an example of such a nonlocal resource. Another example is classical communication [6–10], which is the focus of this paper. In particular, one can ask how many bits of information are needed to reproduce correlations arising from projective measurements on any two-qubit state [6,8,9,11]. For the singlet, it is known that one bit is sufficient (the explicit model is given in Reference [10]); therefore, we are interested in partially entangled states, which are known to be simulable with two bits [10], but not with zero bits [12]. We ask whether one bit also suffices to simulate projective measurements on all two-qubit partially entangled states. It is interesting that such a well-posed binary-answer question for projective measurements on two-qubit pure states has still not been answered, even though several authors have worked on this problem [13,14]. This illustrates the technical difficulty of studying nonlocality. Our strategy is to find Bell-like inequalities that are satisfied by all LHV models supplemented by one bit of communication, and then look for a violation of such inequalities. Although we do not provide an answer to Toner and Bacon's question here, our results already provide a deeper understanding of Bell-like inequalities for scenarios with one bit of communication. Additionally, our work can be of interest to physicists working on alternative causal structures to Bell's theorem (see References [15–17]).

Regular Bell scenarios and Bell scenarios supplemented with one bit of communication sent by Alice to Bob are formally described in Section 2, along with the methods we used to find the main results. In particular, we introduce a useful notation and propose a method to tackle scenarios where direct facet enumeration is difficult. Section 3 gives a proof that all projective measurements on quantum states can be reproduced by one bit of communication, for scenarios where Bob has only two dichotomic measurement settings, despite the fact that we assume the bit to be communicated from Alice to Bob. In Section 4, we discuss the results we obtained for the scenario where both Alice and

Bob have three inputs. Finally, we conclude by discussing the general structure of Bell-like inequalities with one bit of communication, and future directions of research.

2. Bell Inequalities with Auxiliary Communication

2.1. Bell Scenarios

In a bipartite Bell scenario, see Figure 1, the two observers are usually called Alice and Bob. Alice and Bob choose from a set of inputs (measurement settings) and, as a result, get an output (measurement outcome). After they select their inputs, Alice and Bob are not allowed to communicate. Nevertheless, they both have access to the same set of local variables because they share randomness that was generated by a common source at a past time. The observers are allowed to use local variables to produce their outcomes. Alice and Bob both have a number of measurements settings X and Y, respectively, and a number of outputs A, B. This defines the physical setup, or Bell scenario, generally noted XYAB. Since in this article we restrict to binary-outcome measurements, we note Bell scenarios XY22 simply as XY. In the lab, Alice and Bob repetitively perform measurements and record the outcome statistics, which are described by joint probability distribution $p(ab|xy)$. If the correlations allowed by $p(ab|xy)$ are explainable using only common past history and local operations by the observers, physicists say the experiment statistics admit a local hidden variable (LHV model). In such a case, we can write

$$p(ab|xy) = \int q(\lambda) p^A(a|x\lambda) p^B(b|y\lambda) \tag{1}$$

where λ is a local variable (infinite shared randomness), $q(\lambda)$ is its probability distribution, and $p^A(a|x\lambda), p^B(b|y\lambda)$ are, respectively, Alice and Bob's marginal probabilities. If Equation (1) is not satisfied, $p(ab|xy)$ is not local.

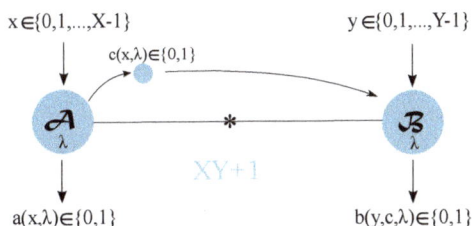

Figure 1. $XY + 1$ scenario where Alice and Bob choose between X and Y binary-outcome measurements, respectively, and share local hidden variables λ (shared randomness). Alice is allowed to send one bit $c(x, \lambda)$ of classical communication to Bob.

If locality is assumed, then deterministic strategies can be defined through the marginals of Alice and Bob [2]. The marginals define their respective local strategies. Set $\mathcal{L}$ of all local strategies $p^{\mathcal{L}}(ab|xy)$ is finite because Alice and Bob choose from a finite set of measurements, and it defines a convex polytope usually called the local polytope. For binary outcomes, there are 2^{X+Y} deterministic strategies, and the local polytope is of dimension $X + Y + XY$. The facets of this polytope define inequalities that are satisfied by any probability distribution in $\mathcal{L}$, but are violated for quantum-probability distributions. These are the famous Bell inequalities, the simplest of which is the CHSH inequality, for binary inputs and outputs on both sides:

$$\begin{aligned} p(00|00) + p(00|01) + p(00|10) - p(00|11) \\ - p^A(0|0) - p^B(0|0) \leq 0 \end{aligned} \tag{2}$$

This inequality is violated by quantum mechanical probability distributions, up to $\frac{1}{\sqrt{2}} - \frac{1}{2} \approx 0.2071$.

2.2. Bell Scenarios Supplemented by One Cbit (Bell + 1)

We are interested in the simulation of projective measurements on qubits through one bit of classical communication. Since quantum correlations are symmetric with respect to Alice and Bob, we specifically consider one-way communication (in one direction; in this case, from Alice to Bob), as two-way communication would have no advantage in a quantum scenario. The protocol goes as follows: Alice and Bob first receive their inputs, then Alice is allowed to send one bit of classical communication to Bob. In this way, Alice and Bob can simulate all $p(ab|xy)$ that satisfy:

$$p(ab|xy) = \int q(\lambda) p^A(a|x\lambda) p^B(b|yc\lambda) \tag{3}$$

where the marginal of Bob now also depends on the value of classical bit $c = c(x, \lambda)$.

One can define all local strategies with one bit of communication analogous to the original Bell scenario. The local strategies can all be written in terms of local deterministic strategies, for which the marginal probabilities of Alice and Bob can only take values 0 and 1. There is a finite number of such strategies and, hence, a finite number of vertices that define a convex polytope. Once we have generated all the vertices, we look for the facets of this polytope: this is the so-called facet-enumeration problem. We call the set of local strategies with one bit of communication $\mathcal{C}$. The inequalities defining these facets are violated only if there exists a two-qubit state and projective measurements yielding correlations that cannot be reproduced using one bit of classical communication.

2.3. Local Strategies for Bell + 1 and Notation

Joint probability distribution $p(ab|xy)$ for each local strategy can be computed in the following way:

$$p(ab|xy) = \sum_{\lambda} q(\lambda) p^A(a|x\lambda) p^B(b|cy\lambda) \tag{4}$$

where $c = c(x, \lambda)$ is the communication function, and can be encoded in multiple ways. In a similar fashion to Bell scenarios, we define such a scenario as $XY + 1$, where we again omit the number of outputs as they are always binary. For a given number of inputs on Alice's side X, the number of communication functions in the case of one cbit is given by the Stirling number of the second kind, denoted $S(X, 2)$ or $\{^X_2\}$, and defined as $\{^X_k\} := \frac{1}{k!} \sum_{j=0}^{k} (-1)^{k-j} \binom{k}{j} j^X$. The Stirling number of the second kind gives the number of distinct ways to divide a set into two nonempty subsets.

By directly generating all local strategies, we obtain $\{^X_2\} \cdot 2^X \cdot 2^{2Y}$ vertices. This method generates repeated vertices because it takes into account the situations where Bob does not use the communication bit. By removing repetitions, we end up with a smaller number of vertices, given by:

$$2^X \left(2^Y + \left\{ ^X_2 \right\} (2^{2Y} - 2^Y) \right) \tag{5}$$

This is a sum of three terms. The first term gives the vertices for the local polytope of the Bell scenario, in which case no communication function is used. The second term accounts for three kinds of strategies: Bell local strategies like the first term, strategies where there is communication but the bit is not used by Bob, and finally strategies for which the bit is used. In order to only keep the Bell local strategies and the strategies for which the bit of communication is useful, we must remove the strategies that do not use the bit, for which the third term accounts. In the second term, the Stirling number gives the number of possible communication functions, and the bit of communication gives a factor of two multiplying Y (the bit is counted as an extra binary input on Bob's side). An interesting consequence of this is that, for different values of (X, Y), one can have the same amount of vertices. In fact, any $XX + 1$ scenario has the same number of vertices as an $(X + 1)(X - 1) + 1$ scenario. Any $X(X + 1) + 1$ scenario also has the same number of vertices as an $(X + 2)(X - 1) + 1$ scenario.

The dimension of the $XY + 1$ polytope is $X + 2XY$. It is easy to see why: Joint probability distribution $p(ab|xy)$ consists of $4XY$ elements, but none of these elements is independent due to normalization and no-signalling constraints. Normalization removes the XY of these elements, and no signalling from Alice to Bob removes $X(Y - 1)$ elements. Therefore, $X + 2XY$ is the minimal number of variables (probability elements) needed to define the polytope. The usual notation for vertices, from Toner and Bacon [11], is given by $\{p(00|xy) \ldots p(10|xy) \ldots p^A(a = 0|x) \ldots\}$. The three dots mean that we run through all the values of x and y, for example, $\{p(00|xy) \ldots\}$ means $\{p(00|00), p(00|01), p(00|10), p(00|11), \text{etc...}\}$. We instead chose to use notation $\{p(00|xy) \ldots p^B(b = 0|xy) \ldots p^A(a = 0|x) \ldots\}$ similarly to Reference [18] because it makes it easier to see what inequalities reduce to when considering probability distributions in the no-signalling (NS) subspace, such as quantum probability distribution (see Table S1, provided as a Supplementary File). This becomes clear when we study the first nontrivial scenarios, $32 + 1$ and $33 + 1$, while $2Y + 1$ is trivial for all Y because Alice can simply send her input as the communication bit; in fact, as we show in Section 3, $X2 + 1$ is also trivial for all X.

A Bell + 1 inequality can be written as:

$$\sum_{xy} d_{xy} p(00|xy) + \sum_{xy} e_{xy} p^B(0|xy) + \sum_x f_x p^A(0|x) \le b \tag{6}$$

We can represent such an inequality as a table (see Table 1) in which elements are the coefficients multiplying each probability element $\{p(00|xy) \ldots p^B(b = 0|xy) \ldots p^A(a = 0|x) \ldots\}$. We denote the coefficients for $p(00|xy)$ elements as d_{xy}, while the coefficients for Bob's marginals are e_{xy}, and for Alice's marginals f_x. Finally, an inequality is also characterized by its bound b.

Table 1. Inequalities notation $33 + 1$. f_x are the weights of Alice's marginals $p_x^A(a = 0|x)$, d_{xy} are the weights of joint probabilities for outcomes $a = b = 0$, and e_{xy} are the coefficients for Bob's marginals $p^B(b = 0|xy)$.

f_0	d_{00}	d_{01}	d_{02}	
	e_{00}	e_{01}	e_{02}	
f_1	d_{10}	d_{11}	d_{12}	$\le b$
	e_{10}	e_{11}	e_{12}	
f_2	d_{20}	d_{21}	d_{22}	
	e_{20}	e_{21}	e_{22}	

Note that a vector of the form

$$\vec{I}^S = (d_{00}, d_{01}, \ldots, d_{XY}, e_{00}, e_{01}, \ldots, e_{XY}, c_0, c_1, \ldots, c_X) \tag{7}$$

belongs to the NS subspace iff e_{xy} is independent of x for all y.

Knowing the vertices, it is possible to compute all facets of a given polytope using dedicated software such as PORTA [19] or PANDA [20].

2.4. Extension of Inequalities from Bell to Bell + 1 Scenarios and Intersection of Bell + 1 Inequalities with NS Subspace

An inequality of a Bell scenario can be extended to the corresponding Bell + 1 scenario. We extend inequalities from the NS space to the one-bit space by choosing the coefficients for Bob's marginals in a clever way. For any Bell inequality, there are infinite such extensions. We chose the one orthogonal to the NS subspace as depicted in Figure 3, i.e., we imposed that the vector characterizing the extension lay within NS subspace. This orthogonal extension is unique. Let us look at the example of $33 + 1$, a scenario where we need to use this technique because a full resolution of the polytope is difficult. In Table 2, we show how to extend an arbitrary 33 inequality to the $33 + 1$ space. We extended the inequality to the $33 + 1$ space by adding coefficients for Bob's marginals, which in this higher-dimensional space dependent on both x and y. We chose the coefficients for Bob's marginals such that e'_y satisfied $e_{0y} = e_{1y} = e_{2y} = e'_y/3$ for all y, where e'_y are coefficients of the 33 inequality for Bob's marginals $p^B(0|y)$. In this way, one can intersect the one-bit inequality with the nonsignalling subspace and map it back to the original Bell inequality that was used for the extension.

Table 2. Orthogonal extension of a Bell inequality to the one-bit communication space (for example, for 33 + 1). The bound in both cases is the local bound.

	e'_0	e'_1	e'_2
f_0	d_{00}	d_{01}	d_{02}
f_1	d_{10}	d_{11}	d_{12}
f_2	d_{20}	d_{21}	d_{22}

$\leq 0 \longrightarrow$

f_0	d_{00}	d_{01}	d_{02}
	$e'_0/3$	$e'_1/3$	$e'_2/3$
f_1	d_{10}	d_{11}	d_{12}
	$e'_0/3$	$e'_1/3$	$e'_2/3$
f_2	d_{20}	d_{21}	d_{22}
	$e'_0/3$	$e'_1/3$	$e'_2/3$

≤ 0

Intersecting a one-bit inequality with NS subspace is also straightforward to do using our choice of notation, as one simply has to sum up the coefficients for Bob's marginals $\sum_x e_{xy} = e'_y$, then

The bound for the NS inequality in Table 3 has to be carefully considered. Indeed, this bound is the one-bit bound for $\vec{I}^S$, a particular extension (not the orthogonal one) of $\vec{I}^{NS}$ of Table 3. Different extensions do not give the same one-bit bound though, see Figure 2. For clarity, we used a simplified scheme. In Figure 2, we represent the signalling space as a plane containing the NS space, represented as a line. Using brackets, we also represent the bounds of the NS polytope that are given by non-negativity condition $p(ab|xy) \geq 0$ for all a, b, x, y. In a similar way, the vertical lines in the NS space delimit the local polytope. The points where those lines are placed represent facets of the local polytope. A facet of the one-bit polytope is a hyperplane I^S; in our representation, it is an interval. In order for probability distribution to not be reproducible by one bit of communication, we need its representative point to be farther to the right than the intersection of I^S with the NS space. For any point in the NS space $\vec{q} \in$ NS, $\vec{I}^S \cdot \vec{q} = \vec{I}^{NS} \cdot \vec{q}$. Therefore, a quantum bound for $\vec{I}^{NS}$ larger than the one-bit bound of $\vec{I}^S$ implies that the distribution attaining the value of the quantum bound cannot be reproduced with one bit of communication. Note that the orthogonal extension's bound is always equal ti or larger than the correct one-bit bound, since having one bit of communication implies leaving the NS subspace.

Table 3. Intersecting one-bit inequality I^S with NS subspace amounts to summing the coefficients for Bob's marginals, characterizing one of his inputs y.

$\vec{I}^S =$

f_0	d_{00}	d_{01}	d_{02}
	e_{00}	e_{01}	e_{02}
f_1	d_{10}	d_{11}	d_{12}
	e_{10}	e_{11}	e_{12}
f_2	d_{20}	d_{21}	d_{22}
	e_{20}	e_{21}	e_{22}

$\leq b \xrightarrow{NS} \vec{I}^{NS} =$

	e'_0	e'_1	e'_2
f_0	d_{00}	d_{01}	d_{02}
f_1	d_{10}	d_{11}	d_{12}
f_2	d_{20}	d_{21}	d_{22}

$\leq b$

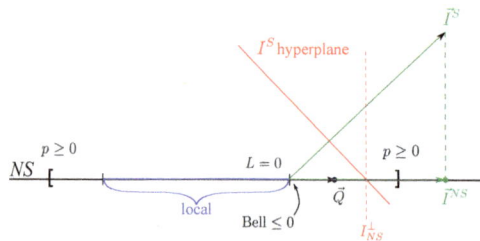

Figure 2. Geometry schematic of one-bit and no-signalling spaces. NS space is represented as a line, while the signalling space is represented as two-dimensional. The non-negativity conditions delimiting the NS polytope are represented by brackets.

2.5. Cutting the Polytope

When direct facet enumeration cannot be done in one or two weeks, we use a trick to find a smaller set of inequalities. The trick consists in enumerating the facets for a subpolytope of $\mathcal{C}$, where $\mathcal{C}$ is the one-bit polytope. The way we select the subpolytope is by taking a Bell scenario inequality, extending it to the one-bit space in an orthogonal way as shown in Figure 3, and removing any vertex that satisfies this new inequality. This amounts to cutting the polytope with a hyperplane.

As previously described, we chose the coefficients for Bob's marginals in the one-bit space to be equal because this corresponds to an orthogonal extension of the facet with respect to the NS space, i.e., $I^S \perp NS$, where I^S is the rightmost inequality in Table 2. We extensively tested the choice of coefficients with the $32 + 1$ scenario, which was already fully solved [14]. In order to generate all relevant facets, it is important that coefficients for Bob's marginals for inputs that give a CHSH inequality are equal. The other coefficients seem completely arbitrary. In the $33 + 1$ example of Table 2, for Bob's input $y = 1$, this means coefficients for $p^B(0|xy)$ for $x = 0, 1$ should be equal, and the coefficient for $x = 2$ is arbitrary.

When we change the choice of coefficients for Bob's marginals, we perform a rotation of the hyperplane used to cut the one-bit polytope. Therefore, one could try different choices of coefficients in order to select different sets of vertices and, therefore, produce several subpolytopes out of the original polytope. Furthermore, each relabelling of the inequality cuts a different region of the polytope, possibly revealing new facets.

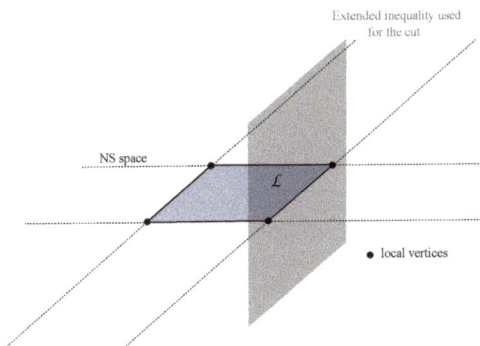

Figure 3. A $\mathcal{C}$ polytope is cut by an extended Bell inequality, which is orthogonal to the NS subspace. The NS subspace is represented as a two-dimensional space. We chose not to represent the $\mathcal{C}$ polytope as we did not know its geometrical form. By keeping all vertices that saturate or violate such an inequality, one obtains a subpolytope for which it is easier to find the facets via direct facet enumeration.

There is another freedom for the cut: one can modify the bound of the inequality used for the cut. This causes a translation of the hyperplane that allows to change the size of the subpolytopes we generate. Therefore, for very hard problems, we can increase the bound to try to solve smaller subpolytopes. This translation technique has been used before (see [21] for further details).

Last but not least, when we cut a polytope and find the facets of the subpolytope, some facets are not facets of the original polytope, but they were created by the cut. In order to keep only the relevant inequalities, we check their rank and whether the vertices of the original polytope exactly saturate the inequalities bound.

3. $X2 + 1$ Scenarios

Scenarios of $2Y + 1$ are trivial because Alice can send her input as the classical bit. Surprisingly, $X2 + 1$ inequalities also cannot be violated by any NS distribution despite the assumption that the classical bit is sent from Alice to Bob. The reason is that every NS vertex of an X2 Bell scenario can be reproduced using a PR box [4,22]. Therefore, one PR box can simulate any quantum state in X2 scenarios, as boxes can be written as convex combinations of the NS vertices. Furthermore, one bit of communication is a strictly stronger nonlocal resource that one PR box [23]. Therefore, one bit of communication can simulate any quantum state in an X2 scenario.

4. 33 + 1 Scenario

In this section, we present our results for the 33 + 1 scenario. For this scenario, facet enumeration is demanding but, by cutting the polytope, we can recover a large list of inequalities. In the corresponding 33 Bell scenario, besides CHSH there is one new inequality, called I_{3322}, which we can also use to cut the 33 + 1 polytope:

$$I_{3322} = \begin{array}{c|ccc} & -1 & 0 & 0 \\ \hline -2 & 1 & 1 & 1 \\ -1 & 1 & 1 & -1 \\ 0 & 1 & -1 & 0 \end{array} \leq 0$$

4.1. Cutting with CHSH

We apply the cut with extended CHSH inequality using the procedure described above.

We then solve the subpolytope. We find 657 inequivalent inequalities, where 179 inequalities have a quantum advantage when intersected with the NS subspace. Note that quantum probabilities do not violate the one-bit bound C, but they can offer, as is the case for the 179 inequalities, an advantage with respect to local bound L in the NS subspace. We can distinguish the inequalities by how close the quantum bound is from the one-bit bound with the following figure of merit:

$$\frac{Q - L}{C - L} \tag{8}$$

This figure of merit also gives a lower bound on the amount of average communication required to reproduce 3322 correlations [24]. The best quantum bound that we obtained with respect to the local bound was halfway between the local and one-bit bounds (see inequalities 195 and 232 in Table S1). This result implies that, to reproduce 3322 correlations, Alice needs to send to Bob one bit at least half of the time on average. By looking inside the NS subspace, we can show that our halfway quantum bound is obtained through a sum of two I_{3322} inequalities (recall that the quantum bound of I_{3322} is equal to 0.25). We also found inequalities that, in the NS subspace, correspond to the sum of two CHSH, and inequalities corresponding to one CHSH or one I_{3322}. In addition, we found violations that correspond to a CHSH or an I_{3322} inequality, plus a term that changes the optimal state/measurement and, therefore, modifies the quantum bound, too. Performing the same analysis in the 32 + 1 scenario, one finds that correlations can be reproduced only if the amount of average communication is higher than 0.4142.

In Table 4, we give an explicit example of a facet that has a larger quantum bound with respect to the local bound (inequality number 232 in Table S1, which can be found in the Appendix). In the nonsignalling subspace, this facet corresponds to a sum of I_{3322}. In order to clarify this, we intersected the facet of the C polytope with the NS subspace.

Table 4. Facet of $33 + 1$, for which the quantum bound is halfway between the local and one-bit bounds. When intersected with the NS space, this inequality reduces to a sum of I_{3322} inequalities. This inequality corresponds to facet number 232 in Table S1.

-3	2	2	2
	-1	-1	-1
-1	2	2	-2
	-1	-1	1
0	2	-2	0
	-1	1	0

$\leq 1 \overset{\text{NS}}{\longrightarrow}$

	-3	-1	0
-3	2	2	2
-1	2	2	-2
0	2	-2	0

≤ 1

The resulting inequality is $I_{3322} + I_{3322}^{\text{perm}}$ with a bound of one instead of zero, where I_{3322}^{perm} is I_{3322} with a relabeling of the parties (permutation of Alice and Bob labels). We found another inequality of the same type, which also includes a sum of I_{3322} and I_{3322}^{perm}, although it is less obvious to see because it also includes some other terms that do not contribute to the quantum bound. The second inequality (number 195 in Table S1) is given in Table 5.

Table 5. Second facet (number 195) of $33 + 1$ for which the quantum bound is halfway between the local and one-bit bounds.

-3	2	2	2
	-1	-1	-1
-1	2	-2	1
	-1	1	0
0	2	1	-2
	-1	-1	1

$\leq 1 \overset{\text{NS}}{\longrightarrow}$

	-3	-1	0
-3	2	2	2
-1	2	-2	1
0	2	1	-2

≤ 1

We give more examples of $33 + 1$ inequalities in the appendix, along with their NS intersections. We also tested the subpolytope method in the $32 + 1$ scenario. By cutting with CHSH, we retrieved 80 inequalities. By removing those that are not true facets of the one-bit polytope, we obtained 17 inequalities. By sorting these inequalities into inequivalence classes, we ended up with nine inequalities, a positivity facet, and the eight new facets that were published in Reference [14]. In this scenario, by cutting the polytope we easily recover the complete list of facet inequalities. Additionally, by intersecting these facets with the NS subspace, we again find that inequalities that have a larger quantum bound than local bound are constructed from CHSH. The best inequality in terms of distance between local and quantum bounds in $32 + 1$ is a sum of two CHSH.

4.2. Cutting with I3322

We repeated the "cutting" procedure using the I_{3322} inequality instead of CHSH. There are two other versions (in fact many more: any relabeling as discussed in Section 2.5) of I_{3322} that we can use. One of them is I_{3322}^{perm}, which we previously introduced. The other is the symmetrized version of I_{3322}:

$$I_{3322}^{\text{sym}} = \begin{array}{c|ccc} & -1 & -1 & 0 \\ \hline -1 & 0 & 1 & 1 \\ -1 & 1 & -1 & 1 \\ 0 & 1 & 1 & -1 \end{array} \leq 0$$

These inequalities are equivalent in the NS subspace, but when extended to the one-bit space they become inequivalent. Therefore, each cut gives a different number of vertices and facets. Cutting with I_{3322}, we obtained 513 inequivalent facets, 151 of them having a larger quantum bound than local bound in the NS subspace. The cut with I_{3322}^{sym} yields 642 inequivalent inequalities, 171 of them having a quantum advantage in the NS subspace. Finally, I_{3322}^{perm} gives 634 facets, 174 with a quantum advantage in the NS subspace.

We grouped all these inequalities together, and removed equivalent inequalities. We ended up with a total of 667 inequalities, 184 of which have a stronger quantum bound than local bound.

We found the same construction as before, and inequalities are constructed out of inequalities of the Bell polytope. For example, we found the same facet inequalities for $33 + 1$ that reduce to the sum of two I_{3322} in the NS subspace.

We also attempted to directly solve the full polytope. At the moment when we extracted the list of inequalities generated with the full polytope, the number of inequalities had not increased in the last two months. We thus conjecture that the list of 668 facet inequalities is complete.

5. Conclusions

We present a method and notation to find facets of Bell scenarios supplemented by one bit of classical communication. The notation we used simplifies the study of one-bit inequalities, especially with respect to their intersection with NS subspace. Even though the one-bit polytope is difficult to directly solve, we were able to find an extensive list of facets that we conjecture to be complete. In the $33 + 1$ scenario, we found no quantum violation of the one-bit bound. Given the structure of $33 + 1$ facets, and assuming our conjecture is correct, we proved that the obtained statistics by choosing between three projective measurements on any two-qubit quantum state can be reproduced by one bit of classical communication between parties. Our results also imply that, in this scenario, Alice must send one bit at least half of the time on average to Bob in order for the two parties to reproduce quantum correlations. These findings constitute a step further toward answering the binary-answer question raised in Section 1. Our results provide a better understanding of the general structure of Bell inequalities supplemented by one bit. Indeed, we found that, by intersecting the facets of the $\mathcal{C}$ polytope with the NS subspace, we derive inequalities that are constructed from Bell inequalities of the corresponding scenario without communication. This can be a starting point to guess new facets for scenarios where Bell inequalities are known.

The next scenarios to tackle are $34 + 1$, $43 + 1$, and $44 + 1$. An important point is that our results show that the best inequalities we found in terms of distance between local and quantum bounds are sums of the same Bell inequality of the corresponding Bell scenario; for example, for $33 + 1$, the best inequality is a sum of two Bell inequalities from 33. If this is a general trend for Bell scenarios supplemented by one bit of communication, in order to find a violation of the one-bit bound we require that Bell inequalities of the corresponding Bell scenario should be:

(1) maximally violated by a partially entangled state; and
(2) have a quantum bound that is more than halfway between local and one-bit bounds.

Only starting from four settings on one side and three on the other do we have partially entangled states maximally violating a facet Bell inequality [25]. In addition, in the $44 + 1$ scenario, states that maximally violate Bell inequalities are, in most cases, very close to maximally entangled [25]. Furthermore, for polytopes of higher dimension than the 44 scenario [26], we still do not know the complete list of facets, which complicates the problem even more. All of these points are quite negative in the perspective of solving the binary-answer question; nevertheless, there are possible avenues to get closer to the solution. An idea is to generate facets from subpolytopes of such complicated scenarios, but one has to be lucky to find optimal inequalities in terms of communication. Another possibility is to guess inequalities using known Bell inequalities, at least up to four settings for each party.

Supplementary Materials: The following are available online at http://www.mdpi.com/1099-4300/21/2/171/s1. Table S1: Conjectured complete list of tight Bell + 1 inequalities with three settings for both parties. Coefficients for each inequality are given in the following order: d_{00} d_{01} d_{02} d_{10} d_{11} d_{12} d_{20} d_{21} d_{22} e_{00} e_{01} e_{02} e_{10} e_{11} e_{12} e_{20} e_{21} e_{22} f_0 f_1 f_2. For each inequality, we give local bound L, two-qubit quantum bound Q, one bit of communication bound C, and quantum state that achieves the largest quantum bound $|\psi(\theta_{max})\rangle = \cos\theta_{max}|00\rangle + \sin\theta_{max}|11\rangle$. All quantities were computed for nondegenerate measurements.

Author Contributions: Both authors contributed equally to this work.

Funding: This research was funded by the Swiss NCCR-QSIT.

Acknowledgments: The authors would like to thank S. Pironio, F. Hirsch, D. Rosset, and N. Brunner for the useful discussions. Financial support by the Swiss NCCR-QSIT is gratefully acknowledged.

Conflicts of Interest: The authors declare no conflict of interest.

Appendix A. Examples of 33 + 1 Facets and Complete List

In this appendix, we give examples of $33 + 1$ facets, along with their NS intersection and connection to Bell inequalities. We start with a $33 + 1$ facet, shown in Table A1, that reduces to the sum of two CHSH inequalities in the NS subspace (inequality 349 in our Table S1).

Table A1. Facet of $33 + 1$, for which the quantum bound is $\sqrt{2} - 1$, for a local bound of zero and a one-bit bound of one. When intersected with the NS space, this inequality reduces to a sum of CHSH inequalities.

0	-1	0	1
	0	0	0
0	0	1	-1
	0	0	0
-2	1	1	2
	0	-1	-1

$\leq 1 \quad \xrightarrow{\text{NS}}$

	0	-1	-1
0	-1	0	1
0	0	1	-1
-2	1	1	2

≤ 1

This inequality corresponds, in the NS subspace, to a sum of one CHSH inequality that uses Alice's inputs $x = 0, 2$ and Bob's inputs $y = 0, 2$, and another CHSH that uses $x = 1, 2$ and $y = 1, 2$. Therefore, the quantum bound of this inequality is $\sqrt{2} - 1 \approx 0.4142$, which is twice the amount of violation for CHSH. The quantum bound is obtained for the maximally entangled state $1/\sqrt{2}(|00\rangle + |11\rangle)$.

One can also have a single I_{3322} contained in the facet, as the example in Table A2 shows (inequality number 529):

Table A2. Facet of $33 + 1$, for which the quantum bound is 0.25, for a local bound of zero and a one-bit bound of one. When intersected with the NS space, this inequality reduces to I_{3322}. In fact, we see that it corresponds to I_{3322}^{sym} if we permute Alice's inputs $x = 1$ and $x = 2$. This inequality is maximally violated by the maximally entangled state, and its quantum bound is the I_{3322} quantum bound.

-1	0	1	1
	0	0	-1
0	1	1	-1
	-1	-1	1
-1	1	-1	1
	0	0	0

$\leq 1 \quad \xrightarrow{\text{NS}}$

	-1	-1	0
-1	0	1	1
0	1	1	-1
-1	1	-1	1

≤ 1

In Table A3, we give an example of a facet which when intersected with the NS space reduces to a CHSH inequality and some other terms. Despite the extra terms, its quantum bound is the maximum violation of CHSH, attained for a maximally entangled state. This inequality is number 380 in Table S1, and it is similar to the inequality of Table 5, in the sense that both are constructed from Bell inequalities, and have some extra terms that do not contribute to the quantum bound. If we remove these extra terms, the quantum and local bounds would therefore not change. Understanding how these extra terms arise could lead to a better understanding of how to construct Bell + 1 inequalities from Bell inequalities.

Table A3. Facet of $33 + 1$, for which the quantum bound is $1/2(\sqrt{2} - 1)$, for a local bound of zero and a one-bit bound of one. When intersected with the NS space, this inequality reduces to a CHSH inequality for two of each party's inputs and some other terms. This inequality is maximally violated by the maximally entangled state, and its quantum bound is the CHSH quantum bound.

0	-1	1	0
	0	-1	1
-2	0	2	2
1	-1	-1	
-1	1	1	0
	-1	0	-1

$\leq 1 \xrightarrow{NS}$

	0	-2	-1
0	-1	1	0
-2	0	2	2
-1	1	1	0

$\leq 1 =$

	0	-1	0
0	-1	1	0
0	0	0	0
-1	1	1	0

$+$

	0	-1	-1
0	0	0	0
-2	0	2	2
0	0	0	0

≤ 1

Most inequalities of $33 + 1$ have a quantum bound that is different than the CHSH bound, I_{3322} or twice their amount. Most inequalities have quantum bounds that do not easily relate to Bell inequalities for binary outcomes, up to three settings. As a final example, we show such an $33 + 1$ facet and how its NS intersection is constructed from CHSH and I_{3322} even if the quantum bound does not directly relate to the maximal violations of the Bell inequalities. One such facet is inequality number 196 in Table S1.

Table A4. Facet of $33 + 1$, for which the quantum bound is 0.4158, for a local bound of zero and a one-bit bound of one. When intersected with the NS space, this inequality reduces to a sum of a CHSH inequality for two of each party's inputs and an I_{3322}. This inequality is maximally violated by the nonmaximally entangled state.

-1	0	1	1
	0	0	0
-2	2	-1	2
	-1	0	-1
0	2	1	-2
	-1	-1	1

$\leq 1 \xrightarrow{NS}$

	-2	-1	0
-1	0	1	1
-2	2	-1	2
0	2	1	-2

$\leq 1 =$

	-1	-1	0
-1	0	1	1
-1	1	-1	1
0	1	1	-1

$+$

	-1	0	0
0	0	0	0
-1	1	0	1
0	1	0	-1

≤ 1

As shown in Table A4, facet number 196 corresponds to a sum of I_{3322}^{sym} and CHSH using inputs $x = 1, 2$ of Alice and $y = 0, 2$ of Bob. maximum violation of 0.4158 is given by a partially entangled state:

$$|\psi\rangle = 0.738|00\rangle + 0.675|11\rangle \tag{A1}$$

and resistance to noise for this inequality is $\lambda = 0.7830$, larger than the resistance to noise of CHSH ($\lambda_{\text{CHSH}} = 0.7071$), but lower than the resistance to noise of I_{3322} ($\lambda_{I_{3322}} = 0.8$).

References and Note

1. Bell, J.S. Physics 1, 195 (1964). *Google Scholar* 1964.
2. Brunner, N.; Cavalcanti, D.; Pironio, S.; Scarani, V.; Wehner, S. Bell nonlocality. *Rev. Mod. Phys.* **2014**, *86*, 419. [CrossRef]
3. Popescu, S.; Rohrlich, D. Quantum nonlocality as an axiom. *Found. Phys.* **1994**, *24*, 379–385. [CrossRef]
4. Barrett, J.; Pironio, S. Popescu-Rohrlich correlations as a unit of nonlocality. *Phys. Rev. Lett.* **2005**, *95*, 140401. [CrossRef]
5. Cerf, N.J.; Gisin, N.; Massar, S.; Popescu, S. Simulating maximal quantum entanglement without communication. *Phys. Rev. Lett.* **2005**, *94*, 220403. [CrossRef] [PubMed]
6. Maudlin, T. Bell's inequality, information transmission, and prism models. *Philos. Sci. Assoc.* **1992**, *1992*, 404–417. [CrossRef]
7. Gisin, N.; Gisin, B. A local hidden variable model of quantum correlation exploiting the detection loophole. *Phys. Lett. A* **1999**, *260*, 323–327. [CrossRef]
8. Brassard, G.; Cleve, R.; Tapp, A. Cost of exactly simulating quantum entanglement with classical communication. *Phys. Rev. Lett.* **1999**, *83*, 1874. [CrossRef]
9. Steiner, M. Towards quantifying non-local information transfer: finite-bit non-locality. *Phys. Lett. A* **2000**, *270*, 239–244. [CrossRef]

10. Toner, B.F.; Bacon, D. Communication cost of simulating Bell correlations. *Phys. Rev. Lett.* **2003**, *91*, 187904. [CrossRef]
11. Bacon, D.; Toner, B.F. Bell inequalities with auxiliary communication. *Phys. Rev. Lett.* **2003**, *90*, 157904. [CrossRef]
12. Gisin, N. Bell's inequality holds for all non-product states. *Phys. Lett. A* **1991**, *154*, 201–202. [CrossRef]
13. Regev, O.; Toner, B. Simulating quantum correlations with finite communication. *SIAM J. Comput.* **2009**, *39*, 1562–1580. [CrossRef]
14. Maxwell, K.; Chitambar, E. Bell inequalities with communication assistance. *Phys. Rev. A* **2014**, *89*, 042108. [CrossRef]
15. Brask, J.B.; Chaves, R. Bell scenarios with communication. *J. Phys. A Math. Theory* **2017**, *50*, 094001. [CrossRef]
16. Van Himbeeck, T.; Brask, J.B.; Pironio, S.; Ramanathan, R.; Sainz, A.B.; Wolfe, E. Quantum violations in the Instrumental scenario and their relations to the Bell scenario. *arXiv* **2018**, arXiv:1804.04119.
17. Chaves, R.; Carvacho, G.; Agresti, I.; Di Giulio, V.; Aolita, L.; Giacomini, S.; Sciarrino, F. Quantum violation of an instrumental test. *Nat. Phys.* **2018**, *14*, 291. [CrossRef]
18. Collins, D.; Gisin, N. A relevant two qubit Bell inequality inequivalent to the CHSH inequality. *J. Phys. A Math. Gen.* **2004**, *37*, 1775. [CrossRef]
19. PORTA Vers. 1.3.2: Sources, Examples, Man-Pages (Tar-File). Available online: https://wwwproxy.iwr.uni-heidelberg.de/groups/comopt/software/PORTA/ (accessed on 7 February 2019).
20. Lörwald, S.; Reinelt, G. PANDA: A software for polyhedral transformations. *EURO J. Comput. Opt.* **2015**, *3*, 297–308. [CrossRef]
21. Pál, K.F.; Vértesi, T. Quantum bounds on Bell inequalities. *Phys. Rev. A* **2009**, *79*, 022120. [CrossRef]
22. Jones, N.S.; Masanes, L. Interconversion of nonlocal correlations. *Phys. Rev. A* **2005**, *72*, 052312. [CrossRef]
23. Gisin, N.; Popescu, S.; Scarani, V.; Wolf, S.; Wullschleger, J. Oblivious transfer and quantum channels as communication resources. *Natural Comput.* **2013**, *12*, 13–17. [CrossRef]
24. Pironio, S. Violations of Bell inequalities as lower bounds on the communication cost of nonlocal correlations. *Phys. Rev. A* **2003**, *68*, 062102. [CrossRef]
25. Brunner, N.; Gisin, N. Partial list of bipartite Bell inequalities with four binary settings. *Phys. Lett. A* **2008**, *372*, 3162–3167. [CrossRef]
26. Cruzeiro, E.Z.; Gisin, N. Complete list of Bell inequalities with four binary settings. *arXiv* **2018**, arXiv:1811.11820.

MDPI

Article

Non-Local Parity Measurements and the Quantum Pigeonhole Effect

G. S. Paraoanu

Department of Applied Physics, Aalto University, P.O. Box 15100, FI-00076 Aalto, Finland;
sorin.paraoanu@aalto.fi

Received: 5 June 2018; Accepted: 8 August 2018; Published: 16 August 2018

check for
updates

Abstract: The pigeonhole principle upholds the idea that by ascribing to three different particles either one of two properties, we necessarily end up in a situation when at least two of the particles have the same property. In quantum physics, this principle is violated in experiments involving postselection of the particles in appropriately-chosen states. Here, we give two explicit constructions using standard gates and measurements that illustrate this fact. Intriguingly, the procedures described are manifestly non-local, which demonstrates that the correlations needed to observe the violation of this principle can be created without direct interactions between particles.

Keywords: non-locality; parity measurements; entanglement; pigeonhole principle; controlled-NOT

1. Introduction

Quantum physics defies our classical intuition in many ways. The founders of this discipline have been keenly aware of this, and further insights obtained over many decades have only deepened this conceptual gap. Amongst the most counterintuitive results, the Einstein-Podolsky-Rosen paradox [1], the quantum Zeno effect [2], the non-cloning theorem [3], interaction-free measurements [4,5], and the no-reflection theorem [6] have challenged the common intuition that properties have a well-defined, pre-existing ontological status.

Recently, Yakir Aharonov et al. [7] have put forward another gedankenexperiment which brings quantum physics in direct conflict with the everyday view of reality. The experiment attemps to establish a quantum version of the well-known pigeonhole principle from mathematical combinatorics. Classically, attempting to place three pigeons in two holes will necessarily result in at least two pigeons being in the same hole. However, in the case of quantum particles, this is no longer the case. Indeed, let us consider that the pigeons are impersonated at the quantum level by particles and the left and right holes from the presentation of Aharonov et. al. correspond to the states $\{|0\rangle, |1\rangle\}$ in a two-dimensional Hilbert space. The three particles are indexed by a, b, c, and they are placed in the labs of Alice, Bob, and Charlie. This allows the problem to be reformulated in the modern quantum-information language of qubits and gates.

The quantum pigeonhole *gedankenexperiment* proceeds as follows: first, each particle (qubit) is prepared in the state $|+\rangle$, where $|+\rangle = \frac{1}{\sqrt{2}}(|0\rangle + |1\rangle)$. Then, a parity measurement is performed on any two of the three particles. The projectors corresponding to the results "same" and "different" are

$$\Pi^{\text{same}} = |00\rangle\langle00| + |11\rangle\langle11|, \tag{1}$$
$$\Pi^{\text{diff}} = |01\rangle\langle01| + |10\rangle\langle10|. \tag{2}$$

After the parity measurement, the two qubits end up either being projected onto the state $\Pi^{\text{same}}|+\rangle|+\rangle$ which, after normalization, is the Bell state $|\Phi^+\rangle = \frac{1}{\sqrt{2}}(|00\rangle + |11\rangle)$ if a "same" result is obtained, or onto the state $\Pi^{\text{diff}}|+\rangle|+\rangle$ which, after normalization, yields the Bell state $|\Psi^+\rangle = \frac{1}{\sqrt{2}}(|01\rangle + |10\rangle)$ in the case of a "different" result. Finally, the protocol ends by applying a measurement with the Pauli-Y operator on each of the two particles. This measurement can have two results ($\pm$), corresponding to single-qubit projection operators:

$$\Pi_{Y\pm} = \frac{1}{2}(1 \pm Y) = |\pm i\rangle\langle\pm i|, \tag{3}$$

where $|\pm i\rangle = \frac{1}{\sqrt{2}}(|0\rangle \pm i|1\rangle)$ are the eigenvalues of the Y operator, $Y|\pm i\rangle = |\pm i\rangle$. Now, we can easily verify that

$$\Pi_{Y+} \otimes \Pi_{Y+}|\Phi^+\rangle = 0. \tag{4}$$

This means that whenever "same" is obtained in the parity measurement of the two qubits, the result of the final measurement of $Y \otimes Y$ cannot be "++" (both qubits in the positive y direction). Thus, a result "++" for the final mesurement implies that a "different" result was obtained in the parity measurement.

Let us now consider the third qubit, which is not involved in the parity measurement. Since this qubit is prepared in the state $|+\rangle$, there is a non-zero probability of $1/2$ to be found in the $|+i\rangle$ state. We now postselect over cases such as the above, with all three qubits starting in the same state ($|+\rangle$) and giving the result (+) under the measurement of the Y operators at the end of the protocol. We now know that, under these conditions, only the result "different" could have been possible in the parity measurement.

Now, a contradiction can be obtained as follows: if we insist that after preparation, each qubit assumes a real (albeit unknow) value of 0 or 1 ("up" or "down" projection for a spin $1/2$), then the complete description of the three-particle system would be $||a_z, ..\rangle\rangle||b_z, ...\rangle\rangle||c_z, ...\rangle\rangle$, where $a_z, b_z, c_z \in \{0,1\}$. Here, using double kets, we denote a fictional representation of the state in terms of unknown "real" values: a_z, b_z, c_z. Thus, the parity operator only reveals if two of these values are equal or not by applying it to the corresponding pair of particles. However, in those situations when the final measured state is $|+i\rangle|+i\rangle|+i\rangle$, the parity operator is always "different" no matter which pair we choose to measure, and therefore, we have $a_z \neq b_z, b_z \neq c_z$, and $c_z \neq a_z$. This cannot be realized if $a_z, b_z, c_z \in \{0,1\}$. These values thus provide an overcomplete description of the state which leads to a logical contradiction. The argument provides a somewhat unexpected refutation of a naive realistic description of the quantum state, which this is achieved purely by logic. Somewhat similar arguments against local realism, involving only logical reasoning, have been presented before [8,9]. In this sense, the quantum pigeonhole effect provides a simple demonstration of contextuality in quantum physics.

There is, however, a conceptual loophole in the argument above. Indeed, a proponent of realism could still invent a mechanism by which the qubits get disturbed during the parity measurement in such a way that the quantum mechanical predictions for this experiment are still correctly reproduced. We show that this loophope can be closed by designing non-local setups, where the parity measurements are realized by local interactions and classical communication, thus avoiding direct physical interactions which could presumably have unknown or uncontrolled effects on the pre-existing values of the qubits. In one of the setups, allowing for such interactions results in a contradiction with a certain symmetry, while we show that the other would require backward causation.

We also noticed that an earlier version of the quantum pigeonhole paradox exists [10], where the protocol starts with the particles prepared in a GHZ state. In this case, the "same" and "diff" results are established by performing standard projection measurements on each qubit separately. Thus, one does not need to use parity measurements and the objection above does not apply.

2. Results

We start by noticing that the argument above hinges on the result that

$$\langle+i|\langle+i|\Pi^{\text{same}}|+\rangle|+\rangle = 0, \tag{5}$$

which is an immediate consequence of Equation (4). To understand why this is a key logical element, let us review the reasoning in a slightly modified formulation. Two qubits are measured with a "different" result and the final measurement also yields ++ in the y direction. Now, let us assume that the third qubit is also measured at the end of the protocol, and it is found in the $|+i\rangle$ state as well. To force a counterfactual reasoning, we ask what would have been the result of a parity measurement of the third qubit and any of the first two that were actually measured. Equation (5) shows that the result could not have been "same"; hence, there is an immediate contradiction with the attempt to assign real parity values.

However, this logic has a weak point—in order to establish the parity, one should perform the measurement which typically implies bringing the particles together in the same region of space and having them interact with a (yet unspecified) apparatus. Suppose, for example, that we believe that the values of the spin along the z and y directions pre-exist the measurement. In this case, the state of the three particles could be written as $||a_z, a_y, \ldots\rangle\rangle, ||b_z, b_y, \ldots\rangle\rangle, ||c_z, c_y, \ldots\rangle\rangle$, where the dots signify the (possible) existence of other variables used to descrive the state. Now, a clever supporter of ontological realism could come up with a model for parity measurement where nothing happens with the y components in case of a "different" measurement; however, a "same" measurement establishing $a_z = b_z$ would imply an interaction of the y components a_y and b_y, such that at the end of this interaction, no matter what their initial values were, a_y and b_y would acquire opposite values. We would thus end up, after a "same" result, with states of the type $||0, \pm\rangle\rangle||0, \mp\rangle\rangle$ and $||1, \pm\rangle\rangle||1, \mp\rangle\rangle$, which will never have a ++ projection on the y axes. Thus, the quantum mechanical result, Equation (5), is reproduced. In this case, if we perform the parity measurement on the first two particles, the cases with $a_z = b_z$ will be eliminated by our postselection. We could still have $a_z = c_z$ or $b_z = c_z$, but these pairs were not measured.

To eliminate these situations, we show that the quantum pigeonhole effect can be realized in completely non-local setups. We present two such realizations, shown in Figures 1 and 2. To simplify the presentation, we assume that the qubits belonging to Alice, Bob, and Charlie are prepared in the $|+\rangle$ state by a standard Hadamard gate applied to $|0\rangle$. On the measurement side, the projection on $|+i\rangle$ can be realized by first rotating the state by $\pi/2$ around the x axis and then performing a standard measurement on the z axis, with the result 0. Indeed, we have $\Pi_{Y+} = R_x^{-\pi/2}|0\rangle\langle 0|R_x^{\pi/2}$.

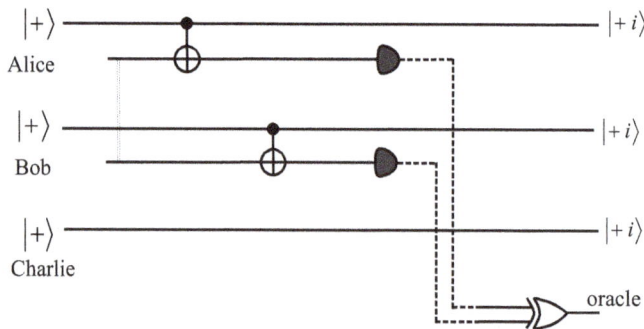

Figure 1. Circuit schematic for the quantum pigeonhole effect based on entanglement distillation. The double dotted line represents the entanglement between Alice's and Bob's ancilla qubits, which were prepared in the Bell state $|\Phi^+\rangle$. The dotted line is a classical channel that transmits the results of the measurements of the ancilla qubits to a classical XOR gate. A parity with the result "same" corresponds to the classical output of XOR having the value 0, while for "different", it takes a value of 1.

In the scheme of Figure 1, two ancilla qubits are used to implement the non-local parity measurement. They are prepared in an entangled state $|\Phi^+\rangle$ and they serve as the target qubits of the local CNOT gates with the $|+\rangle$ states as a control. The CNOT gates are local and could be realized by local interactions between the target qubit and the auxiliary qubits. The results of the measurements on the ancilla qubits are then transmitted in the form of classical bits of information to a coincidence counter which is implemented as a classical XOR gate. It is straigthforward to verify that a value of 0 for the oracle bit at the output of the XOR gate corresponds to the measurement operator Π^{same}, while a value of 1 corresponds to Π^{diff}, see Equations (1) and (2). Interestingly, through the use of CNOT gates—which condition the state of the ancillas on the qubits that we want to measure—this scheme converts a quantum parity measurement into a classical parity evaluation by an XOR gate.

The scheme blocks the counterargument presented above. Indeed, what happens locally (at the site of Alice or Bob) is identical. For the pair of entangled ancilla qubits, we are forced to assume identical local pre-existing values of the spin in all directions (the state $|\Phi^+\rangle$ is invariant under simultaneous rotations from one axis to another). Both Alice and Bob have a qubit in the same $|+\rangle$ state, and in cases of "same" measurements, the two qubits should have had identical pre-existing z component values. No matter what type of interaction we assume for the CNOT gate in this local realistic model, what happened at Alice's and Bob's sites had to result in identical states after the application of these gates. Even if the y-components were affected, they would be affected in the same way. Thus, there is no way to get a zero projection on $|+i\rangle|+i\rangle$.

Finally, let us note that the power of this scheme comes from entanglement distillation [11]. Indeed, while the states of Alice's and Bob's qubits were initially separated, the entanglement of the ancilla qubits was eventually transferred to Alice and Bob. After the measurement of the ancilla qubits had been performed, the state of Alice and Bob was either $|\Phi^+\rangle$ if the ancilla qubits were measured in the same state, or $|\Psi^+\rangle$ if they were measured in a different state.

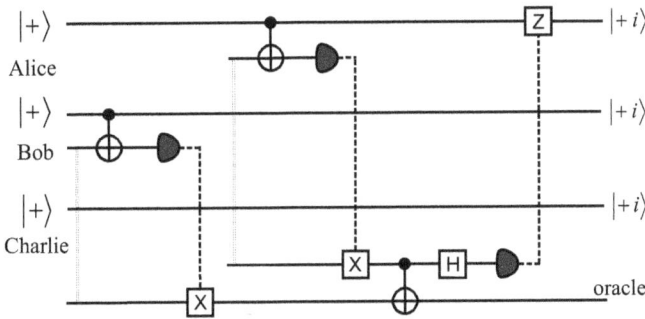

Figure 2. Circuit schematic for the quantum pigeonhole effect based on non-local CNOT gates. As in the previous figure, entanglement is shown with a dotted double line, while classical communication is shown with dashed lines. The result of the measurement is "same" when the oracle qubit is measured to be 0, and "different" when the oracle is measured to be 1.

In Figure 2, we present another possible non-local scheme, this time based on the teleportation of CNOT gates [12]. The idea comes from analyzing a relatively standard construction of parity measurements using two consequtive CNOT gates with Alice's and Bob's qubits as control qubits and one oracle qubit as the common target. We assume, as usual, that the oracle qubit starts in the $|0\rangle$ state. Then, we can check that

$$|+\rangle|+\rangle|0\rangle \xrightarrow{CNOT,CNOT} \frac{1}{\sqrt{2}}|\Phi^+\rangle|0\rangle + \frac{1}{\sqrt{2}}|\Psi^+\rangle|1\rangle. \tag{6}$$

A "0" result on the oracle qubit corresponds to a "same" result, with the Alice-Bob pair projected onto $|\Phi^+\rangle$, while a "1" result corresponds to a "diff" result, with the pair ending up in the $|\Psi^+\rangle$. If these CNOT gates were produced in the standard way by using qubit-qubit interactions, we would need to face the objection that perhaps a physical unknown influence could propagate, say, from Bob's qubit (assuming this one is connected first to its CNOT) to the oracle, and then to Alice's qubit. To avoid this, the construction, shown in Figure 2, makes use of the concept of teleportation of gates. The first part of our scheme, which adresses Bob's qubit, is half of a teleported CNOT gate, while the part that deals with Alices' qubit is a full CNOT gate. If, at the end of the circuit, the oracle qubit is zero, then the parity is "same"; if it is 1, then the parity is "different". This scheme illuminates the paradox in a different way. We assume again that Alice's and Bob's qubits came with pre-defined values a_z and b_z, and that somehow the CNOT gates would affect the y component of the qubit in such a conspiratorial way that whenever the Z components were the same, the y components would be made opposite. Now, in this scheme, the operations for the half teleported CNOT are applied first to Bob's qubit and ancilla, and then to the oracle. At that time, there was no other physical connection or correlation with Alices' qubit (unlike the previous scheme where we had the two ancillas entangled). Yet, the switching or non-switching of the y value of Bob's qubit would have had to be decided at this point. One can still argue that, perhaps, during the half teleported CNOT applied to Bob's qubit, the information about the state of Bob's qubit was transferred to the oracle qubit, and then this would influence the switching at Alice's site. However, this would imply a causation backward in time, sinc the CNOT that connects the oracle to Alice is placed after the operations (CNOT, measurememnt) performed on Alice's site. Note that the last conditional Z gate applied to Alice's qubit cannot produce such a hidden interaction, since it is triggered purely by a classical bit of information.

3. Discussion

In both schemes above, the key is the role of information which is distinct to that of interaction. Without knowedge of the result of the measurements, the effect disappears. Indeed, in both schemes, we are able to trace over the final results of the measurement (of the two ancilla qubits in the first scheme or of the oracle qubit in the second scheme; see also, the final state from Equation (6)), to obtain the mixed state

$$\frac{1}{2}|\Phi^+\rangle\langle\Phi^+| + \frac{1}{2}|\Psi^+\rangle\langle\Psi^+| = \frac{1}{4}\begin{bmatrix} 1 & 0 & 0 & 1 \\ 0 & 1 & 1 & 0 \\ 0 & 1 & 1 & 0 \\ 1 & 0 & 0 & 1 \end{bmatrix}. \tag{7}$$

The second representation is an X-matrix in the $|00\rangle, |01\rangle, |10\rangle, |11\rangle$ basis. The trace of the square of this density matrix is 1/2, showing that it is a mixed state. For this type of density matrices, the concurrence can be calculated with standard expressions (see, e.g., [13,14]), and it yields zero. Thus, in the absence of information on the result of the parity measurements, the entanglement dissappears and the correlations implicit in Equation (5) are no longer established. This underlines the key role played by entanglement in the quantum pigeonhole paradox. In the earliest version presented in [10], the initial state is a GHZ state and entanglement is present from the beginning. In the formulation which uses parity measurement [7], the initial and final states are separable, and so superficially it looks like entanglement plays no role. However, the application of this measurement not only produces the information about parity but also entangles the two qubits in a Bell state. In the non-local version proposed in this work, the entanglement is extracted and transferred, by local measurements and classical communication, from the ancilla qubits.

4. Conclusions

In conclusion, we closed a loophole in the quantum version of the pigeonhole principle by analyzing two manifestly non-local schemes. This eliminated the possibility of unknown local interactions and backaction by the use of non-local parity operators.

Funding: The author acknowledges support from the Academy of Finland under the Centre of Excellence Quantum Technology Finland (QTF), project 312296 and from the Center for Quantum Engineering at Aalto University project QMETRO.

Conflicts of Interest: The author declares no conflict of interest.

References

1. Einstein, A.; Podolsky B.; Rosen, N. Can Quantum-Mechanical Description of Physical Reality be Considered Complete? *Phys. Rev.* **1935**, *47*, 777–780. [CrossRef]
2. Misra, B.; Sudarshan, E.C.G. The Zeno's paradox in quantum theory. *J. Math. Phys.* **1977**, *18*, 756–763. [CrossRef]
3. Wootters, W.; Zurek, W. A Single Quantum Cannot be Cloned. *Nature* **1982**, *299*, 802–803. [CrossRef]
4. Elitzur, A.C.; Vaidman, L. Quantum Mechanical Interaction-Free Measurements. *Found. Phys.* **1993**, *23*, 987–997. [CrossRef]
5. Paraoanu, G.S. Interaction-Free Measurements with Superconducting Qubits. *Phys. Rev. Lett.* **2006**, *97*, 180406, doi:10.1103/PhysRevLett.97.180406. [CrossRef] [PubMed]
6. Kumar, K.S.; Paraoanu, G.S. A quantum no-reflection theorem and the speeding up of Grover's search algorithm. *Europhys. Lett.* **2011**, *93*, 64002. [CrossRef]
7. Aharonov, Y.; Colombo, F.; Popescu, S.; Sabadini, I.; Struppa, D.C.; Tollaksen, J.Q. Quantum violation of the pigeonhole principle and the nature of quantum correlations. *Proc. Natl. Acad. Sci. USA* **2016**, *113*, 532–535. [CrossRef] [PubMed]
8. Hardy, L. Quantum mechanics, local realistic theories, and Lorentz-invariant realistic theories. *Phys. Rev. Lett.* **1992**, *68* 2981–2984. [CrossRef] [PubMed]
9. Paraoanu, G.S. Realism and Single-Quanta Nonlocality. *Found. Phys.* **2011**, *41*, 734. [CrossRef]
10. Aharonov, Y.; Nussinov, S.; Popescu, S.; Vaidman, L. Peculiar features of entangled states with postselection. *Phys. Rev. A* **2013**, *87* 014105. . [CrossRef]
11. Bennett, C.H.; Brassard, G.; Popescu, S.; Schumacher, B.; Smolin, J.A.; Wootters, W.K. Purification of Noisy Entanglement and Faithful Teleportation via Noisy Channels. *Phys. Rev. Lett.* **1996**, *76*, 722–725, doi:10.1103/PhysRevLett.76.722. [CrossRef] [PubMed]
12. Eisert, J.; Jacobs, K.; Papadopoulos, P.; Plenio, M.B. Optimal local implementation of nonlocal quantum gates. *Phys. Rev. A* **2000** *62*, 052317. [CrossRef]
13. Yu, T.; Eberly, J.H. Decay of entanglement in coupled, driven systems with bipartite decoherence. *Quantum Inf. Comput.* **2007**, *7*, 459–468.
14. Li, J.; Paraoanu, G.S. Decay of entanglement in coupled, driven systems. *Eur. Phys. J. D* **2010**, *56*, 255–264, doi:10.1140/epjd/e2009-00247-9. [CrossRef]

entropy

MDPI

Article

Quantum Nonlocality and Quantum Correlations in the Stern–Gerlach Experiment

Alma Elena Piceno Martínez, Ernesto Benítez Rodríguez, Julio Abraham Mendoza Fierro, Marcela Maribel Méndez Otero and Luis Manuel Arévalo Aguilar *

Facultad de Ciencias Físico Matemáticas, Benemérita Universidad Autónoma de Puebla,
18 Sur y Avenida San Claudio, Col. San Manuel, Puebla 72520, Mexico; epic4492@gmail.com (A.E.P.M.);
tortugaleon44@gmail.com (E.B.R.); mfja52@gmail.com (J.A.M.F.); motero@fcfm.buap.mx (M.M.M.O.)
* Correspondence: larevalo@fcfm.buap.mx

Received: 27 February 2018; Accepted: 12 April 2018; Published: 19 April 2018

Abstract: The Stern–Gerlach experiment (SGE) is one of the foundational experiments in quantum physics. It has been used in both the teaching and the development of quantum mechanics. However, for various reasons, some of its quantum features and implications are not fully addressed or comprehended in the current literature. Hence, the main aim of this paper is to demonstrate that the SGE possesses a quantum nonlocal character that has not previously been visualized or presented before. Accordingly, to show the nonlocality into the SGE, we calculate the quantum correlations $C(z, \theta)$ by redefining the Banaszek–Wódkiewicz correlation in terms of the Wigner operator, that is $C(z, \theta) = \langle \Psi | \hat{W}(z, p_z) \hat{\sigma}(\theta) | \Psi \rangle$, where $\hat{W}(z, p_z)$ is the Wigner operator, $\hat{\sigma}(\theta)$ is the Pauli spin operator in an arbitrary direction θ and $|\Psi\rangle$ is the quantum state given by an entangled state of the external degree of freedom and the eigenstates of the spin. We show that this correlation function for the SGE violates the Clauser–Horne–Shimony–Holt Bell inequality. Thus, this feature of the SGE might be interesting for both the teaching of quantum mechanics and to investigate the phenomenon of quantum nonlocality.

Keywords: quantum nonlocality; quantum mechanics; Stern–Gerlach experiment

1. Introduction

The Stern–Gerlach experiment (SGE) [1–4] has played an important role in both the teaching and advancement of quantum mechanics. In quantum physics, this experiment is commonly used to introduce the concept of the internal spin of quantum systems, which has no counterpart in classical systems. Although nowadays, a significant number of new proposals has arisen to enhance our general understanding of the way it works and its possible uses [5–14], nonetheless, after almost one hundred years since its inception, still, there is not a total understanding of how it works yet in terms of an entire quantum mechanical description [4–16]; also, see [17]. In fact, most research analyses and textbooks have regularly focused on an SGE's semiclassical model [5,6,9]. As an example of new approaches, Boustimi et al. gave a step forward by abandoning the usual magnets' configuration producing a quadrupolar static magnetic field (by using just four bars) with the objective of modulating an atomic beam by means of an interference pattern [18]; see also [19]. Additionally, Machluf et al. have produced a field gradient beam splitter to create a coherent momentum superposition for matter-wave interferometry [20]. Of high relevance, the SGE was proposed as a system for implementing quantum roulettes, which are generalized quantum measurements [12]. In this case, a fluctuating magnetic field induces a probability distribution, which is used to implement a positive operator-value measure (POVM), which describes a continuous quantum roulette; for details, see [12].

Recently, we argued that the SGE could easily be used to introduce the concept of entanglement between the external and internal degrees of freedom in the teaching of quantum mechanics [5,6], to exemplify the entanglement generation between discrete and continuous variables and between pure and mixed states, as well. To show these properties of the SGE, we have used the evolution operator method [21–25]; to see an independent test of this method, see [26,27].

As we stated above, one of the main goals of the present paper is to show that the SGE possesses nonlocal correlations between internal and external degrees of freedom. In this way, this finding, of the SGE's nonlocality features, will serve to stress its paramount importance in teaching quantum mechanics, and likewise, it might open a new avenue for investigation and the understanding of this famous experiment. In fact, the physical education research community (PER) is currently undergoing intense research and development regarding the learning and teaching of quantum mechanics, where it is important to highlight the importance of SGE [28–53].

2. Quantum Nonlocality

As was indicated by Clauser and Shimony [54], realism, which is the philosophical conception held by most physicists, claims that the external reality is supposed to have definite properties, i.e., predetermined values, independently of whether or not they are "observed"; this seems to support the objectivity of scientific investigation about nature. However, some of the quantum mechanics implications represent a direct challenge to this conception, e.g., the superposition principle. Particularly, the nonlocal character of quantum mechanics seems to imply that reality is not as direct as it was previously thought; instead, it is "created" by the measurement process. For example, if we get the singlet spin state $|\psi\rangle_{AB} = \frac{1}{\sqrt{2}}(|\uparrow_z\rangle_A |\downarrow_z\rangle_B - |\downarrow_z\rangle_A |\uparrow_z\rangle_B)$, where $|\uparrow_z\rangle_A$ ($|\downarrow_z\rangle_A$) represents spin up (spin down) of system A in the z direction, and we separate each of its two parts very far apart, when measuring the observable $\hat{\sigma}_z$ on part A, then part B acquires a defined z spin component. If, instead of measuring the observable $\hat{\sigma}_z$, we decide to measure the observable $\hat{\sigma}_y$ on A, then part B will acquire a definite y spin component. Therefore, by choosing to measure observable $\hat{\sigma}_z$ or $\hat{\sigma}_y$ on system A, a property of system B is "created"; this is only possible because the nonlocal character of quantum mechanics allows nonlocal correlations [55], since the systems are very far apart.

Hence, quantum nonlocality is a valuable resource, which allows the performance of many non-classical tasks, and at first instance, it was believed to be equivalent to entanglement [56]. However, nowadays, it is understood that nonlocality and entanglement agree with each other when the entangled systems are in pure states only, because it was proven that Bell's inequality holds for all non-product states [57]; that is to say, any entangled state possesses quantum correlations that result in a contradiction with local classical theories [58]. Additionally, there are mixed entangled states whose quantum correlations could be explained by means of local hidden variables theories, and they could still be used to implement probabilistic teleportation protocols [59], i.e., quantum entanglement differs from quantum nonlocality [60,61].

Furthermore, one way of perceiving differences between entanglement and nonlocality is by demonstrating that there exist sets of quantum unentangled states that, however, possess nonlocal correlations in the sense that they may not be reliably distinguished by local measurements on the parts, and neither may the cloning operation be implemented by local operations on them [61].

Historically, the Fifth Physical Conference of the Solvay Institute, held in 1927, was one of the first times when the counterintuitive features of quantum nonlocality were addressed. In this conference, Einstein put forward a thought experiment, now called *Einstein's boxes*, where he uncovers an important facet of the nonlocal character of quantum mechanics [62]. In this experiment, a single particle wave function is diffracted by a single slit, and the ongoing spherical wave function given by $\psi(x,t)$ is spread over a hemisphere screen; then, according to Einstein, $|\psi(x,t)|^2$ expresses the probability that, at a given moment, the particle arrives at an arbitrary point belonging to the hemisphere screen [62]. Therefore, to rule out the possibility of being located at more than one place, there must be an instantaneous action on the entire screen. Jammer's translation of Einstein's words says: "a peculiar action-at-a-distance must be assumed to take place which prevent the continuously distributed wave in space from producing an effect at two places on the screen" [62]. This thought experiment was further stated in terms of two boxes by Einstein in a letter addressed to Schrödinger [63,64]; this letter seems to be the source of Schrödinger's cat paradox [63]. Probably the second occasion where the counterintuitive nonlocal feature of quantum mechanics emerged was in the Einstein et al. paper of 1935, where the Einstein-Podolsky-Rosen paradox was established [65]. Nowadays, quantum nonlocality is believed to be different from entanglement, and it is taken as another quantum resource. Like entanglement, nonlocality cannot be created by local operations and classical communications [66]. Furthermore, the nonlocality resource can be distilled, in a similar way as entanglement [67]. Consequently, it is important to extensively study nonlocality for the sake of a better understanding of its relation with entanglement.

3. The SGE in A Complete Quantum Treatment

The SGE experiment is usually analyzed in most textbooks in a semiclassical way, where the external degrees of freedom are considered as classical variables, and its dynamics is thought of in terms of Newton's second law [5,6,9]. However, in the scientific literature, we can find proposals that treat the external degrees of freedom (EDF) as a quantum variable giving a quantum description of the evolution of the EDF; see [5–11,13]. Nevertheless, in a recent paper, we gave a complete quantum treatment of the SGE [5,6], as it was originally thought of by Scully et al. [9], by using the evolution operator method [21–25] and obtaining the solution to the Schrödinger equation; this allows us to see the quantum features of the EDF. It is worth mentioning that in Figure 1 of [5], we are replacing the usual continuous path by a dotted one to stress the absence of classical paths. Then, one of our aims in this paper is to extract the nonlocal implications of the quantum treatment of the SGE.

For the case of the evolution of an initial superposition state of the spin degree of freedom, we have the initial state:

$$|\psi(0)\rangle = \psi_0 \left(\alpha \,|\!\uparrow_z\rangle + \beta \,|\!\downarrow_z\rangle \right), \tag{1}$$

where α and β are constants obeying $|\alpha|^2 + |\beta|^2 = 1$, and ψ_0 is the Gaussian wave packet:

$$\psi_0 = \frac{1}{(2\pi\sigma_0^2)^{3/4}} \exp\left(-\frac{r^2}{4\sigma_0^2} + i\mathbf{k}\cdot\mathbf{r} \right), \tag{2}$$

where σ_0 is the width of the wave packet, $\mathbf{r}$ is the position and $\mathbf{k}$ is the wave vector. Then, the evolved state is given by:

$$
\begin{aligned}
|\psi(t)\rangle &= e^{-\frac{it}{\hbar}\left(\frac{-\hbar^2}{2m}\nabla^2 + \mu_c(\hat{\sigma}\cdot\mathbf{B}) \right)} \left[\psi_0(\alpha\,|\!\uparrow_z\rangle + \beta\,|\!\downarrow_z\rangle) \right] \\
&= \exp\left(\frac{-it^3\mu_c^2 b^2}{6m\hbar} \right) \left[\frac{\sigma_0}{(2\pi)^{1/2}} \right]^{3/2} \left(\sigma_0^2 + \frac{i\hbar t}{2m} \right)^{-3/2} \exp\left(-\sigma_0^2 k_y^2 \right) \\
&\quad \times\ \exp\left[\frac{i4y\sigma_0^2 k_y}{4(\sigma_0^2 + i\hbar t/2m)} \right] \exp\left[\frac{-(x^2 + y^2 - 4\sigma_0^4 k_y^2)}{4(\sigma_0^2 + i\hbar t/2m)} \right] \\
&\quad \times\ \left\{ \alpha \exp\left[\frac{-it\mu_c}{\hbar}(B_0 + bz) \right] \exp\left[\frac{-1}{4(\sigma_0^2 + i\hbar t/2m)} \left(z + \frac{t^2\mu_c b}{2m} \right)^2 \right] |\!\uparrow_z\rangle \right. \\
&\quad +\ \left. \beta \exp\left[\frac{it\mu_c}{\hbar}(B_0 + bz) \right] \exp\left[\frac{-1}{4(\sigma_0^2 + i\hbar t/2m)} \left(z - \frac{t^2\mu_c b}{2m} \right)^2 \right] |\!\downarrow_z\rangle \right\}.
\end{aligned}
\tag{3}
$$

where ∇ is the vector differential operator, m is the mass of the particle, $\mu_c = g\frac{e\hbar}{4m_e}$ and g is the gyromagnetic ratio; see [5,6] for details. The magnetic field $\mathbf{B}$ of the SGE is an inhomogeneous magnetic field of the form $\mathbf{B} = -bx\hat{i} + (B_0 + bz)\hat{k}$, with B_0 a constant and b the strength of the inhomogeneity of the field.

4. Quantum Correlations and Nonlocality in the Stern–Gerlach Experiment

In this section, we calculate the quantum correlation and nonlocality arising in the SGE. To achieve that, we analytically work out the quantum correlation function $\mathcal{C}(z, \theta)$ in the phase space, by redefining the correlation function proposed by Banaszek and Wódkiewicz [68–70] (see also [71]) in terms of the Wigner operator [72]. In other words, we define the correlation function in the following way:

Firstly, we define a correlation as:

$$c(z, \theta) = \frac{1}{\pi\hbar} \langle \Psi | \hat{W}(z, p_z) \hat{\sigma}(\theta) | \Psi \rangle, \tag{4}$$

where p_z is the momentum in z and $\hat{W}(z, p_z)$ is the Wigner operator given by [72]; see also [73–75]:

$$\hat{W}(z, p_z) = \frac{1}{2} \int_{-\infty}^{\infty} \left| z - \frac{q}{2} \right\rangle \exp\left(-\frac{iqp_z}{\hbar} \right) \left\langle z + \frac{q}{2} \right| dq, \tag{5}$$

$\hat{\sigma}(\theta)$ is the usual Pauli spin operator in an arbitrary direction θ, and q is a parameter. The $1/2$ factor that multiplies the integral in Equation (5) derives from the parity operator defined by Royer [74], which is given by $\Pi_{rp} = \int e^{-i2ips/\hbar} |r - s\rangle\langle r + s| ds$; by changing the variable s for $q/2$, you arrive at Equation (5). A possible path to deduce the Wigner operators is as follows: The definition of the Wigner function given by Wigner and collaborators is $P_w(q, p) = \frac{1}{\pi\hbar} \int e^{2ipy/\hbar} \langle q - y|\hat{\rho}|q + y\rangle dy$ [75]; from $P_w(q, p)$ and setting $\hat{\rho} = |\Psi\rangle\langle\Psi|$, we have $P_w(q, p) = \frac{1}{\pi\hbar} \int e^{2ipy/\hbar} \langle q - y|\Psi\rangle \langle\Psi|q + y\rangle dy = \frac{1}{\pi\hbar} \int e^{2ipy/\hbar} \langle\Psi|q + y\rangle \langle q - y|\Psi\rangle dy = \frac{1}{\pi\hbar} \langle\Psi| \left\{ \int e^{2ipy/\hbar} |q + y\rangle\langle q - y| dy \right\} |\Psi\rangle$, and this is equal to the Royer definition and could explain the $1/2$ factor in front of Equation (5); this renders $\hat{W}^2(z, p_z) = 1$. In addition, to perceive the importance of the Wigner function for the understanding of quantum mechanics, see [75–78].

However, notice that the correlation $c(z, \theta)$ possesses dimensional factors that are originated because it is a correlation between the Wigner function and the Pauli operator. Hence, in order to avoid this dimensional factor and to present a correlation without dimension, we define the correlation $\mathcal{C}(z, \theta)$ in the Stern–Gerlach experiment as:

$$\mathcal{C}(z, \theta) = \pi\hbar c(z, \theta) = \langle \Psi | \hat{W}(z, p_z) \hat{\sigma}(\theta) | \Psi \rangle. \tag{6}$$

Additionally, notice that the Wigner operator is, in fact, the parity operator around the points z and p_z [72,74]. Here, we demonstrate that the SGE's correlation function $\mathcal{C}(z, \theta)$ violates the Bell's inequality [68]. In the case of entangled pure states, this violation of Bell's inequality is usually interpreted as the signature of nonlocality in quantum mechanics [68,69,79,80]; then, based on the preceding, in this paper, we restrict ourselves to this interpretation only. However, see [81] for a different interpretation.

In addition, notice that, as was pointed out by Ferraro and Paris, the amount of violation of Bell's inequality specifically depends on the kind of Bell operators used to test it [82]. Furthermore, the degree of quantum nonlocality depends on the type of entangled state [80]. For example, with regard to the approach of Banazek and Wódkiewicz, the maximal violation attained for a two-mode squeezed state is approximately 2.32 [68,69]; though, for the formalism proposed by Chen et al. [80], the maximal violation attained, for the same two-mode squeezed states, reaches the maximum value, i.e., $\approx 2\sqrt{2}$ [79]. See also [82] for an instructive discussion of nonlocality in continuous variables for two and three modes.

4.1. A Pure State

In this section, we consider a pure state for the position and the spin that traverses an SGE. The effect of the SGE is to produce an entangled state between the internal and the external degrees of freedom, as given in [5,6]. Then, the state coming out from the SGE, given in the previous section, can be written as follows:

$$|\psi(t)\rangle = c_0(x,y,t)\frac{1}{\sqrt{2}}\left(|\varphi_+(t)\rangle\,|\uparrow_z\rangle + |\varphi_-(t)\rangle\,|\downarrow_z\rangle\right), \tag{7}$$

where $c_0(x,y,t)$ are the variables on x and y dimensions that appear in the previous section or in [5,6], which are in concordance with the definition $A_1 = \exp\left(-\frac{it^3\mu_c^2b^2}{mh}\right)\left(\frac{\sigma_0}{\sqrt{2\pi}}\right)^{\frac{1}{2}}\left(\sigma_0^2 + \frac{iht}{2m}\right)^{-\frac{1}{2}}$, that is:

$$
\begin{aligned}
c_0(x,y,t) &= \exp\left(\frac{5it^3\mu_c^2b^2}{6mh}\right)\left[\frac{\sigma_0}{(2\pi)^{1/2}}\right]\left(\sigma_0^2 + \frac{iht}{2m}\right)^{-1}\exp\left(-\sigma_0^2 k_y^2\right)\\
&\times \exp\left[\frac{i4y\sigma_0^2 k_y}{4(\sigma_0^2 + ith/2m)}\right]\exp\left[\frac{-(x^2+y^2-4\sigma_0^4 k_y^2)}{4(\sigma_0^2 + ith/2m)}\right],
\end{aligned}
\tag{8}
$$

we have set the constants α and β of Equation (1) equal to $1/\sqrt{2}$, where the position states $|\varphi_+(t)\rangle$ and $|\varphi_-(t)\rangle$ are such that:

$$\langle z|\varphi_+(t)\rangle = A_1\exp\left[-\frac{it\mu_c}{\hbar}(B_0+bz)\right]\exp\left[-\frac{\left(z+\frac{t^2\mu_cb}{2m}\right)^2}{4\left(\sigma_0^2+\frac{ith}{2m}\right)}\right], \tag{9}$$

$$\langle z|\varphi_-(t)\rangle = A_1\exp\left[\frac{it\mu_c}{\hbar}(B_0+bz)\right]\exp\left[-\frac{\left(z-\frac{t^2\mu_cb}{2m}\right)^2}{4\left(\sigma_0^2+\frac{ith}{2m}\right)}\right]. \tag{10}$$

Notice that $\langle z|\varphi_+(t)\rangle$ and $\langle z|\varphi_-(t)\rangle$ are not orthogonal; however, they are properly normalized in the variable z when taking into account the constant A_1. Henceforward, in order to facilitate this analysis, we focus on the single position dimension z; in other words, we do not take into account the other two dimensions. As a consequence, in the subsequent paragraphs, we will just employ the coordinate z.

Then, using Equation (7), we calculate $\mathcal{C}(z,,\theta)$ as:

$$
\begin{aligned}
\mathcal{C}(z,\theta) =\ & \langle\varphi_+(t)|\,\hat{W}(z,p_z)\,|\varphi_+(t)\rangle\,\langle\uparrow_z|\,\hat{\sigma}(\theta)\,|\uparrow_z\rangle + \langle\varphi_+(t)|\,\hat{W}(z,p_z)\,|\varphi_-(t)\rangle\,\langle\uparrow_z|\,\hat{\sigma}(\theta)\,|\downarrow_z\rangle\\
& + \langle\varphi_-(t)|\,\hat{W}(z,p_z)\,|\varphi_+(t)\rangle\,\langle\downarrow_z|\,\hat{\sigma}(\theta)\,|\uparrow_z\rangle + \langle\varphi_-(t)|\,\hat{W}(z,p_z)\,|\varphi_-(t)\rangle\,\langle\downarrow_z|\,\hat{\sigma}(\theta)\,|\downarrow_z\rangle.
\end{aligned}
\tag{11}
$$

Equation (11) establishes the quantum correlations emerging from the SGE, and after lengthy calculations, we arrive at the next expression:

$$
\begin{aligned}
\mathcal{C}(z,\theta) =\ & \frac{\cos\theta}{2}\left(\exp\left\{-\frac{\sigma_0^2\left(z+\frac{t^2\mu_cb}{2m}\right)^2}{2\left(\sigma_0^4+\frac{\hbar^2t^2}{4m^2}\right)} - \frac{2\left(\sigma_0^4+\frac{\hbar^2t^2}{4m^2}\right)}{\sigma_0^2}\left[\frac{p_z}{\hbar}+\frac{t\mu_cb}{\hbar}-\frac{\hbar t\left(z+\frac{t^2\mu_cb}{2m}\right)}{4m\left(\sigma_0^4+\frac{\hbar^2t^2}{4m^2}\right)}\right]^2\right\}\right.\\
& -\exp\left\{-\frac{\sigma_0^2\left(z-\frac{t^2\mu_cb}{2m}\right)^2}{2\left(\sigma_0^4+\frac{\hbar^2t^2}{4m^2}\right)} - \frac{2\left(\sigma_0^4+\frac{\hbar^2t^2}{4m^2}\right)}{\sigma_0^2}\left[\frac{p_z}{\hbar}-\frac{t\mu_cb}{\hbar}-\frac{\hbar t\left(z-\frac{t^2\mu_cb}{2m}\right)}{4m\left(\sigma_0^4+\frac{\hbar^2t^2}{4m^2}\right)}\right]^2\right\}\right)\\
& +\sin\theta\exp\left[-\frac{\sigma_0^2z^2}{2\left(\sigma_0^4+\frac{\hbar^2t^2}{4m^2}\right)} - \frac{2\left(\sigma_0^4+\frac{\hbar^2t^2}{4m^2}\right)}{\sigma_0^2}\left(\frac{p_z}{\hbar}-\frac{\hbar t z}{4m\left(\sigma_0^4+\frac{\hbar^2t^2}{4m^2}\right)}\right)^2\right]\\
& \cos\left[-\frac{2t\mu_c}{\hbar}(B_0+bz)+\frac{\hbar t^3\mu_cbz}{4m^2\left(\sigma_0^4+\frac{\hbar^2t^2}{4m^2}\right)}+\frac{t^2\mu_cb}{m}\left(\frac{p_z}{\hbar}-\frac{\hbar t z}{4m\left(\sigma_0^4+\frac{\hbar^2t^2}{4m^2}\right)}\right)\right]
\end{aligned}
\tag{12}
$$

In this case, Equation (12) represents the correlation function that arises between the direction of the spin and the z position of the atom; it exhibits the interference and the entanglement of the internal and external degrees of freedom, and it shows that there are values where the correlation is minimum and some values where this correlation is maximum. In essence, this result shows that the measurement outcomes of the z position may depend nonlocally on the measurement outcome of the internal degree and vice versa. In other words, the dichotomic observables in this case are the parity operator given in Equation (5) and the spin of the atom in an arbitrary direction. Moreover, Equation (12) has two terms: the first term has two Wigner functions (if multiplied by $\pi\hbar$), which are displaced by a term $d = t^2 \mu_c b/2m$, and they move in opposite direction in position space with velocity proportional to $\mu_c b/\hbar$, whereas the second one has an oscillating cosine term, which is responsible for the oscillations. Additionally, it is important to mention that Equation (12) depends on six parameters: the time t, the position z, the strength of the divergence of the magnetic field b, the momentum p_z, the initial width of the wave packet σ_0 and the angle θ.

We have plotted Equation (12) on Figure 1, where we take p_z as a constant; we set $\pi\hbar = 1$ and $m = 1$ in a very similar way as is carried out in [11,70]. This figure allows us to see the oscillatory behavior of the correlation function with variables z and θ predicted in Equation (12). Furthermore, there, we can notice how the oscillations in z decay very fast when its values are increased, until this effect is hardly appreciated. In the same way, the negativity of the Wigner function is also perceived, which is commonly associated with some kind of nonclassical behavior, although some care must be taken when using this interpretation of the negativity because it involves the spin variable also. This issue is associated with the definition of the Wigner function for finite dimensional Hilbert space [83–86].

To conclude this section, it is important to remark that the correlation function in Equation (12) can be put in a very similar way to the one of the correlation function for the Schrödinger cat state seen in Wódkiewicz's article [68], so that the same conclusions at which he arrives regarding the displacement D still remain valid.

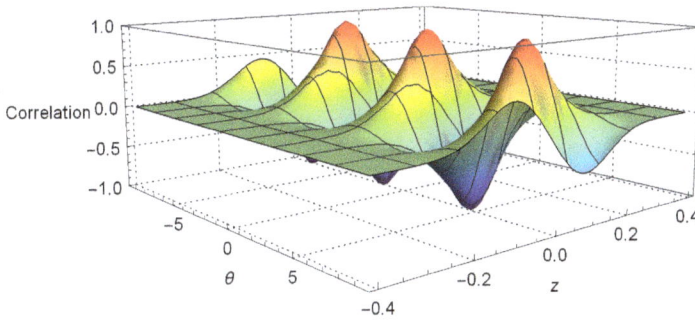

Figure 1. A plot of the correlation function between z and θ given by Equation (12). To obtain this plot, we have employed the following values: first, we set $\pi\hbar = 1$ and $m = 1$; then, we set $\sigma_0 = 0.005$, $\mu_c b/2 = 2.2$, $p_z = 0.01$ and time $t = 0.2$.

4.2. Violation of Bell's Inequalities

In hidden variables theories, quantum correlations are thought to arise from the average of the correlation function with respect to the hidden variables λ over statistical distributions. In particular, for the case of SGE, we apply the analysis implemented by Wódkiewicz where the average is given by [68]; see also [79,80]:

$$C(z,\theta) = \int d\lambda_{ext} \int d\lambda_{int} W(z,\lambda_{ext})\sigma(\theta,\lambda_{int})P(\lambda_{ext},\lambda_{int}), \qquad (13)$$

where λ_{ext} and λ_{int} are hidden variables of the external and internal degrees of freedom, respectively, $W(z,\lambda_{ext}) = \pm 1$ represents the parity operator complemented with the hidden variable λ_{ext} and its values ± 1 are the local realities of the external degree of freedom. On the other hand, $\sigma(\theta, \lambda_{int}) = \pm 1$ represents the Pauli operator complemented with the hidden variable λ_{int}, and its values ± 1 represent the local realities of the internal degree of freedom. Finally, $P(\lambda_{ext}, \lambda_{int})$ is the density distribution of the hidden variables. According to the Clauser–Horne–Shimony–Holt (CHSH) analysis of Bell's inequalities [87], this correlation should obey the following inequalities:

$$-2 \leq C(z', \theta') + C(z', \theta) + C(z, \theta') - C(z, \theta) \leq 2. \tag{14}$$

Thus, a violation of these inequalities by quantum mechanical correlations arises from the nonlocality of quantum phenomena. It is important to emphasize that C represents the correlations that are the product of the hidden variables' average in a hidden variables theory.

On the other hand, from Equation (12), we can calculate the function $\mathcal{B}_z$ for the quantum correlations of the SGE as follows:

$$\mathcal{B}_z = \mathcal{C}(z', \theta') + \mathcal{C}(z', \theta) + \mathcal{C}(z, \theta') - \mathcal{C}(z, \theta). \tag{15}$$

Note that, given the form of Equation (14), we may consider the correlations between z and θ for Equation (15) by taking p_z as a constant in Equation (12) to obtain Equation (15).

Plots of Equation (15), setting $z' = 0.08$ and $\theta' = \pi/2$, are given in Figures 2 and 3. As in the last section, we set the constants by making $\mu_c b/2 = 2.2$, $t = 0.2$, $\sigma_0 = 0.005$ and considering $\pi\hbar = 1$ and $m = 1$. These plots clearly show the violation of Bell's inequalities given by Equation (15). This means that the quantum correlation arising from Equation (12) and shown in Figure 1 cannot be explained by local influences or local causes.

It is important to mention that the violation of the CHSH inequality, shown in Figure 3, does not reach the maximum value $2\sqrt{2}$. This is due to two factors: first, we found it by conjecturing values for the parameters that place us near the violation; then, we varied the values to find the violation. Notice that there could probably exist values where the violation might be higher. Second, as the states of the external degree of freedom are Gaussian, it seems that we can apply the explanation given by Haug et al. [70] stating that the Gaussian form of the correlation function smooths the CHSH-correlation, and therefore, it reduces the maximal possible value of the correlation.

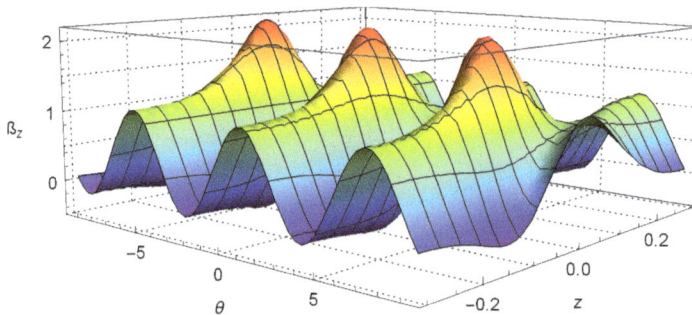

Figure 2. A plot of the function z given by Equation (15) considering $z' = 0.08$ and $\theta' = \pi/2$. Once more, we set $\pi\hbar = 1$, $m = 1$, $\sigma_0 = 0.005$, $\mu_c b/2 = 2.2$, $p_z = 0.01$ and time $t = 0.2$.

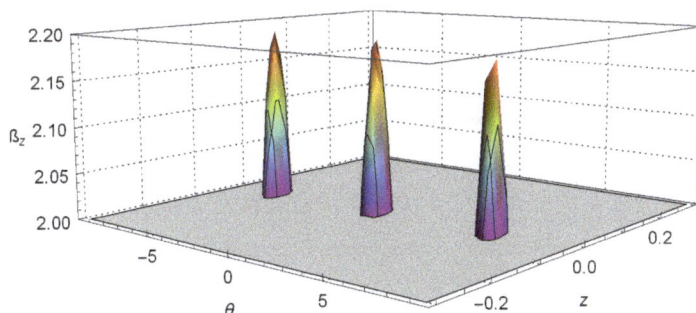

Figure 3. A close up of the region of Figure 2 where the violation of Bell's inequality is perceived.

5. Conclusions

Quantum mechanics is a fascinating field; nonetheless, the core ideas, like quantum nonlocality and "disturbance", are difficult concepts to grasp. In fact, the concept of disturbance, caused by the measurement process, which is responsible for one of the interpretations of the Heisenberg uncertainty principle, is not fully understood yet, and it is under investigation, as well [88]. In this case, disturbance refers to the change and perturbation produced by the measurement process; see [88] and the references therein. Quantum nonlocality is captivating, as well, and the fact of analyzing nonlocality by using the highly significant SGE could lead to understanding this concept better.

With this in mind, in this article, we have studied the correlations arising from the evolution of a pure state in the SGE, using the results given in [5,6] for the quantum mechanical evolution in the SGE and the approach given in [68] for testing nonlocality. In this way, we revealed that the SGE presents a nonlocal behavior, something that has never been thoroughly studied in the literature before. Once the correlations and nonlocality in the SGE have been characterized, we would like to conclude this paper by proposing the SGE as a scaffolding to introduce these concepts to students of quantum mechanics, as its experimental features can make the exposition of these concepts especially intuitive; additionally, this approach could open new investigations in consecutive SGEs or in quantum roulettes.

Acknowledgments: This work was funded by Vicerrectoría de Investigación y Estudios de Posgrado (VIEP), Benemérita Universidad Autónoma de Puebla, through the project titled "Enredamiento Cuántico en aparatos de Stern–Gerlach sucesivos". Ernesto Benítez Rodríguez and Alma Elena Piceno Martínez thank CONACYT for PhD degree fellowship support. Julio Abraham Mendoza Fierro thanks CONACYT for Master's degree fellowship support. Marcela Maribel Mendéz Otero and Luis Manuel Arévalo Aguilar thank Sistema Nacional de Investigadores and PRODEP. We thank Alba Julita Chiyopa Robledo for English editing and proofreading services. Finally, we thank Ricardo Villegas Tovar for helping us with technical bibliographical assistance.

Author Contributions: Alma Elena Piceno Martínez, Ernesto Benítez Rodríguez and Julio Abraham Mendoza Fierro made the plots, did the calculations and discussed the results. Luis Manuel Arévalo Aguilar made the plots, worked on the calculations and discussed and analyzed all the results and data. Marcela Maribel Mendéz Otero discussed the results and provided assistance. Alma Elena Piceno Martínez and Luis Manuel Arévalo Aguilar wrote the first draft. Luis Manuel Arévalo Aguilar supervised the whole project and conceived of it.

Conflicts of Interest: The authors declare no conflict of interest.

References

1. Friedrich, B.; Herschbach, D. Space Quantization: Otto Stern's Lucky Star. *Daedalus* **1998**, *127*, 165–191. [CrossRef]
2. Friedrich, B.; Herschbach, D. Stern and Gerlach: How a bad cigar helped reorient atomic physics. *Phys. Today* **2003**, *56*, 53–59. [CrossRef]
3. Schmidt-Böcking, H.; Schmidt, L.; Lüdde, H.J.; Trageser, W.; Templeton, A.; Sauer, T. The Stern–Gerlach experiment revisited. *Eur. Phys. J. H* **2016**, *41*, 327–364. [CrossRef]

4. Weinert, F. Wrong theory-Right experiment: The significance of the Stern–Gerlach experiments. *Stud. Hist. Phil. Mod. Phys.* **1995**, *26*, 75–86. [CrossRef]

5. Rodríguez, E.B.; Aguilar, L.A.; Martínez, E.P. A full quantum analysis of the Stern–Gerlach experiment using the evolution operator method: Analyzing current issues in teaching quantum mechanics. *Eur. J. Phys.* **2017**, *38*, 025403.

6. Rodríguez, E.B.; Aguilar, L.A.; Martínez, E.P. Corrigendum: A full quantum analysis of the Stern–Gerlach experiment using the evolution operator method: Analysing current issues in teaching quantum mechanics. *Eur. J. Phys.* **2017**, *38*, 069501. [CrossRef]

7. Home, D.; Pan, A.K.; Ali, M.M.; Majumdar, A.S. Aspects of nonideal Stern–Gerlach experiment and testable ramifications. *J. Phys. A: Math. Theor.* **2007**, *40*, 13975. [CrossRef]

8. Roston, G.B.; Casas, M.; Plastino, A.; Plastino, A.R. Quantum entanglement, spin-1/2 and the Stern–Gerlach experiment. *Eur. J. Phys.* **2005**, *26*, 657–672. [CrossRef]

9. Scully, M.O.; Lamb, W.E.; Barut, A. On the theory of the Stern–Gerlach apparatus. *Found. Phys.* **1987**, *17*, 575–583. [CrossRef]

10. Platt, D.E. A modern analysis of the Stern–Gerlach experiment. *Am. J. Phys.* **1992**, *60*, 306–308. [CrossRef]

11. Hsu, B.C.; Berrondo, M.; Van Huele, J.F.S. Stern–Gerlach dynamics with quantum propagators. *Phys. Rev. A* **2011**, *83*, 012109. [CrossRef]

12. Sparaciari, C.; Paris, M.G. Canonical Naimark extension for generalized measurements involving sets of Pauli quantum observables chosen at random. *Phys. Rev. A* **2013**, *87*, 012106. [CrossRef]

13. Potel, G.; Barranco, F.; Cruz-Barrios, S.; Gómez-Camacho, J. Quantum mechanical description of Stern–Gerlach experiments. *Phys. Rev. A* **2005**, *71*, 052106. [CrossRef]

14. Sparaciari, C.; Paris, M.G. Probing qubit by qubit: Properties of the POVM and the information/disturbance tradeoff. *Int. J. Quantum Inf.* **2014**, *12*, 1461012.

15. Fratini, F.; Safari, L. Quantum mechanical evolution operator in the presence of a scalar linear potential: Discussion on the evolved state, phase shift generator and tunneling. *Phys. Scr.* **2014**, *89*, 085004. [CrossRef]

16. Wennerström, H.; Westlund, P.O. A Quantum Description of the Stern–Gerlach Experiment. *Entropy* **2017**, *19*, 186. [CrossRef]

17. Rossi, M.A.; Benedetti, C.; Paris, M.G. Engineering decoherence for two-qubit systems interacting with a classical environment. *Int. J. Quantum Inf.* **2014**, *12*, 1560003. [CrossRef]

18. Boustimi, M.; Bocvarski, V.; de Lesegno, B.V.; Brodsky, K.; Perales, F.; Baudon, J.; Robert, J. Atomic interference patterns in the transverse plane. *Phys. Rev. A* **2000**, *61*, 033602. [CrossRef]

19. Larson, J.; Garraway, B.M.; Stenholm, S. Transient effects on electron spin observation. *Phys. Rev. A* **2004**, *69*, 032103. [CrossRef]

20. Machluf, S.; Japha, Y.; Folman, R. Coherent Stern–Gerlach momentum splitting on an atom chip. *Nat. Commun.* **2013**, *4*, 2424. [CrossRef]

21. Quijas, P.G.; Aguilar, L.A. Factorizing the time evolution operator. *Phys. Scr.* **2007**, *75*, 185–194. [CrossRef]

22. Quijas, P.G.; Aguilar, L.A. Overcoming misconceptions in quantum mechanics with the time evolution operator. *Eur. J. Phys.* **2007**, *28*, 147–159. [CrossRef]

23. Aguilar, L.A.; Quijas, P.G. Reply to Comment on "Overcoming misconceptions in quantum mechanics with the time evolution operator". *Eu. J. Phys.* **2013**, *34*, L77. [CrossRef]

24. Aguilar, L.A.; Luna, F.V.; Robledo-Sánchez, C.; Arroyo-Carrasco, M.L. The infinite square well potential and the evolution operator method for the purpose of overcoming misconceptions in quantum mechanics. *Eur. J. Phys.* **2014**, *35*, 025001, doi:10.1088/0143-0807/35/2/025001.

25. Quijas, P.C.; Aguilar, L.M. A quantum coupler and the harmonic oscillator interacting with a reservoir: Defining the relative phase gate. *Quantum Inf. Comput.* **2010**, *10*, 190–200. [CrossRef]

26. Toyama, F.M.; Nogami, Y. Comment on 'Overcoming misconceptions in quantum mechanics with the time evolution operator'. *Eur. J. Phys.* **2013**, *34*, L73. [CrossRef]

27. Amaku, M.; Coutinho, F.A.; Masafumi Toyama, F. On the definition of the time evolution operator for time-independent Hamiltonians in non-relativistic quantum mechanics. *Am. J. Phys.* **2017**, *85*, 692–697. [CrossRef]

28. Singh, C.; Belloni, M.; Christian, W. Improving students' understanding of quantum mechanics. *Phys. Today* **2006**, *59*, 43–49. [CrossRef]

29. Chhabra, M.; Das, R. Quantum mechanical wavefunction: Visualization at undergraduate level. *Eur. J. Phys.* **2017**, *38*, 015404. [CrossRef]

30. Cataloglu, E.; Robinett, R.W. Testing the development of student conceptual and visualization understanding in quantum mechanics through the undergraduate career. *Am. J. Phys.* **2002**, *70*, 238–251. [CrossRef]

31. Emigh, P.J.; Passante, G.; Shaffer, P.S. Student understanding of time dependence in quantum mechanics. *Phys. Rev. ST Phys. Educ. Res.* **2015**, *11*, 020112. [CrossRef]

32. Dini, V.; Hammer, D. Case study of a successful learner's epistemological framings of quantum mechanics. *Phys. Rev. Phys. Educ. Res.* **2017**, *13*, 010124. [CrossRef]

33. Zhu, G.; Singh, C. Improving students understanding of quantum mechanics via the Stern–Gerlach experiment. *Am. J. Phys.* **2011**, *79*, 499–507. [CrossRef]

34. Carr, L.D.; McKagan, S.B. Graduate quantum mechanics reform. *Am. J. Phys.* **2009**, *77*, 308–319. [CrossRef]

35. Passante, G.; Emigh, P.J.; Shaffer, P.S. Examining student ideas about energy measurements on quantum states across undergraduate and graduate levels. *Phys. Rev. Spec. Top. Phys. Educ. Res.* **2015**, *11*, 020111. [CrossRef]

36. Passante, G.; Emigh, P.J.; Shaffer, P.S. Student ability to distinguish between superposition states and mixed states in quantum mechanics. *Phys. Rev. Spec. Top. Phys. Educ. Res.* **2015**, *11*, 020135.

37. Greca, I.M.; Freire, O. Meeting the Challenge: Quantum Physics in Introductory Physics Courses. In *International Handbook of Research in History, Philosophy and Science Teaching*; Springer: Dordrecht, The Netherlands, 2014; pp. 183–209. [CrossRef]

38. Kohnle, A.; Bozhinova, I.; Browne, D.; Everitt, M.; Fomins, A.; Kok, P.; Kulaitis, G.; Prokopas, M.; Raine, D.; Swinbank, E. A new introductory quantum mechanics curriculum. *Eur. J. Phys.* **2014**, *35*, 015001. [CrossRef]

39. Singh, C. Students understanding of quantum mechanics at the beginning of graduate instruction. *Am. J. Phys.* **2008**, *76*, 277–287. [CrossRef]

40. Singh, C.; Marshman, E. Review of student difficulties in upper-level quantum mechanics. *Phys. Rev. Spec. Top. Phys. Educ. Res.* **2015**, *11*, 020117. [CrossRef]

41. Johansson, A.; Andersson, S.; Salminen-Karlsson, M.; Elmgren, M. "Shut up and calculate": The available discursive positions in quantum physics courses. *Cult. Stud. Sci. Educ.* **2016**, *13*, 205–226. [CrossRef]

42. Greca, I.M.; Freire, O. Teaching introductory quantum physics and chemistry: Caveats from the history of science and science teaching to the training of modern chemists. *Chem. Educ. Res. Pract.* **2014**, *15*, 286–296. [CrossRef]

43. Coto, B.; Arencibia, A.; Suárez, I. Monte Carlo method to explain the probabilistic interpretation of atomic quantum mechanics. *Comput. Appl. Eng. Educ.* **2016**, *24*, 765–774. [CrossRef]

44. Marshman, E.; Singh, C. Investigating and improving student understanding of the expectation values of observables in quantum mechanics. *Eur. J. Phys.* **2017**, *38*, 045701. [CrossRef]

45. Siddiqui, S.; Singh, C. How diverse are physics instructors' attitudes and approaches to teaching undergraduate level quantum mechanics? *Eur. J. Phys.* **2017**, *38*, 035703. [CrossRef]

46. Marshman, E.; Singh, C. Investigating and improving student understanding of quantum mechanical observables and their corresponding operators in Dirac notation. *Eur. J. Phys.* **2018**, *39*, 015707. [CrossRef]

47. Kohnle, A.; Baily, C.; Campbell, A.; Korolkova, N.; Paetkau, M.J. Enhancing student learning of two-level quantum systems with interactive simulations. *Am. J. Phys.* **2015**, *83*, 560–566. [CrossRef]

48. Baily, C.; Finkelstein, N.D. Teaching quantum interpretations: Revisiting the goals and practices of introductory quantum physics courses. *Phys. Rev. Spec. Top. Phys. Educ. Res.* **2015**, *11*, 020124. [CrossRef]

49. McKagan, S.B.; Perkins, K.K.; Wieman, C.E. Design and validation of the Quantum Mechanics Conceptual Survey. *Phys. Rev. Spec. Top. Phys. Educ. Res.* **2010**, *6*, 020121. [CrossRef]

50. Sadaghiani, H.R.; Pollock, S.J. Quantum mechanics concept assessment: Development and validation study. *Phys. Rev. Spec. Top. Phys. Educ. Res.* **2015**, *11*, 010110. [CrossRef]

51. Wuttiprom, S.; Sharma, M.D.; Johnston, I.D.; Chitaree, R.; Soankwan, C. Development and Use of a Conceptual Survey in Introductory Quantum Physics. *Int. J. Sci. Educ.* **2009**, *31*, 631–654. [CrossRef]

52. Bao, L.; Redish, E.F. Understanding probabilistic interpretations of physical systems: A prerequisite to learning quantum physics. *Am. J. Phys.* **2002**, *70*, 210–217.

53. Archer, R.; Bates, S. Asking the right questions: Developing diagnostic tests in undergraduate physics. *New Dir. Teach. Phys. Sci.* **2009**, *5*, 22–25. [CrossRef]

54. Clauser, J.F.; Shimony, A. Bell's theorem. Experimental tests and implications. *Rep. Prog. Phys.* **1978**, *41*, 1881–1927.

55. Gisin, N. *Quantum Chance: Nonlocality, Teleportation and Other Quantum Marvels*; Springer International Publishing: Cham, Switzerland, 2014.[CrossRef]

56. Augusiak, R.; Demianowicz, M.; Acín, A. Local hidden variable models for entangled quantum states. *J. Phys. A Math. Theor.* **2014**, *47*, 424002. [CrossRef]

57. Gisin, N. Bell's inequality holds for all non-product states. *Phys. Lett. A* **1991**, *154*, 201–202. [CrossRef]

58. Popescu, S.; Rohrlich, D. Generic quantum nonlocality. *Phys. Lett. A* **1992**, *166*, 293–297. [CrossRef]

59. Popescu, S. Bell's inequalities versus teleportation: What is nonlocality? *Phys. Rev. Lett.* **1994**, *72*, 797–799. [CrossRef]

60. Brunner, N.; Gisin, N.; Scarani, V. Entanglement and non-locality are different resources. *New J. Phys.* **2005**, *7*, 88. [CrossRef]

61. Bennett, C.H.; DiVincenzo, D.P.; Fuchs, C.A.; Mor, T.; Rains, E.; Shor, P.W.; Smolin, J.A.; Wootters, W.K. Quantum nonlocality without entanglement. *Phys Rev A* **1999**, *59*, 1070–1091, doi:10.1103/PhysRevA.59.1070.

62. Jammer, M. *The Philosophy of Quantum Mechanics*; John Wiley & Sons: New York, NY, USA, 1974.

63. Fine, A. *The Shaky Game*; The University of Chicago Press: London, UK, 1986. [CrossRef]

64. Norsen, T. Einstein's boxes. *Am. J. Phys.* **2005**, *73*, 164–176. [CrossRef]

65. Einstein, A.; Podolsky, B.; Rosen, N. Can Quantum-Mechanical Description of Physical Reality Be Considered Complete? *Phys. Rev.* **1935**, *47*, 777–780. [CrossRef]

66. Gallego, R.; Würflinger, L.E.; Acín, A.; Navascués, M. Operational Framework for Nonlocality. *Phys. Rev. Lett.* **2012**, *109*, 070401. [CrossRef]

67. Forster, M.; Winkler, S.; Wolf, S. Distilling Nonlocality. *Phys. Rev. Lett.* **2009**, *102*, 120401. [CrossRef]

68. Wódkiewicz, K. Nonlocality of the Schrödinger cat. *New J. Phys.* **2000**, *2*, 21. [CrossRef]

69. Banaszek, K.; Wódkiewicz, K. Testing Quantum Nonlocality in Phase Space. *Phys. Rev. Lett.* **1999**, *82*, 2009–2013, [CrossRef]

70. Haug, F.; Freyberger, M.; Wódkiewicz, K. Nonlocality of a free atomic wave packet. *Phys. Lett. A* **2004**, *321*, 6–13. [CrossRef]

71. Agarwal, G.; Home, D.; Schleich, W. Einstein-Podolsky-Rosen correlation—Parallelism between the Wigner function and the local hidden variable approaches. *Phys. Lett. A* **1992**, *170*, 359–362. [CrossRef]

72. Ben-Benjamin, J.S.; Kim, M.B.; Schleich, W.P.; Case, W.B.; Cohen, L. Working in phase-space with Wigner and Weyl. *Fortschr. Phys.* **2017**, *65*, 1600092. [CrossRef]

73. Case, W.B. Wigner functions and Weyl transforms for pedestrians. *Am. J. Phys.* **2008**, *76*, 937–946. [CrossRef]

74. Royer, A. Wigner function as the expectation value of a parity operator. *Phys. Rev. A* **1977**. *15*, 449–450. [CrossRef]

75. Hillery, M.O.S.M.; O'Connell, R.F.; Scully, M.O.; Wigner, E.P. Distribution functions in physics: Fundamentals. *Phys. Rep.* **1984**, *106*, 121–167. [CrossRef]

76. Zurek, W.H. Decoherence and the Transition from Quantum to Classical. *Phys. Today* **1991**, *44*, 36. [CrossRef]

77. Gerry, C.C.; Knight, P.L. Quantum superpositions and Schrödinger cat states in quantum optics. *Am. J. Phys.* **1997**, *65*, 964–974.

78. Ballentine, L.E. *Quantum Mechanics: A Modern Development*; World Scientific Publishing: Singapore, 1998. [CrossRef]

79. Jeong, H.; Son, W.; Kim, M.S.; Ahn, D.; Brukner, Č. Quantum nonlocality test for continuous-variable states with dichotomic observable. *Phys. Rev. A* **2003**, *67*, 012106. [CrossRef]

80. Chen, Z.B.; Pan, J.W.; Hou, G.; Zhang, Y.D. Maximal Violation of Bell's Inequalities for Continuous Variable Systems. *Phys. Rev. Lett.* **2002**, *88*, 040406. [CrossRef]

81. Zukowski, M. Bell's Theorem Tells Us Not What Quantum Mechanics Is, but What Quantum Mechanics Is Not. In *Quantum [Un]Speakables II*; Bertlmann R., Zeilinger A., Eds; Springer: Cham, Switzerland, 2017; pp. 175–185. [CrossRef]

82. Ferraro, A.; Paris, M.G.A. Nonlocality of two- and three-mode continuous variable systems. *J. Opt. B Quantum Semiclassical Opt.* **2005**, *7*, 174–182. [CrossRef]

83. Ferrie, C. Quasi-probability representations of quantum theory with applications to quantum information science. *Rep. Prog. Phys.* **2011**, *74*, 116001. [CrossRef]

84. Vourdas, A. Quantum systems with finite Hilbert space. *Rep. Prog. Phys.* **2004**, *67*, 267–320. [CrossRef]

85. Hinarejos, M.; Bañuls, M.C.; Pérez, A. Wigner formalism for a particle on an infinite lattice: dynamics and spin. *New J. Phys.* **2015**, *17*, 013037.
86. Gomis, P.; Pérez, A. Decoherence effects in the Stern–Gerlach experiment using matrix Wigner Functions. *Phys. Rev. A* **2016**, *94*, 012103. [CrossRef]
87. Clauser, J.F.; Horne, M.A. Shimony A and Holt R A Proposed Experiment to Test Local Hidden-Variable Theories. *Phys. Rev. Lett.* **1969**, *23*, 880–884. [CrossRef]
88. Rodríguez, E.B.; Aguilar, L.A. Disturbance-disturbance uncertainty relation: The statistical distinguishability of quantum states determines disturbance. *Sci. Rep.* **2018**, *8*, 4010, doi:10.1038/s41598-018-22336-3.

entropy

MDPI

Article

Quantum Dynamics and Non-Local Effects Behind Ion Transition States during Permeation in Membrane Channel Proteins

Johann Summhammer [1],*, Georg Sulyok [1] and Gustav Bernroider [2],*,†

[1] Institute of Atomic and Subatomic Physics, TU Wien, Stadionallee 2, 1020 Vienna, Austria; georg.sulyok@tuwien.ac.at

[2] Department of Biosciences, University of Salzburg, 5020 Salzburg, Austria

* Correspondence: johann.summhammer@tuwien.ac.at (J.S.); Gustav.Bernroider@sbg.ac.at (G.B.); Tel.: +43-1-58801-141427 (J.S.)

† Retired.

Received: 30 April 2018; Accepted: 23 July 2018; Published: 27 July 2018

Abstract: We present a comparison of a classical and a quantum mechanical calculation of the motion of K^+ ions in the highly conserved KcsA selectivity filter motive of voltage gated ion channels. We first show that the de Broglie wavelength of thermal ions is not much smaller than the periodic structure of Coulomb potentials in the nano-pore model of the selectivity filter. This implies that an ion may no longer be viewed to be at one exact position at a given time but can better be described by a quantum mechanical wave function. Based on first principle methods, we demonstrate solutions of a non-linear Schrödinger model that provide insight into the role of short-lived (~1 ps) coherent ion transition states and attribute an important role to subsequent decoherence and the associated quantum to classical transition for permeating ions. It is found that short coherences are not just beneficial but also necessary to explain the fast-directed permeation of ions through the potential barriers of the filter. Certain aspects of quantum dynamics and non-local effects appear to be indispensable to resolve the discrepancy between potential barrier height, as reported from classical thermodynamics, and experimentally observed transition rates of ions through channel proteins.

Keywords: ion channels; selectivity filter; quantum mechanics; non-linear Schrödinger model; biological quantum decoherence

1. Introduction

Selective translocation of ions bound to charges across the plasma membrane of cells provides the physical background for the generation and propagation of electrical membrane signals in excitable cells, particularly in nerve cells. The molecules organizing this translocation are provided by membrane-integrated channel proteins, which control the access of ions to permeation ("gating") in response to changes in transmembrane potentials ("voltage-gating") and allow very fast ion conduction without loss of selectivity towards certain ion species [1,2]. An unprecedented series of studies of these proteins has been initiated after the elucidation of the atomic resolution crystal structure of the prototypic *S. lividans* K^+ channel by MacKinnon et al. [3]. It turned out that the critical domain of the protein that can combine fast transduction close to the diffusion limit with selective preference for the intrinsic ion species is provided by the narrow selectivity filter (SF) of the protein [4]. In particular, an evolutionary highly conserved sequence of amino acids, the TVGYG (Thr75, Val76, Gly77, Tyr78, Gly79) motive lining the filter region, allows for an inward orientation of backbone carbonyls with oxygen-bound lone pair electrons interacting with the positively charged alkali ions (see Figure 1). This delicate arrangement involving glycine (Gly79, Gly77) residues serving as "surrogate D-amino acids" [5] can offer a unique "interaction topology", mimicking the ions' hydration shells prior to entering the filter pore. The interactions are realized by short-range attractive (filter atoms) and repulsive (between ion) Coulombic forces [2,6].

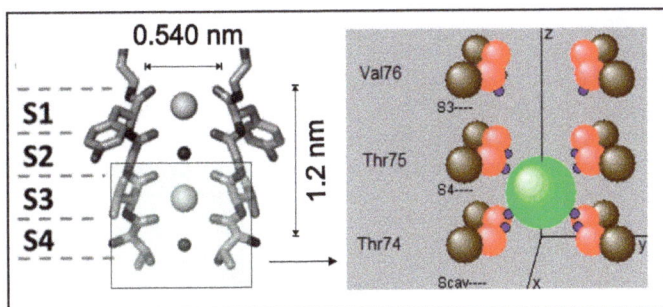

Figure 1. A section through the tetrameric KcsA filter motive, showing a sketch of two transmembrane helices for binding sites S4–S1, with two ions and two waters molecules (left). On the right, a window (insert) for atomic locations of the filter lining during the passage of a K^+ ion (green) from S4 to S3 is sketched. The carbon atoms (brown) of the carbonyl groups are situated at the corners of a square (including all four backbone strands). The charge (blue) of oxygen atoms (in red) is partly contained in the center of the atom and partly within a point location slightly outside the oxygen. As these charges are drawn towards the central axis, they represent the effective charge center of the lone pair electrons (shown in blue). The size of atoms and the K^+ ion on the right are drawn to scale approximately.

The initial picture of ion conduction states was built on an alternating sequence of ions and water molecules (e.g., KwKw) passing through the filter with four equally spaced ion binding sites (labeled as S0–S4, as seen from the extra to the intracellular side) [4]. However, in the course of molecular dynamics (MD) studies this view has become considerably relaxed (e.g., by the observation that different ion permeation mechanisms may coexist energetically [7], pairwise water-ion hopping mechanisms can occur [8], the selectivity filter by itself could play a role in gating [9], and fast permeation involves a direct Coulombic "knock-on" without intermittent water [10]). Even more important are observations from MD studies calculating potentials of mean force (PMF) within the filter, which demonstrate that the potential barriers for ion translocation are simply too high (>5 kT at 300 K) to be in line with experimental conductance as predicted by the Nernst–Planck equation [11]. This seems to be a reflection of an enduring problem within the structure-function relations of purely classical MD simulations at the atomic scale and marks the point where the intention of our present contribution becomes significant.

Generally, the short range Coulombic forces coordinating the atoms in the filter reflect quantum-mechanical effects [2]. As argued before, this requires some quantum dynamics to account for the observed atomic behavior within their molecular environment [12–14]. In our previous work, we have suggested a role for a quantum physical description of ion motion through the selectivity domain of K^+ channel proteins. In these studies, we suggested evidence that at least two important features behind ion permeation and gating dynamics can follow naturally if quantum properties are inserted into the underlying equations of motion. First, inserting quantum interference terms into the canonical version of action potential (AP) initiation can reproduce the fast onset characteristic of APs as seen in experimental recordings of cortical neurons [13]. Second, we have demonstrated evidence for different quantum oscillatory effects within the filter's atomic environment, which discriminate intrinsic (e.g., K^+) from extrinsic (e.g., Na^+) filter occupations in K^+-type channels [14].

In the present paper, we go beyond classical MD simulations by treating the motion of the ion itself in a quantum mechanical (QM) context. Because the de Broglie wavelength of thermal ions at

310 K (~0.025 nm) amounts to up to 10% of the spacing in the periodic structure of Coulomb potentials (~0.3 nm) in the nano-pore model of the selectivity filter, the ions wave packet is found to spread out over a certain region. The associated wave dynamics have a coherent short-lived and significant effect on the Coulomb interaction with the surrounding carbonyl charges. Based on first principle methods and solving a non-linear version of the Schrödinger equation, we find that the quantum trajectory of an ion through the filter is accompanied by different time-dependent phase velocities that can exert a favorable effect on the passage of ions through the confining potential landscape of the filter. We suggest that this observation from a combined QM-MD calculation can possibly explain fast conductance without compromising selectivity in the filter of ion channels. We shall discuss the way this favorable effect is exerted and to what extent it lowers the effective potential barriers for ion translocations. Phrased loosely, it is found that the front part of the particle wave function paves the path for the remaining wave components to "sneak through" the open doors of the confining potentials. Due to the involved barriers and masses, this process is different from "particle tunneling" (although it is naturally considered in the solution of the Schrödinger equation calculated with the Crank–Nicolson formalism [15]). Yet, another finding is remarkable within the context of the enduring debate about quantum coherence times in biological organizations: The quantum characteristic for the present effect not just builds on but also requires very short decoherence times (around 1 ps), a scale that is well within the expected range at biological temperatures [16].

Finally, it should be mentioned why we deal with a "non-local effect", as expressed in the title of this paper: Most frequently, the term "non-locality" in QM states refers to a spatial separation between observables preserving a QM correlation (entanglement) between different modes (i.e., pertaining to different sub-systems behaving as one system). Here, we deal with only one technical QM mode or system (i.e., the ion). However, the present finding, that a short and coherent "spread" or "smear" over space of the particle's mass-bounded charge, according to its QM wave function, can have a strong effect on the dynamic behavior within its environmental potentials implies a functional role for a "non-local property" of a single mode (or system).

2. Methods

Our intention was to observe the K^+ ion during the transition from site S4 to site S3 in the selectivity filter of the KcsA channel (Figure 1). Therefore, the simulation included the carbonyl groups of Thr74, Thr75, and Val76, as shown in Figure 1 (right). The backbone carbons were positioned at the widely used coordinates of Guidoni and Garofoli [17,18]. The carbonyl oxygen–carbon bond was set initially to 0.123 nm bond length and, in the force-free, unperturbed situation, pointed straight to the central axis (the z-axis of the coordinate system). Oxygen atoms were allowed to oscillate within horizontal and vertical bending modes, excluding stretching. The effective spring constant and the damping factor of this oscillation were adjusted to the values of typical thermal frequencies in the range of a few THz and to the expected dissipation of vibrational energy into the protein backbone structure after a few oscillation periods. Although considered in the implementation of the program, the short time interval in the present study did not necessitate setting thermal random kicks from backbone atoms to carbonyl atoms (Table A1). The degrees of freedom for the motion of a single K^+ ion were constrained to the central z-axis of the selectivity filter. This allowed us to implement all calculations on a quad-core computer within a reasonable processing time and does not influence or restrict the conclusions to be drawn from the present results.

The Coulomb type interaction potential between two particles located at r_1, r_2 and charges q_1, q_2, including a repulsion term with a characteristic distance r_{cut} (the distance where the electron shells start to overlap), and ε_0 the vacuum dielectric constant, is as follows:

$$V(r_1, r_2) = \frac{q_1 q_2}{4\pi\varepsilon_0}\left(\frac{1}{|r_1 - r_2|} + \text{sgn}(q_1 q_2)\frac{r_{cut}}{(r_1 - r_2)^2}\right) \qquad (1)$$

where "point-charges" are located at the center of the particle. K$^+$ ions carry unit charge and carbonyl-bound C atoms carry partial charges, usually set to +0.38 units. We assigned two-point charges to oxygen atoms; one at the center of the atom, and the second one representing the effective charge center of the lone pair electrons coordinating the K$^+$ ions and/or water dipoles. The partial charge relocation between the lone pairs and the central O positions was chosen as one of the dynamic variables that determines the depth of the ion-trapping potential (Table A1).

The classical part of the present MD simulation is based on Verlet's algorithm applied to Lennard-Jones molecules [19,20]. The QM model applies to the motional behavior of the K$^+$ ion particle waves and was obtained from a non-linear Schrödinger equation (NLSE) (see Equation (2)), with an initial Gaussian wave packet set to an adjustable width and an adjustable mean ion velocity along the z-axis of the filter. The range of these settings is given in Appendix A in Table A1. It is assumed that the wave packet experiences a potential at every instant of time *t*, which depends on the position of all other particles at this time. Together with the potential term in Equation (1), this can be described by the following NLSE:

$$\left[-\frac{\hbar^2}{2m_K}\frac{d^2}{dz_K^2} + \sum_{i=1}^{36} V(r_i(\psi(r_K,t)),r_K) + gz_K \right] \psi(r_K,t) = i\hbar\frac{\partial}{\partial t}\psi(r_K,t) \tag{2}$$

where m_K denotes the mass of the K$^+$ ion and r_K its position vector along the z-axis of the filter. Due to the tetrameric lining of the observed motive (Figure 1), summation over the potential term runs over 12 backbone atomic positions for the carbon, oxygen, and lone pair centers within the Thr74, Thr75, and Val76 lining amino acids shown in Figure 1 (right). As the atomic positions r_i change in time and are influenced by the position of other atoms, as well as by the probability distribution of the K$^+$ ion, the situation entails non-linearity in the Schrödinger equation (see also Appendix B). In the above Equation (2), this functional dependence of r_i on ψ is explicitly indicated. The linear gradient "*g*" expresses the transmembrane electric potential and z_k the z-component of the vector r_k. The parameters that determine the shape and scaling of interaction potentials (i.e., the geometrical embedding) were adjusted to previous models of the KcsA channel [8,11]. This implied initial values for r_{cut} of 0.13 nm, with the charge separation distance of the lone pair electrons from an oxygen center being 1.4 times the radius of the oxygen atom, leading to an average of 0.0825 nm from the oxygen atom's center. The partial charges of an oxygen atom were split to contain a fraction of 30% at central locations and 70% in the lone pair charge point location (Table A1).

It should be noted that the NLSE given by Equation (2) restricts the motion of the K$^+$ ion to the z-axis and does not include an expansion of the wave packet perpendicular to this axis. This restriction was necessary to keep the computational time within reasonable limits. In addition, the narrow extension and the symmetry of the filter lining in the pore cause sideways forces to mostly cancel each other along the filter's z-axis. For the further formal description, this implies that the position vector r_K of the K$^+$ ion is essentially given by its z-component z_K, because its x- and y-components are always zero.

Prior to solving the SE given by Equation (2), we have to take into account the experience of forces by the surrounding carbonyl C and O atoms due to the interacting K$^+$ wave. At this stage, we assume the backbone C atoms to be rigid (see Section 4) but allow for two bending modes of O atoms while keeping the CO distance constant. The differential *dz* along the z-axis of the wave packet is then found to exert a differential force $d\vec{f}_i$ as follows:

$$d\vec{f}_i = -\vec{\nabla}V_i|\psi(r_K,t)|^2 dz$$
$$= \frac{q_i q_K}{4\pi\varepsilon_0}\left(\frac{1}{(r_i-r_K)^3} + \text{sgn}(q_i q_K)\frac{2r_{cut}}{(r_i-r_K)^4} \right)(r_i - r_K)|\psi(r_K,t)|^2 dz. \tag{3}$$

The total force $\vec{F}_i$ acting from K$^+$ on this O atom is then obtained by integrating over the range defined along z (additional forces acting on this O atom are a restoring force and a decelerating force, as well as attraction/repulsion of the surrounding C and O atoms, see Appendix C.) This will subsequently change the locations r_i of the O atoms and thereby the potential term in the SE acting back on the evolution of the wave packet. The effect of Equation (3) introduces a non-linearity into the SE as shown in Equation (2). The resulting non-linear Schrödinger equation (NLSE) is formally similar but causally different from the description of Bose–Einstein condensation (BEC) at ultra-cold temperatures [21]. This is because under BEC conditions, the probability distribution of the wave function enters into the Hamiltonian itself, while in our case it is the effect integrated over time, as shown in Appendix B. We solved Equation (2) together with Equation (3) in very small-time steps by the Crank-Nicolson method [15] to keep track of the QM phase factor but sampled the positional changes of the O atoms in larger time steps (values given in Appendix A, additional explanations on the NLSE derivation in Appendix B).

We simulated the behavior of classical ensembles to compare with the above-described quantum behavior in Figures 2–5. In these simulations, classical particles were set into motion at 10^2 different starting positions. The positions were generated equidistantly within three times the full 1/e-width of the initial quantum wave packet and weighted with the probability density of this packet. At each initial location 10^2, particles were set into motion with velocities, again sampled equidistantly from within three times the full 1/e-width of the initial Gaussian momentum distribution of the QM wave packet.

The numerical implementation of the present methods was designed to offer an interactive control window that allowed us to change the settings given in the Appendix A for path integration according to Equations (2) and (3), and it delivered the following graphs for time-dependent wave functions and probability plots (Figures 2 and 3).

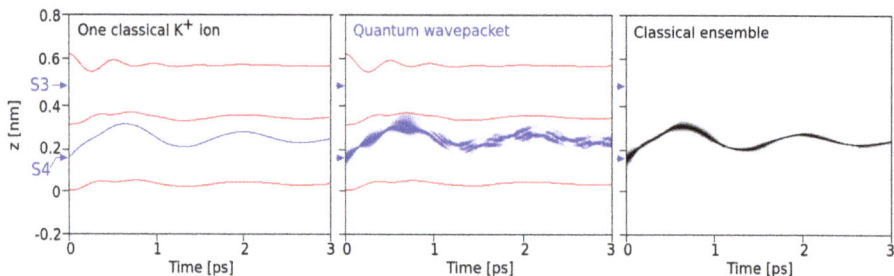

Figure 2. Single ions and the classical ensemble: Comparison of the evolution (within 3 ps time) of a single classical K$^+$ ion (**left**, blue curve) with initial velocity of 300 m/s at the minimum of site S4 with a quantum mechanical wave packet of minimum uncertainty of this ion (**middle**). The red lines are the z-coordinates of the carbonyl oxygen atoms. Middle: Probability density from a quantum mechanical (QM) calculation along the z-axis of the wave packet as a function of time (intensity of blue reflects higher probability densities). The initial full width of this density is 0.05 nm (at 1/e). **Right**: Classical probability density of finding an ion from the ensemble of 10^4 ions at the given z-coordinate as a function of time.

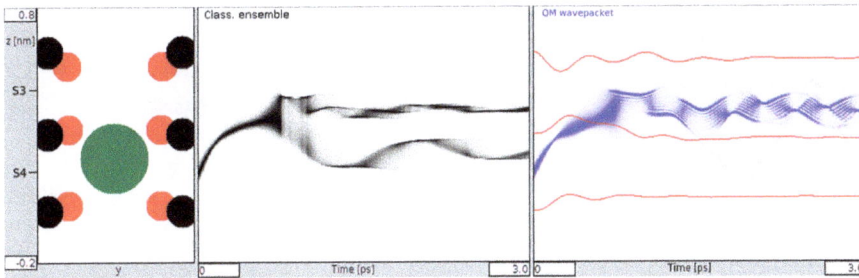

Figure 3. Transition behavior between S4 and S3 (**left** insert) for a classical ensemble (**middle**) and the simulated QM wave packet (**right**), with shades of black and blue coding normalized probability densities for location and time. Red lines (**right**) are again the z-coordinates of carbonyl oxygens. Note: whereas the classical ensemble splits after around 0.8 ps (**middle**), the QM distribution goes beyond the barrier to S3 almost completely (**right**).

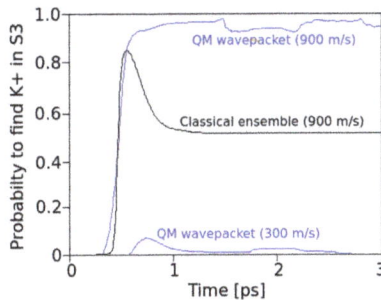

Figure 4. Time-dependent probabilities to find an ion in S3, when the ion was implanted into S4 with different mean onset velocities (900 m/s blue top for the QM wave packet, black for the classical ensemble) and at 300 m/s for the QM wave packet (with some probability <0.1 to cross over to S4). At this initial velocity of 300 m/s, the classical particles do not cross to S3. Note: most classical particles with 900 m/s are in S3 after 0.5 ps but eventually about 45% return to S4 due to oxygen charge derived forces (the spring that returns these ions to equilibrium positions with vibrations around 3 THz, see Figure 2). The QM wave exhibits a small but remaining probability (<10%) of returning to S4.

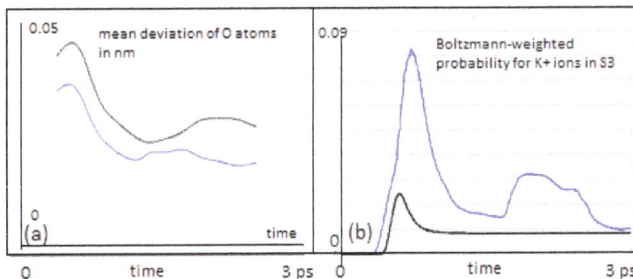

Figure 5. (a) Mean deviation of Tyr75 carbonyl oxygens from their equilibrium positions, while a K^+ is moving past from location S4 to S3 (in nm). The classical particles are in black, QM wave packets in blue; (b) Probability for a K^+ ion to be found in S3, setting out from S4 with mean velocities between 100 m/s and 900 m/s, weighted according a Boltzmann velocity distribution at 310 K (blue QM, black classical).

3. Results

3.1. Classical versus QM Motion

First, our intention focused on the comparison between the classical standard MD setting and the quantum mechanical version under the same interaction potentials between all constituents and the same initial position and velocity distribution of the ion (Figure 2). This yielded comparable results under situations where the K^+ ion is coordinated at a specific site (e.g., S4) and oscillates within this site due to its thermal energy. In the QM version, the wave packet is placed at the minimum of the potential of this site (at $z = 0.15$ nm) and assigned a mean velocity corresponding to a kinetic energy sufficiently below the potential barrier to the next site. For the examples shown, we have chosen $v_0 = 300$ m/s. As can be seen from Figure 2, under identical initial conditions at time $t = 0$, the temporal behavior of a single classical ion (left) is similar to the behavior of the QM wave packet, with the same frequency (around 900 GHz) and amplitudes. This similarity becomes even more striking in a comparison of the single wave packet with a classical ensemble of ions computed for 10^4 particles under an identical initial position and velocity distribution as in the QM version (Figure 2, right).

However, even at the initially relatively low kinetic energy levels of a K^+ ion, the QM wave shows a non-vanishing probability that the ion could make a transition to site S3 in the filter (as seen around <1 ps after onset and in more detail in Figure 4, bottom). This observation is particularly interesting as it occurs within just one picosecond (i.e., well within the expected decoherence time due to thermal noise from protein backbone atoms transmitted to carbonyl atoms).

3.2. Transition Behavior: Classical versus QM Evolution

At the onset of the transition between sites S4 and S3 (i.e., when the initial velocity of the K^+ ion approaches the values needed to cross the potential barrier between these two sites), the QM wave packet and the classical ensemble start to behave quite differently. Figure 3 captures an example for a velocity of $v_0 = 900$ m/s of the ion. At the boundary to S3, the classical ensemble is found to split into roughly $\frac{1}{2}$ (Figure 3, middle), whereas the QM behavior shows that the wave packet manages to cross the barrier with almost all of its location probability (Figure 3, right). Following these initial characteristics, there are also subsequent differences: The splitting in the classical picture remains unchanged during the observed time interval. Ions that have crossed to S3 remain in S3, and ions that did not cross remain in S4.

Despite crossing the potential to S3, the QM wave packet retains a probability of up to 10% to return to location S4 (Figure 4 below). This effect may look inconspicuous at first glance, but in fact, it marks a highly relevant difference between a classical and a quantum mechanical behavior of the moving ion (see Section 4). The passage of "classical ions" through the close coordination distances (in the range of 0.27 nm [17]) provided by the Thr75 oxygen lone pairs, directs these charges to a point towards the center of the ion. Along this view, the ion "drags" a potential valley along its path. In the quantum situation, however, the direction of these forces can split up into different components directed towards a delocalized charge-probability distribution along some distance in the z-direction. One effect to be expected from this interference is that, while the front part of the distribution with the fast-moving momentum components attracts these oxygens towards it, the slower moving "tails" can follow without having to overcome the high barriers.

A signature of this possibility can be seen from the mean deviation of the Tyr75 carbonyl oxygens from their equilibrium as the K^+ ion passes, shown in Figure 5 (left). The deviation depicted in this figure is permanently lower for the QM K^+ ion as compared with the classical ensemble of ions.

3.3. Transition Behavior over a Range of Ion Velocities

The above findings relate to specific initial velocities. From a more general view, one has to take into account that an ion at a specific position (e.g., S4) will oscillate due to its thermal environment and may therefore take on a wide range of velocities. We have therefore performed a series of simulations for both the classical ensemble and the QM version for initial mean velocities between 100 m/s and 900 m/s. For this series of mean velocities (at steps of 100 m/s), the probability to locate the K^+ ion in S3 was calculated (Figure 4) and the resulting curves were corrected by weights obtained from a Boltzmann distribution (which provides the probabilities for each of the initial mean velocities to occur at a temperature of 310 K). The sum of individual probabilities finally provides a Boltzmann-weighted distribution for the probability to find the QM ion at site S3 at a given instant of time (Figure 5b). We have chosen the same procedure for the weights of classical probabilities.

Figure 5b provides a summary of the results for the probabilities of finding a K^+ ion in S3, setting out from a starting location at S4. As mentioned, the probabilities were Boltzmann weighted for 310 K and summed over velocities from 100 to 900 m/s in steps of 100 m/s.

The results shown in Figure 5b clearly demonstrate that the QM wave has an increased probability of crossing from S4 to S3 as compared with the classical particle, throughout the time interval studied and the range of initial velocities. During an early stage of the transitions ($t < 1$ ps), the QM probability for S3 is more than three times the classical probability. After a very short initial time ($t < 0.5$ ps), when the classical ion has hardly reached the barrier, a QM ion has already acquired a significant probability for having crossed this barrier. The QM particle also shows periodic interferences during the 3 ps time interval that are not apparent for a classical behavior with a tendency to return to S4 after 2.5 ps. The classical particle adopts a constant and low probability to transit after about 1 ps. In both cases, in the classical and in the QM situation, the transfer probability to S3 settles to about 10% after 3 ps. This seemingly low value is consistent with a "field-driven" and diffuse transfer through the selectivity filter with alternations between site halts involving oscillations in thermal equilibration and subsequent "hopping" to the next site, a situation that is well predicted by previous MD studies (e.g., [9]). In the "conductive state" of the filter, the ion would naturally transit to the subsequent site S2 along the conduction path ("z" in the filter model). In the "non-conductive filter state", alternating site changes between the configurations 1,3 and 2,4 can occur, which includes a return path from S3 to S4 [3,4,9]. As in the present study, the focus was laid on a single site transition within the conduction cycle from S4 to S3; the coordinating carbonyl groups from subsequent amino acids beyond Val76 were not included in the simulation. Subsequent studies will gradually have to involve the entire conduction path to account for all interaction terms in the guidance of the ion. However, as the force from attractive potentials exerting from distant (e.g., S2) carbonyl cages drops with $1/\Delta r^2$ (Equation (1)) and these distances can be expected to have a lower bound around $\Delta r \geq 0.6$nm, the effect on S4–S3 transitions from more distant oxygens can be expected to be small (in addition, the expected intermittent water dipoles will exert a damping effect on these forces). We provide further comments on this situation in Section 4.

Our observations can shed light on the question of a decoherence function during ion transitions (see Section 4). We find that for a QM wave description of an ion within the potential landscape of the filter atoms, short coherence times are beneficial for the observed high conductivities. If coherence time is short (i.e., not much longer than around 1 ps), the wave packet of the ion can cross the barrier much easier (the peak in Figure 5b) due to its quantum nature as compared to a classical ion. The following loss of coherence, however, is equally important because having crossed the barrier, decoherence will eliminate QM interferences, and the particle starts to adopt classical behavior, which avoids the undesired return to S4. The question of coherence, of course, deserves additional attention, as given in the following section.

4. Discussion

We have investigated the motional behavior of a single K^+ ion between two transition sites in the nano-pore selectivity filter motive of the KcsA channel from two different perspectives: A quantum mechanical simulation implemented by a non-linear version of the Schrödinger equation and a corresponding classical ensemble behavior under identical initial and interaction terms. The non-linear Schrödinger model (NLSE) integrates the solution of the wave equation into its interaction potentials, with all surrounding charges modulating the probability distribution of the wave at a given instance of time. This offers a kind of recursive approach, taking account of mutual interactions of confining Coulomb forces and the QM wave equation, a situation that seems more realistic than the calculation of potentials of mean force (PMF) from classical MD at the atomic scale. The methods were implemented in Java. An executable version is available upon request, as well as a source citation agreement (this requires prior installation of the Java Runtime Environment).

First, our results provide a comparison between classical and QM implementations, which reveal a high degree of similarity in the overall, time-dependent behavior during a several pico-second time interval (Figure 2). The observed similarity is particularly obvious for the ensemble behavior of ions and certainly signals a high level of consistency of the implemented methods. Besides this similarity at the "caged" state, as shown in Figure 2, we also observe small but significant differences in the time evolution of QM and classical ions prior to transitions from S4 to S3 that bear the "seeds" for subsequent QM classical dissociations during the passage from one binding site to the next site. Following this initial situation, the site transition favored by higher onset velocities marks a clear difference between the behavior of the QM wave packet and the classical ensemble of ions. At a critical velocity of ions, the classical probability splits into $\frac{1}{2}$ of its population, with 50% crossings to S3 and the rest remaining in S4 (Figures 3 and 4). Within the same time, and under the same initial conditions, the QM packet, however, can cross to S3 almost completely. An increase in efficiency to cross the barriers of about 50% could be expected to increase permeation rates at larger scales by a similar amount (i.e., within a complete filter occupancy and including intermittent water molecules in the filter domain). Compared with previous reports about the discrepancy of calculated energetic barrier heights from PMF methods (between 5 and 10 kcal/mol) with observed and "effective permeation heights" as required by Nernst–Planck estimates of <~3 kcal/mol at 300 K [11], the ratio of this difference lays well within the range of permeation enhancement as predicted here. In other words, assigning a short (1 ps) coherent quantum mechanical property to the atoms motion interacting with the carbonyl derived forcefields can explain fast conduction speeds, whereas purely classical models cannot.

It must be granted that the present report focuses on a small and short time-scaled view on filter dynamics. This may somehow insufficiently sample the more complete conformational pattern underlying filter conduction states. However, the main findings of the present study strongly suggest that a small scaled and ultra-short-lived quantum state of the permeating ion is not just sufficient but also indispensable to explain fast conduction without compromising selectivity in the filter. The observations demonstrated in Figures 4 and 5 indicate that due to the dispersion of the wave function of a "quantum ion", the coordinating Coulomb forces from surrounding oxygen charges become dissociated to different parts of the wave function. This effect reduces the effective barrier that the ion has to cross. In a metaphorical view, it looks like the faster front part of the particle wave can open the door to the barrier, allowing the slower tail parts to sneak through this door. Due to the engaged height and width of the barrier, this effect is not typical "leap-frogging" (passing over), nor "quantum-tunneling", although there is some formal resemblance to the latter. The resemblance is that the potential energy of the separating barrier is also a function of the state variable of the system. But, as opposed to tunneling, the system actually "manipulates" the barrier to be able to cross it. It is perhaps best described as *"quantum sneaking"* through potential barriers.

The present study takes into account physiological temperatures with respect to ion motion and velocities. At this stage, the effects of these temperatures on thermally induced protein backbone and carbonyl vibrations are not included. Our numerical implementations do offer the extension to thermal carbonyl atomic motions during the temporal evolution of the K$^+$ wave packet. We presently implement these fluctuations by repetitions of K$^+$ evolutions during time-varying random fluctuations of the surrounding O and C atoms. We expect, however, that the main findings reported here about the difference between a classical ensemble and the K$^+$ wave packet will remain largely resistant to these thermal vibrations. The reason is that the main effect found here is due to the spatial dispersion of the QM wave packet, which in turn dynamically spreads the interacting force directions of the surrounding and coordinating charges. In the pure classical case, these forces would permanently be directed towards the center of the moving ion. It is just this distinction that allows for what we called "quantum sneaking" in the discussion above.

Finally, our observations render an important role for decoherence and the quantum to classical transition as predicted in an earlier paper by one of the authors [22]. Fast decoherence of the ions wave function after about 1 ps, a time when almost all of its probability distribution has penetrated the barrier, leads to classical behavior, which can be seen to avoid the return to the previous location. So, decoherence actually "guides" the particle into one direction in the filter, and an oscillation between quantum and classical states cooperate in a directed transport through the potential landscape of the filter.

A possible role of the important and unique QM interaction, the environment-induced, dynamical destruction of quantum coherence deserves some further remarks in the present context. It was not our intention to study decoherence in our simulation explicitly, and we did not include the scattering elements and processes that induce decoherence in the evolving wave packet. In some previous work, we and a co-author of this group (V. Salari) have provided a list of scattering sources and interacting scattering events applicable to the same atomic configuration and dynamics of the KcsA filter model as used in the present study [14,16,23]. The results of these studies suggest that we can expect decoherence times for K^+ ions in the filter model around one or a few pico-seconds at warm temperatures.

The intention of the present simulation was more focused on a potential functional role of decoherence during ion permeation by implementing a comparison between quantum and classical motions. As we can expect that decoherence of the QM wave packet will lead to a certain resemblance with a classical behavior in the course of time, comparisons as those shown in Figures 4 and 5b can give us some inferential information about the time and the role of decoherence. It turns out, that short decoherence times, exactly within the range of the predictions mentioned from the scattering studies above (i.e., around 1 ps), could play a highly beneficial role for the successful transition from S4 to S3. A transition to classical behavior due to decoherence after this time would actually "stabilize" the ion's location at S3 once it has reached this site. The original transition probability would still be at the level of the high QM initial transfer probability. We therefore conjecture that short coherent QM states, in the range of a few ps, are of an advantage for the observed high ion transfer rates without compromising ion coordination. Taken together, we suggest that the quantum dynamics behind the ion motion in the filter open the door through confining potentials, and decoherence guides the moving atoms through the specific path offered by the selectivity filter of channel proteins. To the best of our knowledge, this is one of the first reports about a decisive role of quantum decoherence for an ancient and highly conserved mechanism of membrane signaling in biology.

Author Contributions: J.S. developed the methods and implemented the program environment and numerical calculations. J.S. also analyzed and interpreted the data. G.S. performed numerical experiments and analyzed the quantum mechanical transition mechanism. G.B. raised and conceived the idea, participated in analyzing and interpreting the results, and wrote the manuscript.

Funding: The authors acknowledge the TU Wien University Library for financial support through its Open Access Funding Programme.

Acknowledgments: The authors gratefully acknowledge the very helpful comments and suggestions obtained from three reviewers.

Conflicts of Interest: The authors declare no conflict of interest.

Appendix A.

Table A1. Constants and parameter settings.

Charge of the K⁺ ion	$+1\,q_0$ (q_0 ... unit charge)
Charge of the carbon of a CO-group	$+0.38\,q_0$
Charge of the oxygen of a CO-group	$-0.38\,q_0$
Distance C–O of a CO-group	0.123 nm
z-coordinate of the CO carbons at Thr74	0 nm (by definition)
z-coordinate of the CO carbons at Thr75	0.30 nm
z-coordinate of the CO carbons at Val76	0.62 nm
Distance of the CO carbon atoms from axis of selectivity filter (z-axis)	0.38 nm
Stiffness of bending of the O atom around the C atom in a CO-group	$30°/k_BT$
Damping constant of rotational vibrations of O atoms	1×10^{-13} kg/m
Positions of oxygen atoms at $t = 0$	equilibrium positions [1]
Velocity of oxygen atoms at $t = 0$	0
Distance of lone pairs charge from the center of the O atom	$1.4\,r_O$ (=0.0825 nm)
Percentage of O partial charge in lone pairs	70%
r_{cut}	0.13 nm
Thermal random kicks from backbone to carbonyls	None
Linear potential drop along the axis of the selectivity filter	-100 mV/nm
Initial position of K⁺ ion	0.15 nm [2]
Initial mean velocity of K⁺ ion wavepacket or classical ensemble	varied between 100 m/s and 1200 m/s
Full width of wavepacket (1/e-width)	0.05 nm [3]
Time step for the classical calculations with Verlet algorithm	1 fs
Time step for the quantum mechanical calculations with Crank–Nicolson algorithm	0.003 fs
Time step for sampling positional changes of O atoms due to the K⁺ force	6 fs

[1] C–O perpendicular to axis of selectivity filter and pointing to this axis. [2] This is approximately the middle of site S4. [3] This width entails a velocity spread (1/e-full width) of ±65 m/s. Making the wave packet much narrower would give velocity spreads on the order of the thermal mean velocity of a K⁺ ion. Making it much wider would bring it beyond the width of the ground state of the harmonic oscillator to which a site potential can be approximated.

Appendix B. How the Schrödinger Equation Becomes Nonlinear

The standard Schrödinger equation (SE) of the K⁺ ion has an apparently linear form:

$$\left[-\frac{\hbar^2}{2m_K}\frac{d^2}{dz_K^2} + \sum_{i=1}^{36} V(r_i, r_K) + gz_K \right]\psi(r_K, t) = i\hbar\frac{\partial}{\partial t}\psi(r_K, t)$$

The potential term of this equation contains the position vectors r_i. While the r_i of the carbon atoms of the carbonyl groups are fixed in space, the r_i of the oxygen atoms depend on time. In our calculations, each oxygen atom can rotate around the two orthogonal axes of a polar coordinate system with respect to the C atom of the carbonyl group, but the bond length is kept constant. An O atom is subject to the following forces: a restoring force, which is proportional to the angular elongation of the O atom from its unperturbed position; a dissipative force, which is proportional to its momentary velocity (see Appendix C); Coulomb forces from neighboring O and C atoms; and the force from the K⁺ ion, which depends on the quantum mechanical probability distribution of the K⁺ ion at that moment of time. This latter force introduces a non-linear aspect into the Schrödinger equation:

The time-dependent position vector $r_i(t)$ of the i-th O atom is given by the following relation

$$r_i(t) = r_i(0) + v_i(0)\cdot t + \int_0^t \left[\int_0^{t'} \ddot{r}_i(t'')dt'' \right]dt'$$

where $r_i(0)$ and $v_i(0)$ are the O atom's position and velocity at time $t = 0$ and $\ddot{r}_i(t)$ is the acceleration at time t. By Newton's second law, this acceleration is the quotient of the force acting on the atom, and the atom's mass m_O, as follows:

$$\ddot{r}_i(t) = \frac{\vec{F}_i(t)}{m_O} + \frac{\vec{F}_{other}}{m_O} = \frac{1}{m_O}\int_{z_{min}}^{z_{max}} df_i(z) + \frac{\vec{F}_{other}}{m_O},$$

where $\vec{F}_i(t)$ is the force due to the quantum mechanical probability distribution of the K^+ ion and $\vec{F}_{other}$ subsumes all other forces acting on the atom.

With the differential force element df_i from Equation (3) this acceleration becomes

$$\ddot{r}_i(t) = \frac{q_i q_K}{4\pi\varepsilon_0 m_O} \int_{z_{min}}^{z_{max}} \left(\frac{1}{(r_i(t) - r_K)^3} - \frac{2r_{cut}}{(r_i(t) - r_K)^4} \right) (r_i(t) - r_K) |\psi(r_K, t)|^2 dz + \frac{\vec{F}_{other}}{m_O}.$$

Because the position of the K^+ ion, r_K, is always on the z-axis, the integration extends over the whole range of the wave packet. For the present "cut-off" in our model, the wave packet has appreciable values only between the limits of $z_{min} = -0.2$ nm and $z_{max} = 0.8$ nm. From the above equations, one can see that the position of an O atom at a given moment is determined by the entire history of the spatial spread of the wave packet of the K^+ ion. As this position determines the potential to which the K^+ ion is exposed at that instant of time, it introduces the non-linear term into the Schrödinger equation as given by Equation (2).

Appendix C. Restoring and Dissipating Forces on Carbonyl O Atoms

Generally, carbonyl O atoms of the filter lining can oscillate in two directions similar to a two-dimensional pendulum around the carbonyl C atom. This movement involves a damping term, which leads to a dissipation of energy into the protein backbone and to intermittent water molecules. Here, we model the restoring forces in the following way:

For an O atom out of its equilibrium position, which is defined by its position in the absence of any forces from other atoms and charges, the torque for a return to equilibration in the present model is given by the following:

$$\vec{T}_r = -c\vec{a},$$

where c is an adjustable stiffness constant and $\vec{a}$ is the two-dimensional vector angle on the unit sphere, specifying angular distance and direction from the equilibrium position. This torque can be translated into a restoring force acting on the O atom. Our choice in the present simulation was to adjust the constant "c" so that off-equilibrium angles of 30° are associated with an energy of $k_B T$ (see Table A1), as this gave the typical values for the expected vibrational frequencies.

The damping force was assumed to be characterized by a torque as follows:

$$\vec{T}_d = -b\vec{\omega},$$

where b is a friction constant and $\vec{\omega}$ represents the angular velocity of the O atom at a given instant of time. This torque accounts for a force, which eventually slows down the motion of the O atom and thereby reduces its kinetic energy. In turn, this energy can be assumed to be lost into the protein backbone, water molecules, or other elements not explicitly modeled in the present study.

References

1. Hille, B. *Ion Channels of Excitable Membranes*, 3rd ed.; Sinnauer Associates: Sunderland, MA, USA, 2001.
2. Kuyucak, S.; Andersen, O.S.; Chung, S.H. Models of permeation in ion channels. *Rep. Prog. Phys.* **2001**, *64*, 1427–1471. [CrossRef]
3. Doyle, D.A.; Cabral, J.M.; Pfuetzner, R.A.; Kuo, A.; Gulbis, J.M.; Cohen, S.L.; Chait, B.T.; MacKinnon, R. The structure of the potassium channel: Molecular basis of K^+ conduction and selectivity. *Science* **1998**, *280*, 69–77. [CrossRef] [PubMed]
4. Morais-Cabral, J.H.; Zhou, Y.; MacKinnon, R. Energetic optimization of ion conduction rate by the K^+ selectivity filter. *Nature* **2001**, *414*, 37–42. [CrossRef] [PubMed]
5. Chattopadhyay, A.; Kelkar, D.A. Ion channels and D-amino acids. *J. Biosci.* **2005**, *30*, 147–149. [CrossRef] [PubMed]

6. Bostick, D.L.; Brooks, C.L. III Selectivity in K$^+$ channels is due to topological control of the permeant ion's coordination state. *Proc. Natl. Acad. Sci. USA* **2007**, *104*, 9260–9265. [CrossRef] [PubMed]

7. Furini, S.; Domene, C. Atypical mechanism of conduction in potassium channels. *Proc. Natl. Acad. Sci. USA* **2009**, *106*, 16074–16077. [CrossRef] [PubMed]

8. Gwan, J.F.; Baumgaertner, A. Cooperative transport in a potassium ion channel. *J. Chem. Phys.* **2007**, *127*, 045103. [CrossRef] [PubMed]

9. Berneche, S.; Roux, B. A gate in the selectivity filter of potassium channels. *Structure* **2005**, *13*, 591–600. [CrossRef] [PubMed]

10. Koepfer, D.A.; Song, C.; Gruene, T.; Sheldrick, G.M.; Zachariae, U.; de Groot, B.L. Ion permeation in K$^+$ channels occurs by direct Coulomb knock-on. *Science* **2014**, *346*, 352–355. [CrossRef] [PubMed]

11. Fowler, P.W.; Abad, E.; Beckstein, O.; Sansom, M.S.P. Energetics of multi-ion conduction pathways in potassium ion channels. *J. Chem. Theory Comput.* **2013**, *9*, 5176–5189. [CrossRef] [PubMed]

12. Miller, W.H. Including quantum effects in the dynamics of complex (i.e., large) molecular systems. *J. Chem. Phys.* **2006**, *125*, 132305. [CrossRef] [PubMed]

13. Bernroider, G.; Summhammer, J. Can quantum entanglement between ion-transition states effect action potential initiation? *Cogn. Comput.* **2012**, *4*, 29–37. [CrossRef]

14. Summhammer, J.; Salari, V.; Bernroider, G. A quantum-mechanical description of ion motion within the confining potentials of voltage gated ion channels. *J. Integr. Neurosci.* **2012**, *11*, 123–135. [CrossRef] [PubMed]

15. Crank, J.; Nicolson, P. A practical method for numerical evaluation of solutions of partial differential equations of the heat conduction type. *Math. Proc. Camb. Phil. Soc.* **1947**, *43*, 50–67. [CrossRef]

16. Salari, V.; Moradi, N.; Sajadi, M.; Fazileh, F.; Shahbazi, F. Quantum decoherence time scales for ionic superposition states in ion channels. *Phys. Rev. E* **2015**, *91*, 032704. [CrossRef] [PubMed]

17. Guidoni, L.; Carloni, P. Potassium permeation through the KcsA channel: A density functional study. *Biochim. Biophys. Acta* **2002**, *1563*, 1–6. [CrossRef]

18. Garafoli, G.; Jordan, P.C. Modelling permeation energetics in the KcsA potassium channel. *Biophys. J.* **2003**, *84*, 2814–2830. [CrossRef]

19. Verlet, L. Computer "experiments" on classical fluids. I. Thermodynamical properties of Lennard-Jones molecules. *Phys. Rev.* **1967**, *159*, 98–103. [CrossRef]

20. Allen, M.P.; Tildesley, D.J. *Computer Simulation of Liquids*; Oxford University Press: New York, NY, USA, 2009.

21. Bao, W. The nonlinear Schrödinger equation and applications in Bose-Einstein condensation and Plasma Physics. In *Dynamics in Models of Coarsening, Coagulation, Condensation and Quantization*; Bao, W., Liu, J.G., Eds.; World Scientific: River Edge, NJ, USA, 2007.

22. Bernroider, G.; Roy, S. Quantum entanglement of K$^+$ ions, multiple channels states and the role of noise in the brain. In Proceedings of the Fluctuations and Noise in Biological, Biophysical, and Biomedical Systems III, Austin, TX, USA, 24–26 May 2005.

23. Salari, V.; Tuszynski, J.; Rahnama, M.; Bernroider, G. Plausibility of quantum coherent states in biological systems. *J. Phys. Conf. Ser.* **2011**, *306*, 012075. [CrossRef]

entropy

MDPI

Article

Microscopic Theory of Energy Dissipation and Decoherence in Solid-State Quantum Devices: Need for Nonlocal Scattering Models

Rita Claudia Iotti and Fausto Rossi *

Department of Applied Science and Technology, Politecnico di Torino, Corso Duca degli Abruzzi 24, 10129 Torino, Italy
* Correspondence: fausto.rossi@polito.it; Tel.: +39-011-0907335

Received: 24 July 2018; Accepted: 12 September 2018; Published: 21 September 2018

Abstract: Energy dissipation and decoherence in state-of-the-art quantum nanomaterials and related nanodevices are routinely described and simulated via local scattering models, namely relaxation-time and Boltzmann-like schemes. The incorporation of such local scattering approaches within the Wigner-function formalism may lead to anomalous results, such as suppression of intersubband relaxation, incorrect thermalization dynamics, and violation of probability-density positivity. The primary goal of this article is to investigate a recently proposed quantum-mechanical (nonlocal) generalization (*Phys. Rev. B* **2017**, *96*, 115420) of semiclassical (local) scattering models, extending such treatment to carrier–carrier interaction, and focusing in particular on the nonlocal character of Pauli-blocking contributions. In order to concretely show the intrinsic limitations of local scattering models, a few simulated experiments of energy dissipation and decoherence in a prototypical quantum-well semiconductor nanostructure are also presented.

Keywords: semiconductor nanodevices; quantum transport; density-matrix formalism; Wigner-function simulations; nonlocal dissipation models

1. Introduction

Following the seminal paper by Esaki and Tsu [1], artificially tailored as well as self-assembled solid-state nanostructures form the leading edge of semiconductor science and technology [2]. The design of state-of-the-art optoelectronic nanodevices, in fact, heavily exploits the principles of band-gap engineering [3], achieved by confining charge carriers in spatial regions comparable to their de Broglie wavelengths [4]. This, together with the progressive reduction of the typical time-scales involved, pushes device miniaturization toward limits where, in principle [5], the application of the traditional Boltzmann transport theory [6] becomes questionable, and a comparison with more rigorous quantum-transport approaches [7–13] is desirable; the latter can be qualitatively subdivided into two main classes. On the one hand, so-called double-time approaches based on the nonequilibrium Green's function technique [14] have been proposed and widely employed; an introduction to the theory of nonequilibrium Green's functions with applications to many problems in transport and optics of semiconductors can be found in the books by Haug and Jauho [15], Bonitz [16], and Datta [17]. By employing—and further developing and extending—such nonequilibrium Green's function formalism, a number of groups have recently proposed efficient quantum-transport treatments for the study of various meso- and nanoscale structures as well as of corresponding micro- and optoelectronic devices [18–21]. On the other hand, so-called single-time approaches based on the density-matrix formalism [22,23] have been proposed, including phase-space treatments based on the Wigner-function formalism [7,24]. In spite of the intrinsic validity limits of the semiclassical theory just recalled, during the last few decades, a number of Boltzmann-like Monte Carlo simulation schemes have been successfully employed for the investigation of new-generation semiconductor nanodevices [25–36]. Such modeling strategies—based on the neglect of carrier phase coherence—are, however, unable to properly describe ultrafast phenomena. To this aim, the crucial step is to adopt a quantum-mechanical description of the carrier subsystem; this can be performed at different levels, ranging from phenomenological dissipation and decoherence models [37] to quantum-kinetic treatments [8,10,11]. Indeed, in order to overcome the intrinsic limitations of the semiclassical picture in properly describing ultrafast space-dependent phenomena —e.g., real-space transfer and escape versus capture processes— Jacoboni and co-workers have proposed a quantum Monte Carlo technique [38], while Kuhn and co-workers have proposed a quantum-kinetic treatment [39]; however, due to their high computational cost, these non-Markovian density-matrix approaches are often unsuitable for the design and optimization of new-generation nanodevices.

In order to overcome such limitations, a conceptually simple as well as physically reliable quantum-mechanical generalization of the conventional Boltzmann theory has been recently proposed [40]. The latter, based on the density-matrix formalism, preserves the power and flexibility of the semiclassical picture in describing a large variety of scattering mechanisms; more specifically, employing a microscopic derivation of generalized scattering rates based on a reformulation of the Markov limit [41], a density-matrix equation has been derived, able to properly account for space-dependent ultrafast dynamics in semiconductor nanostructures. Indeed, the density-matrix approach just recalled has been recently applied to the investigation of scattering nonlocality in GaN-based materials [42] and carbon nanotubes [43], as well as to the study of carrier capture processes [44]. It is worth mentioning that a purely phenomenological Lindblad-type approach [45] based on the jump-operator formalism has been recently proposed [46].

In addition to the density-matrix treatments just recalled, quantum-transport phenomena have been extensively investigated via Wigner-function approaches [7,47]. Indeed, the Wigner-function formalism has been adopted in various contexts to study ultrashort space- and/or time-scale phenomena in semiconductor nanomaterials and related nanodevices [48–78]. In view of their formal similarity with the conventional Boltzmann theory, in these Wigner-function treatments, dissipation versus decoherence phenomena are often accounted for in semiclassical terms via local scattering models, such as relaxation-time and Boltzmann-like schemes. It has been recently shown [79] that the use of such local scattering approaches may lead to unphysical results, namely anomalous suppression of intersubband relaxation, incorrect thermalization dynamics, and violation of probability-density positivity. To overcome such severe limitations, in [79], a quantum-mechanical generalization of relaxation-time and Boltzmann-like models has been proposed, resulting in nonlocal electron-phonon scattering superoperators.

The goal of this paper is twofold: on the one hand, we shall elucidate the intimate link between density-matrix and Wigner-function approaches, pointing out intrinsic limitations of semiclassical scattering models within these, apparently different, simulation strategies. On the other hand, we shall extend the carrier–phonon treatment in [79] to carrier–carrier interaction; indeed, the latter has been for a long time to have a dramatic impact both on optical properties [8,10,11] as well as on transport phenomena [80,81], and has more recently been in the spotlight due to the effects of its interplay with spin-orbit coupling [82–85]. Moreover, we shall investigate in more detail the role played by Pauli-blocking terms both within the density matrix formalism (population versus polarization contributions) as well as within the Wigner-function picture (local versus nonlocal action). In order to concretely show the intrinsic limitations of local scattering models, a few simulated experiments of energy dissipation and decoherence in a prototypical quantum-well semiconductor nanostructure are also presented.

The paper is organized as follows: in Section 2, we shall briefly recall the main features of semiclassical scattering models, both for bulk and for nanostructured materials. In Section 3, we shall provide a fully quantum-mechanical treatment of energy-dissipation and decoherence phenomena within the density-matrix formalism, and we shall translate the latter into a nonlocal Wigner-function scattering model for both carrier–phonon and carrier–carrier interaction. In Section 4, we shall analyze the role played by Pauli-blocking contributions, discussing non-classical features, like polarization scattering within the density-matrix formalism, and nonlocal Pauli factors within the Wigner-function picture. Finally, in Section 5, we shall summarize and draw a few conclusions.

2. Semiclassical Scattering Models

To investigate in quantum-mechanical terms the electro-optical response of semiconductor nanomaterials and related nanodevices, it is crucial to study the time evolution of single-particle quantities, e.g., total carrier density, mean kinetic energy, charge current, etc. Such quantities may be conveniently expressed by a suitable (quantum-plus-statistical) average of a corresponding (single-particle) operator in terms of the single-particle density matrix $\rho_{\alpha_1\alpha_2}$ [23] (α denoting the electronic single-particle states of our nanostructure): its diagonal terms $f_\alpha = \rho_{\alpha\alpha}$ describe the population of the generic single-particle state α while the off-diagonal terms describe the quantum-mechanical phase coherence (or polarization) between states α_1 and α_2. More precisely, we may write:

$$\rho_{\alpha_1\alpha_2} = f_{\alpha_1}\delta_{\alpha_1\alpha_2} + p_{\alpha_1\alpha_2}. \tag{1}$$

Here, the first (diagonal) term describes the semiclassical state populations, while the second term

$$p_{\alpha_1\alpha_2} = \rho_{\alpha_1\alpha_2}\left(1 - \delta_{\alpha_1\alpha_2}\right) \tag{2}$$

is the so-called polarization matrix.

Regardless of the specific physical system and related modelling, the time evolution of the single-particle density matrix can be always expressed as the sum of a deterministic (d) and of a scattering (s) contribution:

$$\frac{\partial \rho_{\alpha_1\alpha_2}}{\partial t} = \left.\frac{\partial \rho_{\alpha_1\alpha_2}}{\partial t}\right|_{d} + \left.\frac{\partial \rho_{\alpha_1\alpha_2}}{\partial t}\right|_{s}. \tag{3}$$

Here,

$$\left.\frac{\partial \rho_{\alpha_1\alpha_2}}{\partial t}\right|_{d} = \frac{\epsilon_{\alpha_1} - \epsilon_{\alpha_2}}{i\hbar}\rho_{\alpha_1\alpha_2} \tag{4}$$

(ϵ_α denoting the energy of the single-particle state α), while the explicit form of the scattering contribution depends on our level of description (see Section 3).

As discussed in detail in [13], for quantum nanodevices characterized by a relevant dissipation versus decoherence dynamics and operating in steady-state conditions, it is common practice to adopt the so-called semiclassical picture; this amounts to neglecting the polarization term in (2). Within such semiclassical (or diagonal) approximation ($\rho_{\alpha_1\alpha_2} = f_{\alpha_1}\delta_{\alpha_1\alpha_2}$), the simplest scattering model is given by the well-known relaxation-time approximation (RTA) [23]:

$$\left.\frac{\partial f_\alpha}{\partial t}\right|_{s} = -\Gamma_\alpha\left(f_\alpha - f_\alpha^\circ\right). \tag{5}$$

Here, the relaxation of the state population f_α toward the equilibrium population f_α° is described in terms of a state-dependent relaxation rate Γ_α that purely depends on that state and encodes all relevant scattering processes characterizing the operational conditions of the device.

In order to provide a more accurate description of nonequilibrium phenomena, the RTA model in Equation (5) is usually replaced by a Boltzmann-like scattering model of the form:

$$\left.\frac{\partial f_\alpha}{\partial t}\right|_{s} = \sum_{s}\sum_{\alpha'}\left((1-f_\alpha)P_{\alpha\alpha'}^s f_{\alpha'} - (1-f_{\alpha'})P_{\alpha'\alpha}^s f_\alpha\right). \tag{6}$$

The above collision term exhibits the well-known in- minus out-scattering structure, and allows one to incorporate a number of scattering mechanisms s via corresponding scattering rates $P_{\alpha'\alpha}^s$;

the latter describes the probability per time unit for an electronic transition $\alpha \to \alpha'$ induced by the scattering mechanism s, and are typically derived via the standard Fermi's golden rule; moreover, here the factors $(1 - f_\alpha)$ describe Pauli-blocking effects (see below).

As anticipated in the introductory section, in addition to the density-matrix treatments just recalled, state-of-the-art quantum nanodevices are often modelled via Wigner-function-based simulation schemes [48–78]. Regardless of the specific problem under investigation, the time evolution of the single-particle Wigner function $f(\mathbf{r}, \mathbf{k})$ can be expressed once again as the sum of a deterministic and of a scattering contribution, namely [86]:

$$\frac{\partial f(\mathbf{r}, \mathbf{k})}{\partial t} = \left.\frac{\partial f(\mathbf{r}, \mathbf{k})}{\partial t}\right|_{\mathrm{d}} + \left.\frac{\partial f(\mathbf{r}, \mathbf{k})}{\partial t}\right|_{\mathrm{s}} . \tag{7}$$

Here, the first term is the quantum-mechanical generalization of the deterministic (diffusion-plus-drift) term in the semiclassical theory, and can be conveniently expressed in terms of the well-known Moyal brackets [87], whose explicit form depends on the electron band dispersion and on the electromagnetic gauge [72,79]. The second term, in contrast, describes again energy dissipation and decoherence phenomena induced by various scattering mechanisms. Within a fully quantum-mechanical treatment, such a scattering term is strictly nonlocal, as described in detail in [42], and is of the general form

$$\left.\frac{\partial f(\mathbf{r}, \mathbf{k})}{\partial t}\right|_{\mathrm{s}} = S\left[f(\mathbf{r}', \mathbf{k}')\right](\mathbf{r}, \mathbf{k}) , \tag{8}$$

where, in general, S is a nonlinear scattering superoperator describing a nonlocal action both in $\mathbf{r}$ and $\mathbf{k}$, i.e., the scattering contribution to the generic phase-space point $(\mathbf{r}, \mathbf{k})$ depends on the value of the Wigner function f in any other phase-space point $(\mathbf{r}', \mathbf{k}')$.

Due to the difficulty in dealing with its fully nonlocal character, it is common practice in many quantum-simulation approaches to replace the scattering superoperator in Equation (8) with a local superoperator. The simplest choice is once again the adoption of an RTA model [49,51,66,75] that rewords the semiclassical case, namely:

$$\left.\frac{\partial f(\mathbf{r}, \mathbf{k})}{\partial t}\right|_{\mathrm{s}} = -\Gamma(\mathbf{r}, \mathbf{k}) \left(f(\mathbf{r}, \mathbf{k}) - f^\circ(\mathbf{r}, \mathbf{k})\right) . \tag{9}$$

Here, similar to the RTA model in (5), the relaxation of the Wigner function in the phase-space point $(\mathbf{r}, \mathbf{k})$ toward the equilibrium Wigner function $f^\circ(\mathbf{r}, \mathbf{k})$ is described in terms of a space- and momentum-dependent relaxation rate $\Gamma(\mathbf{r}, \mathbf{k})$; the latter may be extracted from fully microscopic Monte Carlo simulations [6], or modelled via simplified Fermi's Golden-rule treatments.

Another simplified (i.e., local) version of the scattering superoperator in Equation (8) is inspired again by the formal analogy between the Wigner transport equation in (7) and the usual Boltzmann transport theory, and consists of replacing S with a conventional (i.e., semiclassical) Boltzmann collision term [6,23]:

$$\left.\frac{\partial f(\mathbf{r}, \mathbf{k})}{\partial t}\right|_{\mathrm{s}} = \sum_s \int d\mathbf{k}' \left[P^s(\mathbf{r}; \mathbf{k}, \mathbf{k}') f(\mathbf{r}, \mathbf{k}') - P^s(\mathbf{r}; \mathbf{k}', \mathbf{k}) f(\mathbf{r}, \mathbf{k})\right] , \tag{10}$$

where

$$P^s(\mathbf{r}; \mathbf{k}, \mathbf{k}') = (1 - f(\mathbf{r}, \mathbf{k})) \, P_0^s(\mathbf{r}; \mathbf{k}, \mathbf{k}') \tag{11}$$

denotes the low-density scattering rate P_0^s in $\mathbf{r}$ (for the generic transition $\mathbf{k}' \to \mathbf{k}$ induced by the scattering mechanism s) weighted by the usual Pauli-blocking factor, and simply reduces to $P_0^s(\mathbf{r}; \mathbf{k}, \mathbf{k}')$ in the low-density limit ($f(\mathbf{r}, \mathbf{k}) \to 0$).

The Boltzmann collision term in (10) is characterized once again by the well-established in- minus out-scattering structure; indeed, the latter may also be written as

$$\left.\frac{\partial f(\mathbf{r}, \mathbf{k})}{\partial t}\right|_s = \sum_s \int d\mathbf{r}' \, d\mathbf{k}' P^{s,\text{in}}(\mathbf{r}, \mathbf{k}; \mathbf{r}', \mathbf{k}') f(\mathbf{r}', \mathbf{k}')$$
$$- \sum_s \int d\mathbf{r}' \, d\mathbf{k}' P^{s,\text{out}}(\mathbf{r}, \mathbf{k}; \mathbf{r}', \mathbf{k}') f(\mathbf{r}', \mathbf{k}') \tag{12}$$

with

$$P^{s,\text{in}}(\mathbf{r}, \mathbf{k}; \mathbf{r}', \mathbf{k}') = \delta(\mathbf{r} - \mathbf{r}') P^s(\mathbf{r}; \mathbf{k}, \mathbf{k}') \tag{13}$$

and

$$P^{s,\text{out}}(\mathbf{r}, \mathbf{k}; \mathbf{r}', \mathbf{k}') = \delta(\mathbf{r} - \mathbf{r}') \delta(\mathbf{k} - \mathbf{k}') \int d\mathbf{k}'' P^s(\mathbf{r}; \mathbf{k}'', \mathbf{k}) , \tag{14}$$

which shows that, within the conventional Boltzmann theory, both superoperators are local in $\mathbf{r}$, and that the out-scattering one is local in $\mathbf{k}$ as well.

3. Fully Quantum-Mechanical Scattering Models

The quantum-mechanical derivation of effective scattering models within the density-matrix formalism may involve one or more of the following three key steps [88]: (i) mean-field approximation; (ii) adiabatic or Markov limit; and (iii) semiclassical or diagonal limit.

When all of these three approximations are applied, the usual Boltzmann collision term is obtained (see Equation (6)); the latter, if applicable (see above), constitutes a robust/reliable particle-like description in purely stochastic terms, thus providing physically acceptable results.

In contrast, the combination of the first two approximation schemes only, namely mean-field treatment and adiabatic limit, allows one to derive so-called Markovian scattering superoperators, whose action may lead to unphysical results [89]. Indeed, as originally pointed out by Spohn and coworkers [90], the choice of the adiabatic decoupling strategy is definitely not unique and, in general, the positive-definite character of the density-matrix operator may be violated.

To overcome this severe limitation, a few years ago, an alternative and more general Markov procedure has been proposed [41]; the latter allows for a microscopic derivation of Lindblad-type scattering superoperators [45], thus preserving the positive-definite nature of the electronic quantum-mechanical state. More recently [40], such alternative Markov scheme combined with the conventional mean-field approximation just recalled has allowed for the derivation of positive-definite nonlinear scattering superoperators acting on the single-particle density matrix $\rho_{\alpha_1\alpha_2}$; more specifically, as shown in [40], for both carrier–phonon and carrier–carrier interaction, the resulting single-particle equation is given by

$$\left.\frac{d\rho_{\alpha_1\alpha_2}}{dt}\right|_s = \frac{1}{2} \sum_{\alpha'\alpha_1'\alpha_2'} \left(\left(\delta_{\alpha_1\alpha'} - \rho_{\alpha_1\alpha'}\right) \mathcal{P}^s_{\alpha'\alpha_2,\alpha_1'\alpha_2'} \rho_{\alpha_1'\alpha_2'} - \left(\delta_{\alpha'\alpha_1'} - \rho_{\alpha'\alpha_1'}\right) \mathcal{P}^{s*}_{\alpha'\alpha_1',\alpha_1\alpha_2'} \rho_{\alpha_2'\alpha_2} \right) + \text{H.c.} \tag{15}$$

with generalized carrier–phonon scattering rates [91]

$$\mathcal{P}^{\text{cp}}_{\alpha_1\alpha_2,\alpha_1'\alpha_2'} = A^{\text{cp}}_{\alpha_1\alpha_1'} A^{\text{cp}*}_{\alpha_2\alpha_2'} \tag{16}$$

and generalized carrier–carrier scattering rates [92]

$$\mathcal{P}^{\text{cc}}_{\alpha_1\alpha_2,\alpha_1'\alpha_2'} = 2 \sum_{\bar{\alpha}_1\bar{\alpha}_2,\bar{\alpha}_1'\bar{\alpha}_2'} \left(\delta_{\bar{\alpha}_2\bar{\alpha}_1} - \rho_{\bar{\alpha}_2\bar{\alpha}_1}\right) A^{\text{cc}}_{\alpha_1\bar{\alpha}_1,\alpha_1'\bar{\alpha}_1'} A^{\text{cc}*}_{\alpha_2\bar{\alpha}_2,\alpha_2'\bar{\alpha}_2'} \rho_{\bar{\alpha}_1'\bar{\alpha}_2'} . \tag{17}$$

Here, $A^{\text{cp}}_{\alpha\alpha'}$ denotes the matrix element of the corresponding carrier–phonon Lindblad operator for the (one-body) transition $\alpha' \to \alpha$, while $A^{\text{cc}}_{\alpha\bar{\alpha},\alpha'\bar{\alpha}'}$ denotes the matrix element of the corresponding carrier–carrier Lindblad operator for the (two-body) transition $\alpha'\bar{\alpha}' \rightarrow \alpha\bar{\alpha}$. These carrier–phonon and carrier–carrier Lindblad matrix elements can be microscopically derived starting from the corresponding interaction Hamiltonians, as described in [40].

It is worth stressing that, contrary to the generalized carrier–phonon rates in (16), the generalized carrier–carrier rates in (17) are themselves a function of the single-particle density matrix; this is a clear fingerprint of the two-body nature of the carrier–carrier interaction (see below).

The generic single-particle scattering superoperator in (15) is the result of positive-like (in-scattering) and negative-like (out-scattering) contributions, which are nonlinear functions of the single-particle density matrix. Indeed, in the semiclassical limit previously recalled ($\rho_{\alpha_1\alpha_2} = f_{\alpha_1}\delta_{\alpha_1\alpha_2}$), the density-matrix Equation (15) assumes the expected nonlinear Boltzmann-type form

$$\left.\frac{df_\alpha}{dt}\right|_s = \sum_{\alpha'}\left((1-f_\alpha)P^s_{\alpha\alpha'}f_{\alpha'} - (1-f_{\alpha'})P^s_{\alpha'\alpha}f_\alpha\right) \tag{18}$$

with semiclassical carrier–phonon scattering rates

$$P^{cp}_{\alpha\alpha'} = \mathcal{P}^{cp}_{\alpha\alpha,\alpha'\alpha'} = \left|A^{cp}_{\alpha\alpha'}\right|^2 \tag{19}$$

and semiclassical carrier–carrier scattering rates

$$P^{cc}_{\alpha\alpha'} = \mathcal{P}^{cc}_{\alpha\alpha,\alpha'\alpha'} = 2\sum_{\bar{\alpha}\bar{\alpha}'}(1-f_{\bar{\alpha}})\left|A^{cc}_{\alpha\bar{\alpha},\alpha'\bar{\alpha}'}\right|^2 f_{\bar{\alpha}'}. \tag{20}$$

The above semiclassical limit clearly shows that the nonlinearity factors ($\delta_{\alpha_1\alpha_2} - \rho_{\alpha_1\alpha_2}$) in (15) as well as in (17) can be regarded as the quantum-mechanical generalization of the Pauli factors ($1 - f_\alpha$) of the conventional Boltzmann theory (see also Section 4 below).

A closer inspection of Equations (15) and (17)—together with their semiclassical counterparts in (18) and (20)—confirms the two-body nature of the carrier–carrier interaction. Indeed, differently from the carrier–phonon scattering, in this case, the density-matrix equation describes the time evolution of a so-called "main carrier" α interacting with a so-called "partner carrier" $\bar{\alpha}$.

As already pointed out in the introductory section, in addition to the density-matrix treatments just recalled, quantum-transport phenomena in nanomaterials and related nanodevices have been extensively investigated via Wigner-function approaches [48–78]. In view of their formal similarity with the conventional Boltzmann transport theory, in these Wigner-function treatments, dissipation versus decoherence phenomena are often accounted for via local scattering models, such as relaxation-time and Boltzmann-like schemes (see Section 2).

In spite of the fact that density-matrix and Wigner-function treatments have been historically developed and applied independently to the modeling and optimization of various state-of-the-art nanodevices, it is imperative to stress that the single-particle density matrix $\rho_{\alpha_1\alpha_2}$ in (3) is linked to the single-particle Wigner function $f(\mathbf{r}, \mathbf{k})$ in (7) via a one-to-one correspondence provided by the well-known Weyl–Wigner transform [7]. More specifically, adopting the very same notation employed in [72], we have

$$f(\mathbf{r}, \mathbf{k}) = \sum_{\alpha_1\alpha_2} W^*_{\alpha_1\alpha_2}(\mathbf{r}, \mathbf{k})\rho_{\alpha_1\alpha_2}\,, \tag{21}$$

where

$$W_{\alpha_1\alpha_2}(\mathbf{r}, \mathbf{k}) = \int d\mathbf{r}'\phi_{\alpha_1}\left(\mathbf{r} + \frac{\mathbf{r}'}{2}\right)e^{-i\mathbf{k}\cdot\mathbf{r}'}\phi^*_{\alpha_2}\left(\mathbf{r} - \frac{\mathbf{r}'}{2}\right) \tag{22}$$

denotes the Weyl–Wigner transform just recalled, and $\phi_\alpha(\mathbf{r})$ the real-space wavefunction of the single-particle state α.

In view of such one-to-one correspondence, it is thus clear that, given a scattering model within the density-matrix picture, the latter will have a well-defined Wigner-function counterpart, and vice versa. On this basis, the most natural and rigorous approach is to select a reliable/robust model in one picture, and then to translate it into the other one via the Weyl–Wigner transform in (22). This is exactly what has been recently proposed in [79]: applying the nonlinear density-matrix scattering model in (15) to the case of carrier–phonon interaction, a nonlocal scattering superoperator for the Wigner function has been derived. In what follows, we shall extend such nonlocal scattering treatment to the case of carrier–carrier interaction as well.

In order to get the desired Wigner-function version of the density-matrix scattering superoperator in (15), the crucial step is to apply to the latter the Weyl–Wigner transform (21) together with its inverse, namely [93]

$$\rho_{\alpha_1\alpha_2} = \int \frac{d\mathbf{r}\,d\mathbf{k}}{(2\pi)^3} W_{\alpha_1\alpha_2}(\mathbf{r},\mathbf{k}) f(\mathbf{r},\mathbf{k}). \tag{23}$$

The resulting Wigner-function scattering superoperator is given by

$$\left.\frac{\partial f(\mathbf{r},\mathbf{k})}{\partial t}\right|_s = \int d\mathbf{r}'\,d\mathbf{k}'\,P^{s,\text{in}}(\mathbf{r},\mathbf{k};\mathbf{r}',\mathbf{k}') f(\mathbf{r}',\mathbf{k}')$$
$$- \int d\mathbf{r}'\,d\mathbf{k}'\,P^{s,\text{out}}(\mathbf{r},\mathbf{k};\mathbf{r}',\mathbf{k}') f(\mathbf{r}',\mathbf{k}'), \tag{24}$$

where

$$P^{s,\text{in/out}}(\mathbf{r},\mathbf{k};\mathbf{r}',\mathbf{k}') = \int \frac{d\mathbf{r}''\,d\mathbf{k}''}{(2\pi)^3}\,(1 - f(\mathbf{r}'',\mathbf{k}''))\,\tilde{P}^{s,\text{in/out}}(\mathbf{r}'',\mathbf{k}'';\mathbf{r},\mathbf{k};\mathbf{r}',\mathbf{k}') \tag{25}$$

with

$$\tilde{P}^{s,\text{in}}(\mathbf{r}'',\mathbf{k}'';\mathbf{r},\mathbf{k};\mathbf{r}',\mathbf{k}') = \frac{1}{(2\pi)^3} \sum_{\alpha_1\alpha_2\alpha'\alpha_1'\alpha_2'} \Re\left\{ W_{\alpha_1\alpha_2}(\mathbf{r},\mathbf{k}) W^*_{\alpha_1\alpha'}(\mathbf{r}'',\mathbf{k}'') P^s_{\alpha'\alpha_2,\alpha_1'\alpha_2'} W^*_{\alpha_1'\alpha_2'}(\mathbf{r}',\mathbf{k}') \right\} \tag{26}$$

and

$$\tilde{P}^{s,\text{out}}(\mathbf{r}'',\mathbf{k}'';\mathbf{r},\mathbf{k};\mathbf{r}',\mathbf{k}') = \frac{1}{(2\pi)^3} \sum_{\alpha_1\alpha_2\alpha'\alpha_1'\alpha_2'} \Re\left\{ W_{\alpha_1\alpha_2}(\mathbf{r},\mathbf{k}) W^*_{\alpha'\alpha_1'}(\mathbf{r}'',\mathbf{k}'') P^{s,*}_{\alpha'\alpha_1',\alpha_1\alpha_2'} W^*_{\alpha_2'\alpha_2}(\mathbf{r}',\mathbf{k}') \right\}. \tag{27}$$

As expected, for both carrier–phonon and carrier–carrier interaction, the proposed quantum-mechanical generalization of the standard Boltzmann collision term in (10) is thus intrinsically nonlocal. In particular, comparing Equation (25) with its semiclassical counterpart in (11), we realize that the action of the Pauli exclusion principle within the Wigner phase-space is itself nonlocal: the generalized in and out scattering rates for a given transition $\mathbf{r},\mathbf{k} \to \mathbf{r}',\mathbf{k}'$ depend on the value of the Wigner function in any other phase-space point $\mathbf{r}'',\mathbf{k}''$ via the Pauli factor $1 - f(\mathbf{r}'',\mathbf{k}'')$. Such Pauli-blocking nonlocality will be discussed in more detail at the end of Section 4.

In the low-density limit ($f(\mathbf{r},\mathbf{k}) \to 0$), the proposed scattering model in (25) reduces to:

$$P^{s,\text{in}}(\mathbf{r},\mathbf{k};\mathbf{r}',\mathbf{k}') = \frac{1}{(2\pi)^3} \sum_{\alpha_1\alpha_2\alpha_1'\alpha_2'} \Re\left\{ W_{\alpha_1\alpha_2}(\mathbf{r},\mathbf{k}) P^s_{\alpha_1\alpha_2,\alpha_1'\alpha_2'} W^*_{\alpha_1'\alpha_2'}(\mathbf{r}',\mathbf{k}') \right\} \tag{28}$$

and

$$P^{s,\text{out}}(\mathbf{r},\mathbf{k};\mathbf{r}',\mathbf{k}') = \frac{1}{(2\pi)^3} \sum_{\alpha_1\alpha_2\alpha_1'\alpha_2'} \Re\left\{ W_{\alpha_1\alpha_2}(\mathbf{r},\mathbf{k}) P^{s,*}_{\alpha_1'\alpha_1',\alpha_1\alpha_2'} W^*_{\alpha_2'\alpha_2}(\mathbf{r}',\mathbf{k}') \right\}. \tag{29}$$

It is however worth stressing that, while for carrier–phonon interaction the above low-density scattering rates are different from zero, for carrier–carrier interaction, the latter vanish; this is due to the fact that, in the low-density limit ($\rho_{\alpha_1\alpha_2} \to 0$), the generalized carrier–carrier scattering rates in (17) tend to zero.

For the case of carrier–phonon interaction, we may easily derive the explicit form of the corresponding Wigner-function scattering rates. By inserting Equation (16) into Equations (26) and (27), we get

$$\tilde{P}^{\text{cp,in}}(\mathbf{r}'',\mathbf{k}'';\mathbf{r},\mathbf{k};\mathbf{r}',\mathbf{k}') = \frac{1}{(2\pi)^3}\sum_{\alpha_1\alpha_2\alpha_1'\alpha_2'}\Re\left\{W_{\alpha_1\alpha_2}(\mathbf{r},\mathbf{k})W^*_{\alpha_1'\alpha'}(\mathbf{r}'',\mathbf{k}'')A^{\text{cp}}_{\alpha'\alpha_1'}A^{\text{cp}*}_{\alpha_2\alpha_2'}W^*_{\alpha_1'\alpha_2'}(\mathbf{r}',\mathbf{k}')\right\} \quad (30)$$

and

$$\tilde{P}^{\text{cp,out}}(\mathbf{r}'',\mathbf{k}'';\mathbf{r},\mathbf{k};\mathbf{r}',\mathbf{k}') = \frac{1}{(2\pi)^3}\sum_{\alpha_1\alpha_2\alpha_1'\alpha_2'}\Re\left\{W_{\alpha_1\alpha_2}(\mathbf{r},\mathbf{k})W^*_{\alpha'\alpha_1'}(\mathbf{r}'',\mathbf{k}'')A^{\text{cp}*}_{\alpha'\alpha_1}A^{\text{cp}}_{\alpha_1'\alpha_2'}W^*_{\alpha_2\alpha_2}(\mathbf{r}',\mathbf{k}')\right\}. \quad (31)$$

In contrast, for the case of carrier–carrier interaction, getting the explicit form of the corresponding Wigner-function scattering rates is not so straightforward. To this aim, the first step is to rewrite the generalized carrier–carrier rates in (17) in terms of the Wigner function $f(\mathbf{r},\mathbf{k})$. More specifically, by inserting into Equation (17) the inverse Weyl–Wigner transform (23), we get:

$$P^{\text{cc}}_{\alpha_1\alpha_2,\alpha_1'\alpha_2'} = 2\sum_{\bar{\alpha}_1\bar{\alpha}_2,\bar{\alpha}_1'\bar{\alpha}_2'}\int\frac{d\bar{\mathbf{r}}''d\bar{\mathbf{k}}''\,d\bar{\mathbf{r}}'d\bar{\mathbf{k}}'}{(2\pi)^6}\cdot$$

$$\left(1-f(\bar{\mathbf{r}}'',\bar{\mathbf{k}}'')\right)W_{\bar{\alpha}_2\bar{\alpha}_1}(\bar{\mathbf{r}}'',\bar{\mathbf{k}}'')A^{\text{cc}}_{\alpha_1\bar{\alpha}_1,\alpha_1'\bar{\alpha}_1'}A^{\text{cc}*}_{\alpha_2\bar{\alpha}_2,\alpha_2'\bar{\alpha}_2'}W_{\bar{\alpha}_1'\bar{\alpha}_2'}(\bar{\mathbf{r}}',\bar{\mathbf{k}}')f(\bar{\mathbf{r}}',\bar{\mathbf{k}}'). \quad (32)$$

By inserting this last expression into Equations (26) and (27), the latter can be compactly rewritten as

$$\tilde{P}^{\text{cc,in/out}}(\mathbf{r}'',\mathbf{k}'';\mathbf{r},\mathbf{k};\mathbf{r}',\mathbf{k}')=\int\frac{d\bar{\mathbf{r}}''d\bar{\mathbf{k}}''\,d\bar{\mathbf{r}}'d\bar{\mathbf{k}}'}{(2\pi)^6}\left(1-f(\bar{\mathbf{r}}'',\bar{\mathbf{k}}'')\right)\tilde{p}^{\text{in/out}}(\bar{\mathbf{r}}'',\bar{\mathbf{k}}'';\mathbf{r}'',\mathbf{k}'';\mathbf{r},\mathbf{k};\mathbf{r}',\mathbf{k}';\bar{\mathbf{r}}',\bar{\mathbf{k}}')f(\bar{\mathbf{r}}',\bar{\mathbf{k}}') \quad (33)$$

with

$$\tilde{p}^{\text{in}}(\bar{\mathbf{r}}'',\bar{\mathbf{k}}'';\mathbf{r}'',\mathbf{k}'';\mathbf{r},\mathbf{k};\mathbf{r}',\mathbf{k}';\bar{\mathbf{r}}',\bar{\mathbf{k}}') = \frac{1}{4\pi^3}\sum_{\alpha_1\alpha_2\alpha_1'\alpha_2'}\sum_{\bar{\alpha}_1\bar{\alpha}_2,\bar{\alpha}_1'\bar{\alpha}_2'}\cdot$$

$$\Re\left\{W_{\bar{\alpha}_2\bar{\alpha}_1}(\bar{\mathbf{r}}'',\bar{\mathbf{k}}'')W^*_{\alpha_1\alpha'}(\mathbf{r}'',\mathbf{k}'')A^{\text{cc}}_{\alpha'\bar{\alpha}_1,\alpha_1'\bar{\alpha}_1'}W_{\alpha_1\alpha_2}(\mathbf{r},\mathbf{k})A^{\text{cc}*}_{\alpha_2\bar{\alpha}_2,\alpha_2'\bar{\alpha}_2'}W^*_{\alpha_1'\alpha_2'}(\mathbf{r}',\mathbf{k}')W_{\bar{\alpha}_1'\bar{\alpha}_2'}(\bar{\mathbf{r}}',\bar{\mathbf{k}}')\right\} \quad (34)$$

and

$$\tilde{p}^{\text{out}}(\bar{\mathbf{r}}'',\bar{\mathbf{k}}'';\mathbf{r}'',\mathbf{k}'';\mathbf{r},\mathbf{k};\mathbf{r}',\mathbf{k}';\bar{\mathbf{r}}',\bar{\mathbf{k}}') = \frac{1}{4\pi^3}\sum_{\alpha_1\alpha_2\alpha_1'\alpha_2'}\sum_{\bar{\alpha}_1\bar{\alpha}_2,\bar{\alpha}_1'\bar{\alpha}_2'}\cdot$$

$$\Re\left\{W^*_{\bar{\alpha}_2\bar{\alpha}_1}(\bar{\mathbf{r}}'',\bar{\mathbf{k}}'')W^*_{\alpha'\alpha_1'}(\mathbf{r}'',\mathbf{k}'')A^{\text{cc}*}_{\alpha'\bar{\alpha}_1,\alpha_1\bar{\alpha}_1'}W_{\alpha_1\alpha_2}(\mathbf{r},\mathbf{k})A^{\text{cc}}_{\alpha_1\bar{\alpha}_2,\alpha_2'\bar{\alpha}_2'}W^*_{\alpha_2'\alpha_2}(\mathbf{r}',\mathbf{k}')W^*_{\bar{\alpha}_1'\bar{\alpha}_2'}(\bar{\mathbf{r}}',\bar{\mathbf{k}}')\right\}. \quad (35)$$

Exactly as for the density-matrix treatment previously considered, the Wigner-function version of the corresponding carrier–carrier scattering superoperator reveals again its two-body nature. Indeed, combining the general in- minus-out structure in (24) with the explicit form of the carrier–carrier scattering rates in (33) and adopting the compact notation $\xi \equiv \mathbf{r},\mathbf{k}$, it is easy to realize that the carrier–carrier scattering superoperator is always of the form:

$$\left.\frac{\partial f(\xi)}{\partial t}\right|_s = \int d\bar{\xi}''d\xi''d\xi'd\bar{\xi}'\left(1-f(\bar{\xi}'')\right)\left(1-f(\xi'')\right)K\left(\bar{\xi}'',\xi'',\xi,\xi',\bar{\xi}'\right)f(\xi')f(\bar{\xi}'). \quad (36)$$

As we can see, the scattering contribution to the Wigner function in $\xi = \mathbf{r},\mathbf{k}$ is the result of a fully nonlocal two-body transition: while the "main carrier" performs the generic transition $\xi' = \mathbf{r}',\mathbf{k}' \to \xi'' = \mathbf{r}'',\mathbf{k}''$, the "partner carrier" performs the generic transition $\bar{\xi}' = \bar{\mathbf{r}}',\bar{\mathbf{k}}' \to \bar{\xi}'' = \bar{\mathbf{r}}'',\bar{\mathbf{k}}''$.

4. Nonlocal Character of Pauli-Blocking Contributions

The aim of this section is to further investigate—both within the density-matrix formalism and within the Wigner-function picture—the role played by Pauli-blocking terms.

As discussed in [89], the time evolution of the single-particle density matrix is always characterized by a highly non-trivial coupling between diagonal (population) and non-diagonal (polarization) terms; indeed, starting from the density-matrix-based nonlinear scattering model in (15), the equation of motion for the diagonal elements $f_\alpha = \rho_{\alpha\alpha}$ of the semiclassical theory (see Section 2) is given by:

$$\left.\frac{df_\alpha}{dt}\right|_s = \frac{1}{2} \sum_{\alpha'\alpha'_1\alpha'_2} \left((\delta_{\alpha\alpha'} - \rho_{\alpha\alpha'}) \, \mathcal{P}^s_{\alpha'\alpha,\alpha'_1\alpha'_2} \rho_{\alpha'_1\alpha'_2} - \left(\delta_{\alpha'\alpha'_1} - \rho_{\alpha'\alpha'_1}\right) \mathcal{P}^{s*}_{\alpha'\alpha'_1,\alpha\alpha'_2} \rho_{\alpha'_2\alpha} \right) + \text{c.c.} \qquad (37)$$

This shows that the time evolution of the carrier population involves, in general, diagonal as well as non-diagonal elements; this is different from the semiclassical Boltzmann-like scattering model in (6), where all non-diagonal (polarization) terms are neglected.

In order to better compare the semiclassical scattering model in (6) with the fully quantum-mechanical result in (37), let us insert into Equation (37) the separation between population and polarization terms introduced in (1):

$$\begin{aligned}
\left.\frac{df_\alpha}{dt}\right|_s = \ & \sum_{\alpha'} ((1-f_\alpha)\mathcal{P}^s_{\alpha\alpha'}f_{\alpha'} - (1-f_{\alpha'})\mathcal{P}^s_{\alpha'\alpha}f_\alpha) \\
& + \frac{1}{2}\sum_{\alpha'_1\alpha'_2} \left((1-f_\alpha)\,\mathcal{P}^s_{\alpha\alpha,\alpha'_1\alpha'_2}\rho_{\alpha'_1\alpha'_2} - \left(1-f_{\alpha'_1}\right)\mathcal{P}^{s*}_{\alpha'_1\alpha'_1,\alpha\alpha'_2}\rho_{\alpha'_2\alpha} \right) + \text{c.c.} \\
& - \frac{1}{2}\sum_{\alpha'\alpha'_1} \left(\rho_{\alpha\alpha'}\,\mathcal{P}^s_{\alpha'\alpha,\alpha'_1\alpha'_1}\,f_{\alpha'_1} - \rho_{\alpha'\alpha'_1}\,\mathcal{P}^{s*}_{\alpha'\alpha'_1,\alpha\alpha}\,f_\alpha \right) + \text{c.c.} \\
& - \frac{1}{2}\sum_{\alpha'\alpha'_1\alpha'_2} \left(\rho_{\alpha\alpha'}\,\mathcal{P}^s_{\alpha'\alpha,\alpha'_1\alpha'_2}\,\rho_{\alpha'_1\alpha'_2} - \rho_{\alpha'\alpha'_1}\,\mathcal{P}^{s*}_{\alpha'\alpha'_1,\alpha\alpha'_2}\,\rho_{\alpha'_2\alpha} \right) + \text{c.c.} , \qquad (38)
\end{aligned}$$

where $\mathcal{P}^s_{\alpha\alpha'} = \mathcal{P}^s_{\alpha\alpha,\alpha'\alpha'}$ denote the diagonal terms of our generalized scattering rates, which coincide with the standard semiclassical rates of the Boltzmann theory provided by the usual Fermi's-golden-rule-prescription (see Equation (6)).

As we can see, the original scattering contribution in (37) splits into four different terms: the first one describes population–population contributions and coincides with the semiclassical model in (6), the second and third term describe, respectively, population–polarization and polarization–population contributions, while the last one describes polarization-polarization contributions, also referred to as "polarization scattering" [10]. At high carrier densities and in the presence of electronic phase coherence, these last three (polarization-induced) contributions may lead to significant modifications compared to the semiclassical case; it is however hard to draw conclusions about the impact of such quantum-mechanical corrections, since the sign of these three extra-terms depend strongly on the specific problem under investigation as well as on the device operational conditions; in contrast, in the low-density limit, the last two (polarization–population and polarization-polarization) terms vanish, and the quantum-mechanical correction with respect to the semiclassical contribution is given by the second (population–polarization) term only.

The density-matrix analysis presented so far shows that, at high carrier concentrations, the Pauli blocking factors $(\delta_{\alpha_1\alpha_2} - \rho_{\alpha_1\alpha_2})$ may lead to significant modifications to the dissipation versus decoherence process via its diagonal (population) contributions as well as via its non-diagonal (polarization) ones.

Employing once again the Weyl–Wigner transform in (21), the above density-matrix Pauli factors are straightforwardly translated into the corresponding Pauli factors of the Wigner-function formulation (see Equations (25) and (33)):

$$\sum_{\alpha_1\alpha_2} W^*_{\alpha_1\alpha_2}(\mathbf{r}, \mathbf{k}) \left(\delta_{\alpha_1\alpha_2} - \rho_{\alpha_1\alpha_2}\right) = \left(1 - f(\mathbf{r}, \mathbf{k})\right). \tag{39}$$

As shown in the previous section, within our fully quantum-mechanical Wigner-function treatment, the action of these Pauli factors is always nonlocal; this can be clearly seen in Equation (25), where the generic scattering process from $\mathbf{r}', \mathbf{k}' \to \mathbf{r}, \mathbf{k}$ is "weighted" by a corresponding Pauli factor $(1 - f(\mathbf{r}'', \mathbf{k}''))$ and integrated over its phase-space coordinates $\mathbf{r}'', \mathbf{k}''$; this implies that the impact of such nonlocal Pauli factor may be relevant, also if the value of the Wigner function in $\mathbf{r}, \mathbf{k}$ is equal to zero.

We finally stress that, similar to the density-matrix case previously considered, it is difficult to evaluate the real impact of nonlocal Pauli factors within the Wigner-function picture. Indeed, as for the case of the population–polarization, polarization–population and polarization–polarization terms in (38), it is hard to draw general conclusions about the overall impact (scattering increase versus suppression) induced by such nonlocal Pauli factors. Indeed, opposite to the case of a semiclassical carrier distribution, it is imperative to recall that the Wigner function is a real quantity which may take negative values as well as values greater than one (see Figure 1c below). This implies that phase-space regions with a positive Wigner function will lead to a local suppression of dissipation versus decoherence phenomena, while phase-space regions characterized by a negative Wigner function will correspond to a Pauli factor larger than one, thus leading to a local increase of the scattering dynamics; moreover, for phase-space regions characterized by a Wigner function greater than one, the Pauli factor is negative, leading again to a scattering suppression. In a similar way, it is also important to recall that the Wigner-function scattering probabilities $\tilde{P}^{s,\mathrm{in/out}}$ in (25) are pseudoprobabilities, i.e., real functions which, in general, are not positive-definite. This implies that, for phase-space regions where the latter are negative, the two regimes of Pauli-induced scattering suppression versus increase just discussed are simply interchanged.

As a result of the non-positive-definite character of both the Wigner function and of the corresponding scattering probabilities, we are then forced to conclude that the nonlocal Pauli blocking factors previously discussed do not necessarily lead to an overall scattering suppression; we stress that such a conclusion is in clear contrast with the behaviour predicted by semiclassical models (see Equation (6)), where the presence of local Pauli factors leads in any case to a suppression of the scattering dynamics.

In order to concretely show the intrinsic limitations of local scattering models, we shall now present a few simulated experiments of phonon-induced energy dissipation for the prototypical nanosystem depicted in Figure 1a: it consists of a $l = 20\,\mathrm{nm}$ thick GaAs quantum well (QW) surrounded by (Al,Ga)As barriers with band offset $V_\circ = 0.3\,\mathrm{eV}$; its three-dimensional electronic states exhibit the usual subband structure due to confinement along the growth direction (z). To simplify our analysis, we shall neglect in-plane phase-space coordinates and adopt an effective one-dimensional (1D) description of the QW nanosystem, i.e., $\mathbf{r}, \mathbf{k} \to z, k$. This implies that, within such simplified treatment, the set of single-particle quantum numbers of our nanostructure coincides with the partially discrete index of our 1D states only: $\alpha \equiv n$. Moreover, since in the low-temperature simulated experiments discussed below the only electronic states involved in the dissipation process are the ground ($n = 1$) and first excited state ($n = 2$), our QW nanostructure may be described as a two-level system, whose energy levels and electronic wave functions, depicted in Figure 1a, may be safely described via the following infinite-barrier model:

$$k_1 = \frac{\pi}{l}, \qquad \epsilon_1 = \frac{\hbar^2 k_1^2}{2m^*}, \qquad \phi_1(z) = \sqrt{\frac{2}{l}}\cos(k_1 z),$$

$$k_2 = \frac{2\pi}{l}, \qquad \epsilon_2 = \frac{\hbar^2 k_2^2}{2m^*}, \qquad \phi_2(z) = -\sqrt{\frac{2}{l}}\sin(k_2 z) \tag{40}$$

(m^* denoting the GaAs effective mass). The prototypical QW nanostructure in Figure 1a has been optimized in order to maximize the intersubband carrier–phonon coupling; indeed, for $l = 20\,$nm, the interlevel splitting ($\epsilon_2 - \epsilon_1 \simeq 40\,$meV) matches with the GaAs LO-phonon energy [6].

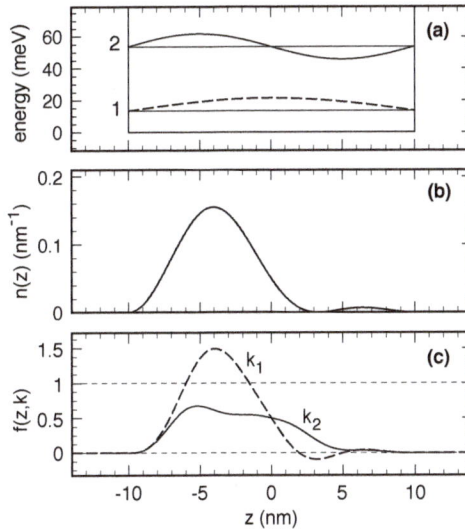

Figure 1. (a) conduction band profile along the growth (z) direction for the prototypical GaAs/(Al,Ga)As QW nanostructure considered in our simulated experiments. Energy levels of the first two confined states (ϵ_1 and ϵ_2) are shown, together with the corresponding wavefunctions ($\phi_1(z)$ and $\phi_2(z)$); (b) probability density ($n(z) = |\psi(z)|^2$) corresponding to the coherent state in (41); (c) Wigner function (see Equation (43)) of the coherent state in (41) plotted for the two relevant values k_1 and k_2 corresponding to the two QW basis states in (40) (see also panel (a)).

In order to better emphasize the intrinsic limitations of local scattering models, let us consider an electronic state given by a coherent and equally weighted superposition [94] of the two QW basis states in (40), namely

$$\psi(z) = c_1\phi_1(z) + c_2\phi_2(z), \qquad c_1 = c_2 = \frac{1}{\sqrt{2}}, \tag{41}$$

whose probability density $n(z) = |\psi(z)|^2$ is depicted in Figure 1b. It is easy to show that the coherent electronic state in (41) corresponds to the following (two-by-two) single-particle density matrix [95]:

$$\begin{pmatrix} \rho_{11} & \rho_{12} \\ \rho_{21} & \rho_{22} \end{pmatrix} = \begin{pmatrix} |c_1|^2 & c_1 c_2^* \\ c_2 c_1^* & |c_2|^2 \end{pmatrix} = \begin{pmatrix} \frac{1}{2} & \frac{1}{2} \\ \frac{1}{2} & \frac{1}{2} \end{pmatrix}. \tag{42}$$

As for any pure state $\psi(z)$, the corresponding Wigner function is simply given by:

$$f(z,k) = \int dz' \psi \left(z + \frac{z'}{2} \right) e^{-ikz'} \psi^* \left(z - \frac{z'}{2} \right). \tag{43}$$

Figure 1c shows the above Wigner function for the two relevant values k_1 and k_2 corresponding to the two QW basis states in (40) (see also Figure 1a). In addition to the strongly asymmetric nature of both the probability density $n(z)$ in Figure 1b and of the corresponding Wigner function profiles in Figure 1c, the latter exhibit negative values as well as values significantly greater than one (see dashed curve).

Combining Equations (24) and (25), the 1D version $(\mathbf{r}, \mathbf{k} \to z, k)$ of the nonlocal scattering model in (24) for the case of carrier–phonon interaction comes out to be:

$$\left. \frac{\partial f(z,k)}{\partial t} \right|_s = \int \frac{dz'' dk'' \, dz' dk'}{2\pi} \left(1 - f(z'',k'') \right) \Delta \tilde{P}^{cp}(z'',k'';z,k;z',k') f(z',k') \tag{44}$$

with

$$\Delta \tilde{P}^{cp}(z'',k'';z,k;z',k') = \tilde{P}^{cp,\text{in}}(z'',k'';z,k;z',k') - \tilde{P}^{cp,\text{out}}(z'',k'';z,k;z',k'). \tag{45}$$

Here, $\tilde{P}^{cp,\text{in/out}}$ are the 1D version of the fully nonlocal scattering rates in (30)–(31), and, for the case of our simplified QW model, the generalized carrier–phonon scattering rates in (16) acquire the diagonal form: $\mathcal{P}^{cp}_{\alpha_1 \alpha_2, \alpha_1' \alpha_2'} = P_{\alpha_1 \alpha_1'} \delta_{\alpha_1 \alpha_1', \alpha_2 \alpha_2'}$. In particular, in the low-temperature limit, the only active relaxation channel is the $2 \to 1$ transition induced by LO-phonon emission, namely

$$\begin{pmatrix} P_{11} & P_{12} \\ P_{21} & P_{22} \end{pmatrix} = \begin{pmatrix} 0 & P^\circ \\ 0 & 0 \end{pmatrix}, \tag{46}$$

where for our GaAs-based QW nanostructure the $2 \to 1$ phonon-emission rate P° is of the order of $5\,\text{ps}^{-1}$.

In order to compare the fully nonlocal QW scattering model described so far with its local counterpart, we shall describe energy relaxation via an effective Boltzmann-like equation coupling the two energy levels of the QW nanostructure depicted in Figure 1a. According to such a local scattering model, the phonon-induced time evolution of the upper-level Wigner function (see solid curve in Figure 1c) is given by:

$$\left. \frac{\partial f(z,k_2)}{\partial t} \right|_s = (1 - f(z,k_2)) P_{21} f(z,k_1) - (1 - f(z,k_1)) P_{12} f(z,k_2), \tag{47}$$

and in the low-temperature limit (see Equation (46)), the latter reduces to:

$$\left. \frac{\partial f(z,k_2)}{\partial t} \right|_s = - (1 - f(z,k_1)) P^\circ f(z,k_2). \tag{48}$$

In Figure 2, we show the time derivative

$$g(z) = \left. \frac{\partial f(z,k_2)}{\partial t} \right|_s \tag{49}$$

of the upper-level Wigner-function profile (see solid curve in Figure 1c) comparing the nonlocal model in (44) (solid curves) with its local counterpart in (47) (dash-dotted curves) in the absence (a) and presence (b) of Pauli-blocking terms.

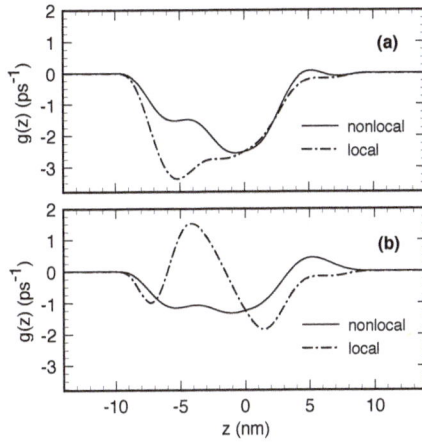

Figure 2. Time derivative of the upper-level Wigner-function profile (see Equation (49)): comparison between the nonlocal model in (44) (solid curves) and its local counterpart in (47) (dash-dotted curves) in the absence (**a**) and presence (**b**) of Pauli-blocking terms (see text).

As we can see, already neglecting Pauli-blocking factors, the nonlocal and local scattering models exhibit qualitatively different behaviours. Indeed, while the local result (dash-dotted curve in (a)) is always negative and simply proportional to the Wigner function $f(z, k_2)$ (see Equation (48) and solid curve in Figure 1c), the nonlocal one (solid curve in (a)) comes out to be significantly different. This is due to the nonlocal nature of the carrier–phonon scattering model in (44) present also in the absence of the Pauli factor $(1 - f(z'', k''))$ and ascribed to the spatial integration with respect to z'.

In the presence of Pauli-blocking factors, the discrepancies between nonlocal and local models are strongly amplified. Indeed, while for the nonlocal model the presence of the Pauli factors leads basically to an overall suppression of the time derivative (solid curve in (b)), with respect to the Pauli-free case (solid curve in (a)), the local result (dash-dotted curve in (b)) exhibits significant positive-definite regions, due to negative values of the Pauli factor $(1 - f(z, k_1))$.

As a confirmation of the intrinsic limitations of the local scattering model pointed out so far, it is easy to show that the Wigner function of the QW ground state $\phi_1(z)$—corresponding to the zero-temperature equilibrium state of our nanostructure—is not a steady-state solution of the local scattering model in (48).

5. Conclusions

Thanks to their simple physical interpretation as well as to their straightforward implementation within various quantum-mechanical simulation schemes, semiclassical scattering models have been widely employed in the design and optimization of new-generation quantum nanomaterials and related nanodevices. In particular, during the last few decades, two different classes of semiclassical treatments have been independently used, namely density-matrix and Wigner-function schemes. The first class is based on the so-called diagonal approximation, i.e., the neglect of non-diagonal density-matrix elements (i.e., polarization terms). The second class includes local scattering models borrowed from the conventional Boltzmann transport theory, namely relaxation-time schemes as well as Boltzmann collision terms; it has been recently shown [79] that the use of such local scattering approaches within the Wigner-function formalism may lead to unphysical results, namely anomalous suppression of intersubband relaxation, incorrect thermalization dynamics, and violation of probability-density positivity. To overcome such severe limitations, a quantum-mechanical generalization of relaxation-time and Boltzmann-like models has been recently proposed [79], resulting in nonlocal electron-phonon scattering superoperators.

The primary goal of this paper is twofold: on the one hand, we have investigated the intimate link between density-matrix and Wigner-function approaches, pointing out intrinsic limitations of semiclassical scattering models within these, apparently different, simulation strategies. On the other hand, we have extended the carrier–phonon treatment in [79] to carrier–carrier interaction, deriving the explicit form of the corresponding two-body scattering superoperator.

The main result of our investigation is that, for both carrier–phonon and carrier–carrier interaction, it is hard to evaluate the impact (scattering suppression or increase) of Pauli-blocking factors. More specifically, within the density-matrix picture, such terms give rise to quantum corrections (with respect to the semiclassical case), namely population–polarization, polarization–population, and polarization-polarization terms, often referred to as "polarization scattering". At the same time, within the Wigner-function picture, the action of the corresponding Pauli factors comes out to always be nonlocal. Combining such nonlocal character with the non-positive-definite nature of both the Wigner function and of the corresponding scattering probabilities, it is again hard to draw general conclusions on the overall impact of Pauli blocking terms on energy dissipation and decoherence processes.

In order to concretely show the intrinsic limitations of local scattering models, a few simulated experiments of energy dissipation and decoherence in a QW semiconductor nanostructure have also been presented. The latter show that, already in the low-density limit (i.e., neglecting Pauli-blocking terms), one deals with significant nonlocal corrections, and that, at high carrier densities, these corrections are strongly amplified.

Author Contributions: Investigation, R.C.I.; supervision, F.R.

Funding: This research received no external funding.

Conflicts of Interest: The authors declare no conflict of interest.

References and Notes

1. Esaki, L.; Tsu, R. Superlattice and negative differential conductivity in semiconductors. *IBM J. Res. Dev.* **1970**, *14*, 61–65. [CrossRef]
2. Ihn, T. *Semiconductor Nanostructures: Quantum States and Electronic Transport*; OUP Oxford: Hong Kong, China, 2010.
3. Capasso, F. *Physics of Quantum Electron Devices*; Springer Series in Electronics and Photonics; Springer: London, UK, 2011.
4. Bastard, G. *Wave Mechanics Applied to Semiconductor Heterostructures*; Monographies de Physique, Les Éditions de Physique; John Wiley and Sons Inc.: New York, NY, USA, 1988.
5. It is imperative to stress that, in spite of the intrinsic limitations of the semiclassical theory, in transport experiments—Characterized by strong energy dissipation and decoherence—It is hard to find clear indications of quantum-transport corrections.
6. Jacoboni, C.; Lugli, P. *The Monte Carlo Method for Semiconductor Device Simulation*; Springer: Berlin, Germany, 1989.
7. Frensley, W.R. Boundary-conditions for open quantum-systems driven far from equilibrium. *Rev. Mod. Phys.* **1990**, *62*, 745–791. [CrossRef]
8. Axt, V.M.; Mukamel, S. Nonlinear optics of semiconductor and molecular nanostructures; A common perspective. *Rev. Mod. Phys.* **1998**, *70*, 145–174. [CrossRef]
9. Datta, S. Nanoscale device modeling: The Green's function method. *Superlattices Microstruct.* **2000**, *28*, 253–278. [CrossRef]
10. Rossi, F.; Kuhn, T. Theory of ultrafast phenomena in photoexcited semiconductors. *Rev. Mod. Phys.* **2002**, *74*, 895–950. [CrossRef]
11. Axt, V.M.; Kuhn, T. Femtosecond spectroscopy in semiconductors: A key to coherences, correlations and quantum kinetics. *Rep. Prog. Phys.* **2004**, *67*, 433–512. [CrossRef]
12. Pecchia, A.; Di Carlo, A. Atomistic theory of transport in organic and inorganic nanostructures. *Rep. Prog. Phys.* **2004**, *67*, 1497–1561. [CrossRef]

13. Iotti, R.C.; Rossi, F. Microscopic theory of semiconductor-based optoelectronic devices. *Rep. Prog. Phys.* **2005**, *68*, 2533–2571. [CrossRef]

14. Kadanoff, L.; Baym, G. Quantum statistical mechanics: Green's function methods in equilibrium and nonequilibrium problems. In *Frontiers in Physics*; W.A. Benjamin: San Francisco, CA, USA, 1962.

15. Haug, H.; Jauho, A. *Quantum Kinetics in Transport and Optics of Semiconductors*; Springer: Berlin, Germany, 2007.

16. Bonitz, M. *Quantum Kinetic Theory*; Teubner-Texte zur Physik, Teubner; Springer: Berlin, Germany, 1998.

17. Datta, S. *Quantum Transport: Atom to Transistor*; Cambridge University Press: Cambridge, UK, 2005.

18. Taylor, J.; Guo, H.; Wang, J. *Ab initio* modeling of quantum transport properties of molecular electronic devices. *Phys. Rev. B* **2001**, *63*, 245407. [CrossRef]

19. Faleev, S.V.; Léonard, F.M.C.; Stewart, D.A.; van Schilfgaarde, M. *Ab initio* tight-binding LMTO method for nonequilibrium electron transport in nanosystems. *Phys. Rev. B* **2005**, *71*, 195422. [CrossRef]

20. Luisier, M.; Klimeck, G. Atomistic full-band simulations of silicon nanowire transistors: Effects of electron-phonon scattering. *Phys. Rev. B* **2009**, *80*, 15543. [CrossRef]

21. Zhang, L.; Xing, Y.; Wang, J. First-principles investigation of transient dynamics of molecular devices. *Phys. Rev. B* **2012**, *86*, 155438. [CrossRef]

22. Haug, H.; Koch, S. *Quantum Theory of the Optical and Electronic Properties of Semiconductors*; World Scientific: Singapore, 2004.

23. Rossi, F. *Theory of Semiconductor Quantum Devices: Microscopic Modeling and Simulation Strategies*; Springer: Berlin, Germany, 2011.

24. Buot, F. *Nonequilibrium Quantum Transport Physics in Nanosystems: Foundation of Computational Nonequilibrium Physics in Nanoscience and Nanotechnology*; World Scientific: Singapore, 2009.

25. Ryzhii, M.; Ryzhii, V. Monte Carlo analysis of ultrafast electron transport in quantum well infrared photodetectors. *Appl. Phys. Lett.* **1998**, *72*, 842–844. [CrossRef]

26. Iotti, R.C.; Rossi, F. Nature of charge transport in quantum-cascade lasers. *Phys. Rev. Lett.* **2001**, *87*, 146603. [CrossRef] [PubMed]

27. Köhler, R.; Tredicucci, A.; Beltram, F.; Beere, H.E.; Linfield, E.H.; Davies, A.G.; Ritchie, D.A.; Iotti, R.C.; Rossi, F. Terahertz semiconductor-heterostructure laser. *Nature* **2002**, *417*, 156–159. [CrossRef] [PubMed]

28. Callebaut, H.; Kumar, S.; Williams, B.S.; Hu, Q.; Reno, J.L. Importance of electron-impurity scattering for electron transport in terahertz quantum-cascade lasers. *Appl. Phys. Lett.* **2004**, *84*, 645–647. [CrossRef]

29. Lu, J.T.; Cao, J.C. Coulomb scattering in the Monte Carlo simulation of terahertz quantum-cascade lasers. *Appl. Phys. Lett.* **2006**, *89*, 211115. [CrossRef]

30. Bellotti, E.; Driscoll, K.; Moustakas, T.D.; Paiella, R. Monte Carlo study of GaN versus GaAs terahertz quantum cascade structures. *Appl. Phys. Lett.* **2008**, *92*, 101112. [CrossRef]

31. Jirauschek, C. Monte Carlo study of carrier-light coupling in terahertz quantum cascade lasers. *Appl. Phys. Lett.* **2010**, *96*, 011103. doi:10.1063/1.3284523. [CrossRef]

32. Matyas, A.; Belkin, M.A.; Lugli, P.; Jirauschek, C. Temperature performance analysis of terahertz quantum cascade lasers: Vertical versus diagonal designs. *Appl. Phys. Lett.* **2010**, *96*, 201110. [CrossRef]

33. Iotti, R.C.; Rossi, F.; Vitiello, M.S.; Scamarcio, G.; Mahler, L.; Tredicucci, A. Impact of nonequilibrium phonons on the electron dynamics in terahertz quantum cascade lasers. *Appl. Phys. Lett.* **2010**, *97*, 033110. [CrossRef]

34. Vitiello, M.S.; Iotti, R.C.; Rossi, F.; Mahler, L.; Tredicucci, A.; Beere, H.E.; Ritchie, D.A.; Hu, Q.; Scamarcio, G. Non-equilibrium longitudinal and transverse optical phonons in terahertz quantum cascade lasers. *Appl. Phys. Lett.* **2012**, *100*, 091101. [CrossRef]

35. Matyas, A.; Lugli, P.; Jirauschek, C. Role of collisional broadening in Monte Carlo simulations of terahertz quantum cascade lasers. *Appl. Phys. Lett.* **2013**, *102*, 011101. [CrossRef]

36. Iotti, R.C.; Rossi, F. Coupled carrier–phonon nonequilibrium dynamics in terahertz quantum cascade lasers: A Monte Carlo analysis. *New J. Phys.* **2013**, *15*, 075027. [CrossRef]

37. Gmachl, C.; Capasso, F.; Sivco, D.L.; Cho, A.Y. Recent progress in quantum cascade lasers and applications. *Rep. Prog. Phys.* **2001**, *64*, 1533–1601. [CrossRef]

38. Brunetti, R.; Jacoboni, C.; Price, P.J. Quantum-mechanical evolution of real-space transfer. *Phys. Rev. B* **1994**, *50*, 11872–11878. [CrossRef]

39. Reiter, D.; Glanemann, M.; Axt, V.M.; Kuhn, T. Spatiotemporal dynamics in optically excited quantum wire-dot systems: Capture, escape, and wave-front dynamics. *Phys. Rev. B* **2007**, *75*, 205327. [CrossRef]

40. Rosati, R.; Iotti, R.C.; Dolcini, F.; Rossi, F. Derivation of nonlinear single-particle equations via many-body Lindblad superoperators: A density-matrix approach. *Phys. Rev. B* **2014**, *90*, 125140. [CrossRef]

41. Taj, D.; Iotti, R.C.; Rossi, F. Microscopic modeling of energy relaxation and decoherence in quantum optoelectronic devices at the nanoscale. *Eur. Phys. J. B* **2009**, *72*, 305–322. [CrossRef]

42. Rosati, R.; Rossi, F. Scattering nonlocality in quantum charge transport: Application to semiconductor nanostructures. *Phys. Rev. B* **2014**, *89*, 205415. [CrossRef]

43. Rosati, R.; Dolcini, F.; Rossi, F. Electron-phonon coupling in metallic carbon nanotubes: Dispersionless electron propagation despite dissipation. *Phys. Rev. B* **2015**, *92*, 235423. [CrossRef]

44. Rosati, R.; Reiter, D.E.; Kuhn, T. Lindblad approach to spatiotemporal quantum dynamics of phonon-induced carrier capture processes. *Phys. Rev. B* **2017**, *95*, 165302. [CrossRef]

45. Lindblad, G. Generators of quantum dynamical semigroups. *Commun. Math. Phys.* **1976**, *48*, 119–130. [CrossRef]

46. Kiršanskas, G.; Franckié, M.; Wacker, A. Phenomenological position and energy resolving Lindblad approach to quantum kinetics. *Phys. Rev. B* **2018**, *97*, 035432. [CrossRef]

47. Jacoboni, C.; Bordone, P. The Wigner-function approach to non-equilibrium electron transport. *Rep. Prog. Phys.* **2004**, *67*, 1033. [CrossRef]

48. Frensley, W.R. Transient Response of a Tunneling Device Obtained from the Wigner Function. *Phys. Rev. Lett.* **1986**, *57*, 2853–2856. [CrossRef] [PubMed]

49. Kluksdahl, N.C.; Kriman, A.M.; Ferry, D.K.; Ringhofer, C. Self-consistent study of the resonant-tunneling diode. *Phys. Rev. B* **1989**, *39*, 7720–7735. [CrossRef]

50. Buot, F.A.; Jensen, K.L. Lattice Weyl–Wigner formulation of exact many-body quantum-transport theory and applications to novel solid-state quantum-based devices. *Phys. Rev. B* **1990**, *42*, 9429–9457. [CrossRef]

51. Jensen, K.; Buot, F. The effects of scattering on current-voltage characteristics, transient response, and particle trajectories in the numerical simulation of resonant tunneling diodes. *J. Appl. Phys.* **1990**, *67*, 7602–7607. [CrossRef]

52. Miller, D.R.; Neikirk, D.P. Simulation of intervalley mixing in double-barrier diodes using the lattice Wigner function. *Appl. Phys. Lett.* **1991**, *58*, 2803–2805. [CrossRef]

53. McLennan, M.J.; Lee, Y.; Datta, S. Voltage drop in mesoscopic systems: A numerical study using a quantum kinetic equation. *Phys. Rev. B* **1991**, *43*, 13846–13884. [CrossRef]

54. Tso, H.C.; Horing, N.J.M. Wigner-function formulation of nonlinear electron-hole transport in a quantum well and analysis of the linear transient and steady state. *Phys. Rev. B* **1991**, *44*, 11358–11380. [CrossRef]

55. Gullapalli, K.K.; Miller, D.R.; Neikirk, D.P. Simulation of quantum transport in memory-switching double-barrier quantum-well diodes. *Phys. Rev. B* **1994**, *49*, 2622–2628. [CrossRef]

56. Fernando, C.L.; Frensley, W.R. Intrinsic high-frequency characteristics of tunneling heterostructure devices. *Phys. Rev. B* **1995**, *52*, 5092–5104. [CrossRef]

57. El Sayed, K.; Kenrow, J.A.; Stanton, C.J. Femtosecond relaxation kinetics of highly excited electronic wave packets in semiconductors. *Phys. Rev. B* **1998**, *57*, 12369–12377. [CrossRef]

58. Pascoli, M.; Bordone, P.; Brunetti, R.; Jacoboni, C. Wigner paths for electrons interacting with phonons. *Phys. Rev. B* **1998**, *58*, 3503–3506. [CrossRef]

59. Kim, K.Y.; Lee, B. Wigner-function formulation in anisotropic semiconductor quantum wells. *Phys. Rev. B* **2001**, *64*, 115304. [CrossRef]

60. Nedjalkov, M.; Kosina, H.; Selberherr, S.; Ringhofer, C.; Ferry, D.K. Unified particle approach to Wigner-Boltzmann transport in small semiconductor devices. *Phys. Rev. B* **2004**, *70*, 115319. [CrossRef]

61. Nedjalkov, M.; Vasileska, D.; Ferry, D.K.; Jacoboni, C.; Ringhofer, C.; Dimov, I.; Palankovski, V. Wigner transport models of the electron-phonon kinetics in quantum wires. *Phys. Rev. B* **2006**, *74*, 035311. [CrossRef]

62. Taj, D.; Genovese, L.; Rossi, F. Quantum-transport simulations with the Wigner-function formalism: Failure of conventional boundary-condition schemes. *Europhys. Lett.* **2006**, *74*, 1060–1066. [CrossRef]

63. Weetman, P.; Wartak, M.S. Self-consistent model of a nanoscale semiconductor laser using Green and Wigner functions in two bases. *Phys. Rev. B* **2007**, *76*, 035332. [CrossRef]

64. Querlioz, D.; Saint-Martin, J.; Bournel, A.; Dollfus, P. Wigner Monte Carlo simulation of phonon-induced electron decoherence in semiconductor nanodevices. *Phys. Rev. B* **2008**, *78*, 165306. [CrossRef]

65. Morandi, O. Multiband Wigner-function formalism applied to the Zener band transition in a semiconductor. *Phys. Rev. B* **2009**, *80*, 024301. [CrossRef]

66. Wójcik, P.; Spisak, B.; Wołoszyn, M.; Adamowski, J. Self-consistent Wigner distribution function study of gate-voltage controlled triple-barrier resonant tunnelling diode. *Semicond. Sci. Technol.* **2009**, *24*, 095012. [CrossRef]

67. Barraud, S. Phase-coherent quantum transport in silicon nanowires based on Wigner transport equation: Comparison with the nonequilibrium-Green-function formalism. *J. Appl. Phys.* **2009**, *106*, 063714. [CrossRef]

68. Yoder, P.D.; Grupen, M.; Smith, R. Demonstration of Intrinsic Tristability in Double-Barrier Resonant Tunneling Diodes With the Wigner Transport Equation. *IEEE Trans. Electron Devices* **2010**, *57*, 3265–3274. [CrossRef]

69. Álvaro, M.; Bonilla, L.L. Two miniband model for self-sustained oscillations of the current through resonant-tunneling semiconductor superlattices. *Phys. Rev. B* **2010**, *82*, 035305. [CrossRef]

70. Savio, A.; Poncet, A. Study of the Wigner function at the device boundaries in one-dimensional single- and double-barrier structures. *J. Appl. Phys.* **2011**, *109*, 033713. [CrossRef]

71. Trovato, M.; Reggiani, L. Quantum maximum-entropy principle for closed quantum hydrodynamic transport within a Wigner function formalism. *Phys. Rev. E* **2011**, *84*, 061147. [CrossRef] [PubMed]

72. Rosati, R.; Dolcini, F.; Iotti, R.C.; Rossi, F. Wigner-function formalism applied to semiconductor quantum devices: Failure of the conventional boundary condition scheme. *Phys. Rev. B* **2013**, *88*, 035401. [CrossRef]

73. Sellier, J.; Amoroso, S.; Nedjalkov, M.; Selberherr, S.; Asenov, A.; Dimov, I. Electron dynamics in nanoscale transistors by means of Wigner and Boltzmann approaches. *Physica A* **2014**, *398*, 194–198. [CrossRef]

74. Sellier, J.; Dimov, I. A Wigner approach to the study of wave packets in ordered and disordered arrays of dopants. *Physica A* **2014**, *406*, 185–190. [CrossRef]

75. Jonasson, O.; Knezevic, I. Dissipative transport in superlattices within the Wigner function formalism. *J. Comput. Electron.* **2015**, *14*, 879–887. [CrossRef]

76. Hamerly, R.; Mabuchi, H. Quantum noise of free-carrier dispersion in semiconductor optical cavities. *Phys. Rev. A* **2015**, *92*, 023819. [CrossRef]

77. Cabrera, R.; Bondar, D.I.; Jacobs, K.; Rabitz, H.A. Efficient method to generate time evolution of the Wigner function for open quantum systems. *Phys. Rev. A* **2015**, *92*, 042122. [CrossRef]

78. Kim, K.Y.; Kim, S. Effect of uncertainty principle on the Wigner function-based simulation of quantum transport. *Solid State Electron.* **2015**, *111*, 22–26. [CrossRef]

79. Iotti, R.C.; Dolcini, F.; Rossi, F. Wigner-function formalism applied to semiconductor quantum devices: Need for nonlocal scattering models. *Phys. Rev. B* **2017**, *96*, 115420. [CrossRef]

80. Beenakker, C.W.J. Theory of Coulomb-blockade oscillations in the conductance of a quantum dot. *Phys. Rev. B* **1991**, *44*, 1646–1656. [CrossRef]

81. Schoeller, H.; Schön, G. Mesoscopic quantum transport: Resonant tunneling in the presence of a strong Coulomb interaction. *Phys. Rev. B* **1994**, *50*, 18436–18452, doi:10.1103/PhysRevB.50.18436. [CrossRef]

82. Dolcini, F.; Dell'Anna, L. Multiple Andreev reflections in a quantum dot coupled to superconducting leads: Effect of spin-orbit coupling. *Phys. Rev. B* **2008**, *78*, 024518. [CrossRef]

83. Secchi, A.; Rontani, M. Coulomb versus spin-orbit interaction in few-electron carbon-nanotube quantum dots. *Phys. Rev. B* **2009**, *80*, 041404. [CrossRef]

84. Ström, A.; Johannesson, H.; Japaridze, G.I. Edge Dynamics in a Quantum Spin Hall State: Effects from Rashba Spin-Orbit Interaction. *Phys. Rev. Lett.* **2010**, *104*, 256804. [CrossRef] [PubMed]

85. Dolcini, F. Signature of interaction in dc transport of ac-gated quantum spin Hall edge states. *Phys. Rev. B* **2012**, *85*, 033306. [CrossRef]

86. The Wigner transport equation in Equation (7) is formally reminiscent of the Boltzmann transport one for the semiclassical distribution function. Such basic link has also stimulated the development of so-called Wigner Monte Carlo schemes [58,60], namely simulation techniques based on a Monte Carlo solution of the Wigner transport equation.

87. Moyal, J.E. Quantum mechanics as a statistical theory. *Math. Proc. Camb. Philos. Soc.* **1949**, *45*, 99–124. [CrossRef]

88. A relevant exception is the so-called "dynamics controlled truncation" introduced by Axt and Stahl [8], based on an expansion in powers of the exciting laser field.
89. Iotti, R.C.; Ciancio, E.; Rossi, F. Quantum transport theory for semiconductor nanostructures: A density-matrix formulation. *Phys. Rev. B* **2005**, *72*, 125347. [CrossRef]
90. Spohn, H. Kinetic equations from Hamiltonian dynamics: Markovian limits. *Rev. Mod. Phys.* **1980**, *52*, 569–615. [CrossRef]
91. It is worth stressing that this treatment is based on the assumption of thermal-equilibrium phonons with a uniform effective temperature. In the presence of significant hot-phonon effects [36], additional nonlocal contributions due to the spatial modulation of the phonon population may arise; however, the latter are expected to play a minor role on the nanometric scale.
92. As usual, the two-body carrier–carrier coupling considered here describes the short-range Coulomb contribution only. The long-range contribution may be accounted for via coupled Wigner-Poisson schemes [7].
93. The fact that Equation (23) is the inverse of the Weyl–Wigner transform in (21) can be easily checked noting that:
$$(2\pi)^{-3} \sum_{\alpha_1\alpha_2} W^*_{\alpha_1\alpha_2}(\mathbf{r},\mathbf{k}) W_{\alpha_1\alpha_2}(\mathbf{r}',\mathbf{k}') = \delta(\mathbf{r}-\mathbf{r}')\,\delta(\mathbf{k}-\mathbf{k}').$$
94. Such a quantum-mechanical state superposition may be realized via ultrafast coherent laser excitation in the infrared spectral range [10].
95. We stress that such a pure state constitutes the building block for the generation of maximally entangled electronic Bell states in semiconductors [23].

MDPI
St. Alban-Anlage 66
4052 Basel
Switzerland
Tel. +41 61 683 77 34
Fax +41 61 302 89 18
www.mdpi.com

Entropy Editorial Office
E-mail: entropy@mdpi.com
www.mdpi.com/journal/entropy

www.ingramcontent.com/pod-product-compliance
Lightning Source LLC
Chambersburg PA
CBHW051837210326

41597CB00033B/5683